Lecture Notes in Mathematics 2052

Editors:
J.-M. Morel, Cachan
B. Teissier, Paris

T0207332

For further volumes:
http://www.springer.com/series/304

Fabien Morel

$\mathbb{A}^1$-Algebraic Topology over a Field

 Springer

Fabien Morel
Mathematisches Institut der LMU
München, Germany

ISBN 978-3-642-29513-3   ISBN 978-3-642-29514-0 (eBook)
DOI 10.1007/978-3-642-29514-0
Springer Heidelberg New York Dordrecht London

Lecture Notes in Mathematics ISSN print edition: 0075-8434
                              ISSN electronic edition: 1617-9692

Library of Congress Control Number: 2012942079

Mathematics Subject Classification (2010): 14-XX, 19-XX, 55-XX

Printed on acid-free paper

Springer is part of Springer Science+Business Media (www.springer.com)

Preface

This work should be considered as a natural sequel to the foundational paper [59] where the $\mathbb{A}^1$-homotopy category of smooth schemes over a base scheme was defined and its first properties studied. In this text the base scheme will always be the spectrum of a perfect field k.

One of our first motivations is to emphasize that, contrary to the first impression, the relationship between the $\mathbb{A}^1$-homotopy theory over k and the category Sm_k of smooth k-schemes is of the same nature as the relationship between the classical homotopy theory and the category of differentiable manifolds. This explains the title of this work; we hope to convince the reader in this matter. This slogan was already discussed in [55], see also [4].

This text is the result of the compilation of two preprints "$\mathbb{A}^1$-algebraic topology over a field" and "$\mathbb{A}^1$-homotopy classification of vector bundles over smooth affine schemes" to which we added some recent new stuff, consisting of the two sections: "Geometric *versus* canonical transfers" (Chap. 4) and "The Rost–Schmid complex of a strongly $\mathbb{A}^1$-invariant sheaf" (Chap. 5).

The main objective of these new sections was primarily to correctly establish the equivalence between the notions of "strongly $\mathbb{A}^1$-invariant" and "strictly $\mathbb{A}^1$-invariant" for sheaves of abelian groups, see Theorem 1.16. These new sections appear also to be interesting in their own. As the reader will notice, the introduction of the notion of *strictly $\mathbb{A}^1$-invariant sheaves with generalized transfers* in our work on the Friedlander–Milnor conjecture [56–58] is directly influenced from these.

Our treatment of transfers in Chap. 4, which is an adaptation of the original one of Bass and Tate for Milnor K-theory [9], is, we think, clarifying. Besides reaching the structure of strictly $\mathbb{A}^1$-invariant sheaves with generalized transfers, we obtain on the way a new proof of the fact [38] that the transfers in Milnor K-theory do not depend on choices of a sequence of generators, see Theorem 4.27 and Remark 4.32. Our proof is in spirit different from the one of Kato [38], as we prove directly by geometric means the independence in the case of two generators.

The construction in Chap. 5 for a general strongly $\mathbb{A}^1$-invariant sheaf of abelian groups of its "Rost–Schmid" complex is directly influenced by the work of Rost [68] and its adaptation to Witt groups in [70]. Our philosophy on transfers in this work was to use them as less as possible, and to only use them when they really show up by themselves. For instance we will define in Chap. 3 the sheaves of unramified Milnor K-theory (as well as Milnor–Witt K-theory) without using any transfers. The proof of Lemma 5.24 contains the geometric explanation of the formula for the differential of the "Gersten–Rost–Schmid" complex (Corollary 5.44) and the appearance of transfers in it.

The results and ideas in this work have been used and extended in several different directions. In [16] some very concrete computations on $\mathbb{A}^1$-homotopy classes of rational fractions give a nice interpretation of our sheaf theoretic computations of $\pi_1^{\mathbb{A}^1}(\mathbb{P}^1)$. The structure and property of the $\mathbb{A}^1$-fundamental group sheaves as well as the associated theory of $\mathbb{A}^1$-coverings have been used in [2,3,86]. It is also the starting point of [4]. Our result (Chap. 9) concerning the Suslin–Voevodsky construction $Sing_\bullet^{\mathbb{A}^1}(SL_n)$, $n \geq 3$, has been generalized to a general split semi-simple group of type not SL_2 in [87] and to the case of SL_2 in [60].

The present work plays also a central role in our approach to the Friedlander–Milnor conjecture [56–58].

Conventions and notations. Everywhere in this work, k denotes a fixed perfect field and Sm_k denotes the category of smooth finite type k-schemes.

We denote by $\mathcal{F}_k$ the category of field extensions $k \subset F$ of k such that F is of finite transcendence degree over k. By a discrete valuation v on $F \in \mathcal{F}_k$ we will always mean one which is trivial on k. We let $\mathcal{O}_v \subset F$ denote its valuation ring, $m_v \subset \mathcal{O}_v$ its maximal ideal, and $\kappa(v)$ its residue field.

For any scheme X and any integer i we let $X^{(i)}$ denote the set of points in X of codimension i. Given a point $x \in X$, $\kappa(x)$ will denote its residue field, and $\mathcal{M}_x$ will denote the maximal ideal of the local ring $\mathcal{O}_{X,x}$.

The category Sm_k is always endowed with the Nisnevich topology [59,62], unless otherwise explicitly stated. Thus for us "sheaf" always means sheaf in the Nisnevich topology.

We will let Set denote the category of sets, Ab that of abelian groups. A *space* is a simplicial object in the category of sheaves of sets on Sm_k [59]. We will also assume the reader is familiar with the notions and results of *loc. cit.*

We denote by Sm_k' the category of *essentially smooth k-schemes*. For us, an *essentially smooth k-scheme* is a noetherian k-scheme X which is the inverse limit of a left filtering system $(X_\alpha)_\alpha$ with each transition morphism $X_\beta \to X_\alpha$ being an étale affine morphism between smooth k-schemes (see [29]).

For any $F \in \mathcal{F}_k$ the k-scheme $Spec(F)$ is essentially k-smooth. For each point $x \in X \in Sm_k$, the local scheme $X_x := Spec(\mathcal{O}_{X,x})$ of X at x as well as its henselization $X_x^h := Spec(\mathcal{O}_{X,x}^h)$ are essentially smooth k-schemes.

In the same way the complement of the closed point in $Spec(\mathcal{O}_{X,x})$ or X_x^h is essentially smooth over k. We will sometime make the abuse of writing "smooth k-scheme" instead of "essentially smooth k-scheme", if no confusion can arise.

Given a presheaf of sets on Sm_k, that is to say a functor $F : (Sm_k)^{op} \to Sets$, and an essentially smooth k-scheme $X = lim_\alpha X_\alpha$ we set $F(X) := colimit_\alpha F(X_\alpha)$. From the results of [29] this is well defined. When $X = Spec(A)$ is affine we will also simply denote this set by $F(A)$.

Acknowledgements I want to take this opportunity to warmly thank Mike Hopkins and Marc Levine for their interest in this work during the past years when I conceived it; their remarks, comments, and some discussions on and around this subject helped me to improve it very much.

Contents

Chapter 1
Introduction

Let k be a perfect field. Our aim in this work is to address in the $\mathbb{A}^1$-homotopy theory of smooth k-schemes considered in [49, 59, 81] the analogues of the following classical theorems:

Theorem 1.1 (Brouwer degree). *Let $n > 0$ be an integer and let S^n denote the n-sphere. Then for an integer i*

$$\pi_i(S^n) = \begin{cases} 0 \ if \ i < n \\ \mathbb{Z} \ if \ i = n \end{cases}$$

Theorem 1.2 (Hurewicz Theorem). *For any pointed connected topological space X and any integer $n \geq 1$ the Hurewicz morphism*

$$\pi_n(X) \to H_n(X)$$

is the abelianization if $n = 1$, is an isomorphism if $n \geq 2$ and X is $(n-1)$-connected, and is an epimorphism if $n \geq 3$ and X is $(n-2)$-connected.

Theorem 1.3 (Coverings and π_1). *Any "reasonable" pointed connected space X admits a universal pointed covering*

$$\tilde{X} \to X$$

It is (up to unique isomorphism) the only pointed simply connected covering of X. Its automorphism group (as unpointed covering) is $\pi_1(X)$ and it is a $\pi_1(X)$-Galois covering.

Theorem 1.4. $\pi_1(\mathbb{P}^1(\mathbb{R})) = \mathbb{Z}$ *and* $\pi_1(\mathbb{P}^n(\mathbb{R})) = \mathbb{Z}/2$ *for* $n \geq 2$, $\pi_1(SL_2(\mathbb{R})) = \mathbb{Z}$ *and* $\pi_1(SL_{n+1}(\mathbb{R})) = \mathbb{Z}/2$ *for* $n \geq 2$.
  The corresponding complex spaces are simply connected: for $n \geq 1$ one has $\pi_1(\mathbb{P}^n(\mathbb{C})) = \pi_1(SL_n(\mathbb{C})) = *$

F. Morel, $\mathbb{A}^1$-*Algebraic Topology over a Field*, Lecture Notes in Mathematics 2052,
DOI 10.1007/978-3-642-29514-0_1, © Springer-Verlag Berlin Heidelberg 2012

Theorem 1.5. *Given a C.W. complex X and r an integer we let $\Phi_r(X)$ be the set of isomorphism classes of rank r real vector bundles on X. Then there exists a natural bijection $\Phi_r(X) \cong [X, B\mathbb{GL}_r(\mathbb{R})]$, where $[-, -]$ means the set of homotopy classes of continuous maps.*

Theorem 1.6. *If X is a C.W. complex of dimension $\leq r$, the obstruction to split off a trivial line bundle of a given rank r vector bundle ξ over X is its Euler class $e(\xi) \in H^n(X; \mathbb{Z})$.*

Let us now explain more in details the structure of this work. The main technical achievement of this work is the understanding of the precise nature of the $\mathbb{A}^1$-homotopy sheaves of $\mathbb{A}^1$-connected pointed spaces. This will occupy us in the first four chapters. Once this is done, we will introduce the $\mathbb{A}^1$-homology sheaves of a space (see also [52]) and then, the analogues of the previous Theorems 1.1–1.4 become almost straightforward adaptations of the classical cases. This is done in Chaps. 6 and 7.

The analogues of the previous last two Theorems requires more work and will take the Chaps. 7 and 8. We adapt classical techniques and results concerning locally free modules of finite rank. These two sections might be considered as the sequel to some of the results in [49]. The two Appendices A and B concern technical results used in Chaps. 7 and 8.

To explain our results, recall that given a space $\mathcal{X}$, we denote by $\pi_0^{\mathbb{A}^1}(\mathcal{X})$ the associated sheaf to the presheaf $U \mapsto Hom_{\mathcal{H}(k)}(U, \mathcal{X})$ where $\mathcal{H}(k)$ denotes the $\mathbb{A}^1$-homotopy category of smooth k-schemes defined in [59, 81]. Sometimes, when no confusion can arise, we will simply denote by $[\mathcal{X}, \mathcal{Y}]$ the set of morphisms $Hom_{\mathcal{H}(k)}(\mathcal{X}, \mathcal{Y})$ in the $\mathbb{A}^1$-homotopy category $\mathcal{H}(k)$ between two spaces $\mathcal{X}$ and $\mathcal{Y}$.

Let $\mathcal{H}_\bullet(k)$ denote the pointed $\mathbb{A}^1$-homotopy category of pointed spaces. For two pointed spaces we will sometimes denote by $[\mathcal{X}, \mathcal{Y}]_\bullet$ the (pointed) set of morphisms $Hom_{\mathcal{H}_\bullet(k)}(\mathcal{X}, \mathcal{Y})$ in $\mathcal{H}_\bullet(k)$ between two pointed spaces $\mathcal{X}$ and $\mathcal{Y}$.

If $\mathcal{X}$ is a pointed space and n is an integer ≥ 1, we denote by $\pi_n^{\mathbb{A}^1}(\mathcal{X})$ the associated sheaf of groups (in the Nisnevich topology) to the presheaf of groups $U \mapsto Hom_{\mathcal{H}_\bullet(k)}(\Sigma^n(U_+), \mathcal{X}) = [\Sigma^n(U_+), \mathcal{X}]_\bullet$. Here Σ is the simplicial suspension. $\pi_n^{\mathbb{A}^1}(\mathcal{X})$ is a sheaf of abelian groups for $n \geq 2$.

In classical topology, the underlying structure to the corresponding homotopy "sheaves" is quite simple: π_0 is a discrete set, π_1 is a discrete group and the π_n's, $n \geq 2$, are discrete abelian groups.

The following notions are the analogues of being "$\mathbb{A}^1$-discrete", or "discrete" in the $\mathbb{A}^1$-homotopy theory; we will justify these in Theorem 1.9 below.

Definition 1.7. 1) A sheaf of sets $\mathcal{S}$ on Sm_k (in the Nisnevich topology) is said to be $\mathbb{A}^1$-*invariant* if for any $X \in Sm_k$, the map

$$\mathcal{S}(X) \to \mathcal{S}(\mathbb{A}^1 \times X)$$

induced by the projection $\mathbb{A}^1 \times X \to X$, is a bijection.

2) A sheaf of groups $\mathcal{G}$ on Sm_k (in the Nisnevich topology) is said to be *strongly $\mathbb{A}^1$-invariant* if for any $X \in Sm_k$, the map

$$H^i_{Nis}(X; G) \to H^i_{Nis}(\mathbb{A}^1 \times X; G)$$

induced by the projection $\mathbb{A}^1 \times X \to X$, is a bijection for $i \in \{0, 1\}$.

3) A sheaf M of abelian groups on Sm_k (in the Nisnevich topology) is said to be *strictly $\mathbb{A}^1$-invariant* if for any $X \in Sm_k$, the map

$$H^i_{Nis}(X; M) \to H^i_{Nis}(\mathbb{A}^1 \times X; M)$$

induced by the projection $\mathbb{A}^1 \times X \to X$, is a bijection for any $i \in \mathbb{N}$.

Remark 1.8. By the very definitions of [59], it is straightforward to check that:

1) A sheaf of set $\mathcal{S}$ is $\mathbb{A}^1$-invariant if and only if it is $\mathbb{A}^1$-local as a space;
2) A sheaf of groups G is strongly $\mathbb{A}^1$-invariant if and only if the classifying space $B(G) = K(G, 1)$ is an $\mathbb{A}^1$-local space;
3) A sheaf of abelian groups M is strictly $\mathbb{A}^1$-invariant if and only if for any $n \in \mathbb{N}$ the Eilenberg–MacLane space $K(M, n)$ is $\mathbb{A}^1$-local.   □

The notion of strict $\mathbb{A}^1$-invariance is directly taken from [79,80]; indeed the $\mathbb{A}^1$-invariant sheaves with transfers of Voevodsky are the basic examples of strictly $\mathbb{A}^1$-invariant sheaves. The Rost's cycle modules [68] give also examples of strictly $\mathbb{A}^1$-invariant sheaves, in fact more precisely of $\mathbb{A}^1$-invariant sheaves with transfers by [21]. Other important examples of strictly $\mathbb{A}^1$-invariant sheaves, which are not of the previous type, are the sheaf $\underline{\mathbf{W}}$ of unramified Witt groups (this is the associated sheaf to the presheaf of Witt groups $X \mapsto W(X)$) in characteristic $\neq 2$ (this fact is proven in [64]), or the sheaves $\underline{\mathbf{I}}^n$ of unramified powers of the fundamental ideal used in [53] (still in characteristic $\neq 2$). These sheaves will be defined below in characteristic 2 as well, by considering the Witt groups of inner product spaces (instead of quadratic forms) studied in [48].

The notion of strong $\mathbb{A}^1$-invariance is new and we will meet important examples of genuine *non commutative* strongly $\mathbb{A}^1$-invariant sheaves of groups. For instance the $\mathbb{A}^1$-fundamental group of $\mathbb{P}^1_k$:

$$\pi^{\mathbb{A}^1}_1(\mathbb{P}^1_k)$$

is non abelian, see Sect. 7.3 below. More generally the $\pi^{\mathbb{A}^1}_1$ of the blow-up of $\mathbb{P}^n$, $n \geq 2$, at several points is highly non-commutative; see [4].

We believe that the $\mathbb{A}^1$-fundamental group sheaf should play a fundamental role in the understanding of $\mathbb{A}^1$-connected[1] projective smooth k-varieties in very much the same way as the usual fundamental group plays a fundamental role in the classification of compact connected differentiable manifolds; for a precise discussion of this idea, see [4].

One of our main global result, which justifies *a posteriori* the previous definition, is:

Theorem 1.9. *Let $\mathcal{X}$ be a pointed space. Then the sheaf of groups $\pi_1^{\mathbb{A}^1}(\mathcal{X})$ is strongly $\mathbb{A}^1$-invariant and for any $n \geq 2$ the sheaf of abelian groups $\pi_n^{\mathbb{A}^1}(\mathcal{X})$ is strictly $\mathbb{A}^1$-invariant.*

Remark 1.10. The proof relies very much on the fact that the base is a field, as we will use Gabber's geometric presentation Lemma [19, 25], see also Lemma 1.15 below, at several places.   □

Remark 1.11. Recall from [59] that any space $\mathcal{X}$ is the homotopy inverse limit of its Postnikov tower $\{P^n(\mathcal{X})\}_n$ and that if $\mathcal{X}$ is pointed and connected, for each $n \geq 1$ the homotopy fiber of the morphism $P^n(\mathcal{X}) \to P^{n-1}(\mathcal{X})$ is $\mathbb{A}^1$-equivalent to the Eilenberg–MacLane space $K(\pi_n^{\mathbb{A}^1}(\mathcal{X}), n)$. Thus the strongly or strictly $\mathbb{A}^1$-invariant sheaves and their cohomology play exactly the same role as the usual homotopy groups and singular cohomology groups play in classical algebraic topology.   □

We are unfortunately unable to prove the analogue structure result for the $\pi_0^{\mathbb{A}^1}$ which appears to be the most difficult case:

Conjecture 1.12. For any space $\mathcal{X}$ the sheaf $\pi_0^{\mathbb{A}^1}(\mathcal{X})$ is $\mathbb{A}^1$-invariant.

Remark 1.13. This conjecture is easy to check for smooth k-schemes of dimension ≤ 1. The case of smooth surfaces is already very interesting and non trivial. Assume that k is algebraically closed and that X is a projective smooth k-surface which is birationally ruled over the smooth projective curve C of genus g then:

$$\pi_0^{\mathbb{A}^1}(X) = \begin{cases} * \text{ if } g = 0 \\ C \text{ if } g > 0 \end{cases}$$

This confirms the conjecture. However the conjecture is unknown in the case of arbitrary smooth surfaces.   □

Remark 1.14. Another interesting general test, which is still open, though known in several cases, is the geometric classifying space $B_{gm}G$ of a smooth affine algebraic k-group G [59, 77]. From [59], there exists a natural transformation in $X \in Sm_k$: $\theta_G : H^1_{et}(X; G) \to Hom_{\mathcal{H}(k)}(X, B_{gm}G)$ which induces a natural morphism

---

[1] A space is said to be $\mathbb{A}^1$-connected if its $\pi_0^{\mathbb{A}^1}$ is the point.

$$\mathcal{H}_{et}^1(G) \to \pi_0^{\mathbb{A}^1}(B_{gm}G)$$

on associated sheaves of sets. From [59, Corollary 3.2 p. 137] it follows that for G finite of order prime to the characteristic of k this morphism is an isomorphism. We can prove more generally that the above morphism induces an isomorphism on sections on perfect fields (at least if the group of irreducible components of G is of order prime to the characteristic of k). It also seems to be possible to prove that the sheaf $\mathcal{H}_{et}^1(G)$ is $\mathbb{A}^1$-invariant (in progress) which would imply the conjecture in that case as well. □

The previous results on $\mathbb{A}^1$-homotopy sheaves rely on the detailed analysis of *unramified sheaves* of groups given in Chap. 2. Our analysis is done very much in the spirit of a "non-abelian variant" of Rost's cycle modules [68]. These unramified sheaves can be described in terms of their sections on function fields in $\mathcal{F}_k$ plus extra structures like "residues" maps and "specializations" morphisms. This analysis is entirely elementary, and contrary to Rost's approach [68], the structure involved in our description does not use any notion of transfers. Instead we use the Gabber's geometric presentation Lemma[2] in the case of one single point, which is equivalent to the following:

Lemma 1.15 (Gabber [19, 25]). *Let F a field, Let X be the localization of a smooth F-scheme of dimension n at a point z and let $Y \subset X$ be a closed subscheme everywhere of codimension ≥ 1. Then there is a point $t \in \mathbb{A}_F^{n-1}$, and an étale morphism $X \to \mathbb{A}_S^1$, where $S = Spec(\mathcal{O}_{\mathbb{A}_{F,t}^{n-1}})$, such that the composition $Y \subset X \to \mathbb{A}_S^1$ is still a closed immersion and Y is finite over S.*

Using the techniques developed in Chap. 2, we will be able in Chap. 3 to define a bunch of universal unramified sheaves, like the unramified sheaves $\underline{\mathbf{K}}_n^M$ of Milnor K-theory in weight n, $n \in \mathbb{N}$, as well as the unramified Milnor–Witt K-theory sheaves $\underline{\mathbf{K}}_n^{MW}$ in weight n, $n \in \mathbb{Z}$, without using any transfers.

Besides the techniques of Chaps. 2 and 3, one of the most important technical result that we prove in this work, and which is essential to prove Theorem 1.9, is the following one (see Theorem 5.46 below).

Theorem 1.16. *Let M be a sheaf of abelian groups on Sm_k. Then: M is strongly $\mathbb{A}^1$-invariant $\Leftrightarrow$ M is strictly $\mathbb{A}^1$-invariant.*

It means that the notions of strong $\mathbb{A}^1$-invariance and strict $\mathbb{A}^1$-invariance are in fact the "same" (when they both make sense at the same time, that is to say for sheaves of abelian groups). The proof is rather non trivial and occupies almost entirely the Chaps. 4 and 5. As we previously explained however, these sections also contain new point of views and ideas on Transfers, and are preparatory to our notion of "strictly $\mathbb{A}^1$-invariant sheaves with

---

[2]This is proven in [19, 25] only when k is infinite, but O. Gabber proved also the statement of the Lemma in the above form when k is finite, private communication.

generalized transfers" used in [56–58]. We also observe that we treat the case of characteristic 2 as well.

The idea of the proof of the previous theorem is roughly speaking the following. Given a strongly A^1-invariant sheaf of abelian groups M, one may write explicitly its "Rost–Schmid" complex $C^*(X; M)$ for any (essentially) k-smooth scheme X, following [68,70]. This requires first to construct canonical (or rather absolute) transfers morphisms on the sheaves of the form M_{-2}: this is done in Chaps. 4 and 5. We use the approach of Bass–Tate [9] for Milnor K-theory conveniently adapted. The proof of the independence of the choices is new. There are two further problems to be solved: prove that the Rost–Schmid differential is indeed a differential and prove the homotopy invariance property for this complex. This is again done mostly following the ideas from Rost [68] and Schmid [70] conveniently adapted. Once this is done, from Gabber's presentation Lemma above, one gets acyclicity for local smooth ring by induction on the dimension, from which the theorem follows.

Remark 1.17. Marc Levine found a gap in a previous argument to prove the Theorem 1.16 above. The new argument is clearly much more elaborated. To write a proof on a non perfect base field seems also possible, but it would require much more work and technicalities. We decided not to try to do it, to keep this text have a reasonable length. This is the only reason why we assume that the base field is perfect from the beginning.   □

The Hurewicz Theorem (see Theorems 6.35 and 6.57) is a straightforward consequence of Theorem 1.9, at least once the notion of A^1-chain complex and the corresponding notion of A^1-homology sheaves $H_*^{A^1}(\mathcal{X})$ of a space $\mathcal{X}$ are conveniently introduced; see Sect. 6.3 and also [52]. An important non trivial consequence of the Hurewicz Theorem is the following *unstable A^1-connectivity theorem* (see Theorem 6.38 in Sect. 6.3 below):

Theorem 1.18. *Let $\mathcal{X}$ be a pointed space and $n \geq 0$ be an integer. If $\mathcal{X}$ is simplicially n-connected then it is n-A^1-connected, which means that $\pi_i^{A^1}(\mathcal{X})$ is trivial for $i \leq n$.*

A stable and thus weaker version was obtained in [52]. The unstable case was previously known before only for $n = 0$ [59] but over a general base scheme.

Remark 1.19. The proof relies very much again on the fact that the base is a field. Over a general base the situation is definitely false, at least when the base scheme has dimension at least 2 as pointed out by J. Ayoub [5]. It could be that the previous statement still holds in case the base scheme is the spectrum of a Dedekind ring, it would follow that most of this work can be extended to the base scheme being the spectrum of a Dedekind ring. □

An example of simplicially $(n - 1)$-connected pointed space is the n-fold simplicial suspension $\Sigma^n(\mathcal{X})$ of a pointed space $\mathcal{X}$. As $A^n - \{0\}$ is

$\mathbb{A}^1$-equivalent to the simplicial $(n-1)$-suspension $\Sigma^{n-1}(\mathbb{G}m^{\wedge n})$ (see [59]), it is thus $(n-2)$-$\mathbb{A}^1$-connected. In the same way the n-th smash power $(\mathbb{P}^1)^{\wedge n}$, which is $\mathbb{A}^1$-equivalent to the simplicial suspension of the previous one [59], is $(n-1)$-$\mathbb{A}^1$-connected.

Remark 1.20. In general, the "correct" $\mathbb{A}^1$-connectivity is given by the connectivity of the "corresponding" topological space of real points, through a real embedding of k, rather than the connectivity of its topological space of complex points through a complex embedding. This principle[3] has been a fundamental guide to our work. For instance the pointed algebraic "sphere" $(\mathbb{G}_m)^{\wedge n}$ is not $\mathbb{A}^1$-connected: it must be considered as a "twisted 0-dimensional sphere". Observe that its space of real points has the homotopy type of the 0-dimensional sphere $S^0 = \{+1, -1\}$.   □

Brouwer degree and Milnor–Witt K-theory. The Hurewicz Theorem implies formally that for $n \geq 2$, the first non-trivial $\mathbb{A}^1$-homotopy sheaf of the $(n-2)$-$\mathbb{A}^1$-connected sphere

$$\mathbb{A}^n - \{0\} \cong_{\mathbb{A}^1} \Sigma^{n-1}(\mathbb{G}_m{}^{\wedge n})$$

is the free strongly $\mathbb{A}^1$-invariant sheaf of abelian groups (or equivalently free strictly $\mathbb{A}^1$-invariant sheaf of abelian groups by Theorem 1.16) generated by the pointed 0-dimensional sphere $(\mathbb{G}_m)^{\wedge n}$. This computation is the analogue of the fact that in classical topology the first non-trivial homotopy group of the n-dimensional sphere S^n is the free "discrete abelian group" generated by the pointed 0-dimensional sphere S^0.

To understand the analogue of Theorem 1.1, it remains thus to understand the free strongly $\mathbb{A}^1$-invariant sheaf on $(\mathbb{G}_m)^{\wedge n}$. We will denote it by $\underline{\mathbf{K}}_n^{MW}$ and will call it the sheaf of Milnor–Witt K-theory in weight n; there is a canonical map of sheaves of pointed sets, the *universal symbol on n units*:

$$(\mathbb{G}_m)^{\wedge n} \to \underline{\mathbf{K}}_n^{MW}$$

$$(U_1, \ldots, U_n) \mapsto [U_1, \ldots, U_n]$$

As $\underline{\mathbf{K}}_n^{MW}$ will be shown below to be automatically *unramified*, as any strongly $\mathbb{A}^1$-invariant sheaf, in particular for any irreducible smooth k-scheme X with function field F, the homomorphism of abelian groups $\underline{\mathbf{K}}_n^{MW}(X) \to \underline{\mathbf{K}}_n^{MW}(F)$ is injective. To describe $\underline{\mathbf{K}}_n^{MW}$ thus, we first need to describe the abelian group $\underline{\mathbf{K}}_n^{MW}(F) =: K_n^{MW}(F)$ for any field F.

Definition 1.21. Let F be a commutative field. The *Milnor–Witt K-theory* of F is the graded associative ring $K_*^{MW}(F)$ generated by the symbols $[u]$,

---

[3]Which owns much to conversations with V. Voevodsky.

for each unit $u \in F^\times$, of degree $+1$, and one symbol η of degree -1 subject to the following relations:

1 (Steinberg relation) For each $a \in F^\times - \{1\} : [a].[1 - a] = 0$
2 For each pair $(a, b) \in (F^\times)^2 : [ab] = [a] + [b] + \eta.[a].[b]$
3 For each $u \in F^\times : [u].\eta = \eta.[u]$
4 Set $h := \eta.[-1] + 2$. Then $\eta . h = 0$

This object was introduced in a complicated way by the author, until the above very simple and natural description was found in collaboration with Mike Hopkins. As we will see later on, each generator and relation has a natural $\mathbb{A}^1$-homotopic interpretation. For instance, the class of the algebraic Hopf map $\eta \in [\mathbb{A}^2 - \{0\}, \mathbb{P}^1]_{\mathcal{H}_\bullet(k)}$ is closely related to the element η of the Milnor–Witt K-theory.

The quotient ring $K_*^{MW}(F)/\eta$ is clearly the Milnor K-theory ring $K_*^M(F)$ of F introduced by Milnor in [47]. It is not hard to prove (see Sect. 3.1) that the ring $K_*^{MW}(F)[\eta^{-1}]$ obtained by inverting η is the ring of Laurent polynomials $W(F)[\eta, \eta^{-1}]$ with coefficients in the Witt ring $W(F)$ of non-degenerate symmetric bilinear forms on F (see [48], and [69] in characteristic $\neq 2$). More precisely, $K_0^{MW}(F)$ is the Grothendieck–Witt ring $GW(F)$ of non-degenerate symmetric bilinear forms on F. The isomorphism sends the 1-dimensional form $(X, Y) \mapsto uXY$ on F to $< u >:= \eta[u] + 1$ (see Sect. 2.1). For $n > 0$, the homomorphism multiplication by η^n: $K_0^{MW}(F) \to K_{-n}^{MW}(F)$ is surjective and defines through relation **(4)** an isomorphism $W(F) \cong K_{-n}^{MW}(F)$.

Using residue morphisms in Milnor–Witt K-theory defined as in [47] and the above Gabber's Lemma, we define in Sect. 3.2 the sheaf $X \mapsto \underline{\mathbf{K}}_n^{MW}(X)$. This approach is new as we do not at all use any transfer at this level. For X irreducible with function field F, $\underline{\mathbf{K}}_n^{MW}(X) \subset K_n^{MW}(F)$ denotes the intersection of the kernel of all the residue maps at points in X of codimension 1. Then:

Theorem 1.22. *For $n \geq 1$, the above morphism of sheaves*

$$(\mathbb{G}_m)^{\wedge n} \to \underline{\mathbf{K}}_n^{MW} \quad , \quad (U_1, \ldots, U_n) \mapsto [U_1, \ldots, U_n]$$

is the universal one to a strongly $\mathbb{A}^1$-invariant (or strictly $\mathbb{A}^1$-invariant) sheaf of abelian groups: any morphism of pointed sheaves $(\mathbb{G}_m)^{\wedge n} \to M$ to a strongly $\mathbb{A}^1$-invariant sheaf of abelian groups induces a unique morphism of sheaves of abelian groups

$$\underline{\mathbf{K}}_n^{MW} \to M$$

The Hurewicz Theorem gives now for free the analogue of Theorem 1.1 we had in mind:

Theorem 1.23. *For $n \geq 2$ one has a canonical isomorphisms of sheaves*

$$\pi_{n-1}^{\mathbb{A}^1}(\mathbb{A}^n - \{0\}) \cong \pi_n^{\mathbb{A}^1}((\mathbb{P}^1)^{\wedge n}) \cong \underline{\mathbf{K}}_n^{MW}$$

It is not hard to compute for (n, m) a pair of integers the abelian group of morphisms of sheaves of abelian groups from $\underline{\mathbf{K}}_n^{MW}$ to $\underline{\mathbf{K}}_m^{MW}$: it is $\underline{\mathbf{K}}_{m-n}^{MW}(k)$, the isomorphism being induced by the product $\underline{\mathbf{K}}_n^{MW} \times \underline{\mathbf{K}}_{m-n}^{MW} \to \underline{\mathbf{K}}_m^{MW}$. This implies in particular for $n = m$:

Corollary 1.24 (Brouwer degree). *For $n \geq 2$, the canonical morphism*

$$[\mathbb{A}^n - \{0\}, \mathbb{A}^n - \{0\}]_\bullet \cong [(\mathbb{P}^1)^{\wedge n}, (\mathbb{P}^1)^{\wedge n}]_\bullet \to K_0^{MW}(k) = GW(k)$$

is an isomorphism.

The Theory of the Brouwer degree in $\mathbb{A}^1$-homotopy theory thus assigns to an $\mathbb{A}^1$-homotopy class of maps from an algebraic sphere to itself an element in $GW(k)$; see [55] for a heuristic discussion in case $n = 1$, and also [18] for a concrete point of view. We observe that in case $n = 1$ the morphism $[\mathbb{P}^1, \mathbb{P}^1]_{\mathcal{H}_\bullet(k)} \to GW(k)$ is only an epimorphism.

Our computations in Theorem 1.23 stabilize as follows:

Corollary 1.25. *[50, 51] Let $\mathcal{SH}(k)$ be the stable $\mathbb{A}^1$-homotopy category of $\mathbb{P}^1$-spectra (or T-spectra) over k [50, 51, 81]. Let $\mathbb{S}^0$ be the sphere spectrum, $(\mathbb{G}_m)$ be the suspension spectrum of the pointed $\mathbb{G}_m$, let $\eta : (\mathbb{G}_m) \to \mathbb{S}^0$ be the (suspension of the) Hopf map and let MGL be the Thom spectrum [81]. For any integer $n \in \mathbb{Z}$ one has a commutative diagram in which the verticals are canonical isomorphisms:*

$$
\begin{array}{ccccc}
[\mathbb{S}^0, (\mathbb{G}_m)^{\wedge n}]_{\mathcal{SH}(k)} & \to & [\mathbb{S}^0, Cone(\eta) \wedge (\mathbb{G}_m)^{\wedge n}]_{\mathcal{SH}(k)} & \cong & [\mathbb{S}^0, MGL \wedge (\mathbb{G}_m)^{\wedge n}]_{\mathcal{SH}(k)} \\
\downarrow \wr & & \downarrow \wr & & \downarrow \wr \\
K_n^{MW}(k) & \twoheadrightarrow & K_n^{MW}(k)/\eta & = & K_n^M(k)
\end{array}
$$

Remark 1.26. A general base change argument as in [52] implies that the previous computations are still valid over an arbitrary field.

Observe that the proof we give here is completely elementary, as opposed to [50, 51], which at that time used the Milnor conjectures.   □

$\mathbb{A}^1$-*coverings and* $\pi_1^{\mathbb{A}^1}$. Another natural consequence of our work is the theory of $\mathbb{A}^1$-coverings and their relation to $\mathbb{A}^1$-fundamental sheaves of groups (see Sect. 7.1). The notion of $\mathbb{A}^1$-covering is quite natural: it is a morphism of spaces having the unique left lifting property with respect to "trivial $\mathbb{A}^1$-cofibrations". The Galois étale coverings of order prime to $char(k)$, or the $\mathbb{G}_m$-torsors are the basic examples of $\mathbb{A}^1$-coverings. We will prove the existence of a universal $\mathbb{A}^1$-covering for any pointed $\mathbb{A}^1$-connected space $\mathcal{X}$, and more precisely the exact analogue of Theorem 1.3.

This theory is in some sense orthogonal, or "complementary", to the étale theory of the fundamental group: an $\mathbb{A}^1$-connected space $\mathcal{X}$ has no nontrivial pointed Galois étale coverings [4, 55]. In case $\mathcal{X}$ is not $\mathbb{A}^1$-connected, the étale finite coverings are "captured" by the sheaf $\pi_0^{\mathbb{A}^1}(\mathcal{X})$ which may have non trivial étale covering, like the case of abelian varieties shows; these are equal to their own $\pi_0^{\mathbb{A}^1}$.

The universal $\mathbb{A}^1$-covering and the sheaf $\pi_1^{\mathbb{A}^1}$ encode a much more combinatorial and geometrical information than the arithmetical information of the étale one. As we already mentioned, we hope that this combinatorial object will play a central role in the "$\mathbb{A}^1$-surgery classification" approach to projective smooth $\mathbb{A}^1$-connected k-varieties [4, 55].

In Sect. 7.2 we compute the $\pi_1^{\mathbb{A}^1}$ of $\mathbb{P}^n$, $n \geq 2$ and of SL_n, $n \geq 3$:

Theorem 1.27. *1) For $n \geq 2$, the canonical $\mathbb{G}_m$-torsor $(\mathbb{A}^{n+1} - \{0\}) \to \mathbb{P}^n$ is the universal $\mathbb{A}^1$-covering of $\mathbb{P}^n$, and thus yields an isomorphism*

$$\pi_1^{\mathbb{A}^1}(\mathbb{P}^n) \cong \mathbb{G}_m$$

2) One has a canonical isomorphism

$$\pi_1^{\mathbb{A}^1}(SL_2) \cong \underline{\mathbf{K}}_2^{MW}$$

and the inclusions $SL_2 \to SL_n$, $n \geq 3$, induce an isomorphism

$$\underline{\mathbf{K}}_2^{MW}/\eta = \underline{\mathbf{K}}_2^{M} \cong \pi_1^{\mathbb{A}^1}(SL_n)$$

Remark 1.28. 1) It is interesting to determine the $\pi_1^{\mathbb{A}^1}$ of split semi-simple groups which are simply connected (in the sense of algebraic group theory). It is possible using [85] and [43]. One finds that for every group which is not of symplectic type the $\pi_1^{\mathbb{A}^1}(G)$ is isomorphic to $\underline{\mathbf{K}}_2^{M}$ and for each group of symplectic type the $\pi_1^{\mathbb{A}^1}(G)$ is $\underline{\mathbf{K}}_2^{MW}$. This fits with the previous result.

2) Our computations make clear that the $\mathbb{Z}$ or $\mathbb{Z}/2$ in the statement of Theorem 1.4 have different "motivic" natures, the fundamental groups of projective spaces being of "weight one" and that of special linear groups of "weight two". □

We finish in Sect. 7.3 this computational part with a very explicit description of the sheaf $\pi_1^{\mathbb{A}^1}(\mathbb{P}^1)$. The sheaf of groups $\pi_1^{\mathbb{A}^1}(\mathbb{P}^1)$ will be shown to be non-commutative, and is thus not equal to $\underline{\mathbf{K}}_1^{MW}$. The algebraic Hopf map $\eta : \mathbb{A}^2 - \{0\} \to \mathbb{P}^1$ fits into a fiber sequence of the form:

$$\mathbb{A}^2 - \{0\} \to \mathbb{P}^1 \to \mathbb{P}^\infty$$

which gives, using the isomorphisms $\pi_1^{\mathbb{A}^1}(\mathbb{A}^2 - \{0\}) \cong \underline{\mathbf{K}}_2^{MW}$ and $\pi_1^{\mathbb{A}^1}(\mathbb{P}^\infty) \cong \mathbb{G}_m$, a central extension of sheaves of groups

$$0 \to \underline{\mathbf{K}}_2^{MW} \to \pi_1^{\mathbb{A}^1}(\mathbb{P}^1) \to \mathbb{G}_m \to 1$$

which is explicitly described in term of cocycles. In particular we will see it is non commutative.

Homotopy classification of algebraic vector bundles. Let us now come to the analogues of Theorems 1.5 and 1.6 above. First denote by $\Phi_r(X) :=$ $H^1_{Zar}(X; GL_r) \cong H^1_{Nis}(X; GL_r)$ the set of isomorphism classes of rank r vector bundles over a smooth k-scheme X. It follows from [59] that there is a natural transformation in $X \in Sm_k$

$$\Phi_r(X) \to Hom_{\mathcal{H}(k)}(X, BGL_r)$$

Theorem 1.29. *Let r be an integer different from 2. Then for X a smooth affine k-scheme, the natural map*

$$\Phi_r(X) \cong Hom_{\mathcal{H}(k)}(X, BGL_r)$$

is a bijection.

Remark 1.30. The proof will be given in Chaps. 8 and 9. It is rather technical and relies in an essential way on the solution of the so-called *generalized Serre problem* given by Lindel [41], after Quillen [66] and Suslin [73] for the case of polynomial rings over fields, as well as some works of Suslin [74] and Vorst [82,83] on the "analogue" of the Serre problem for the general linear group. The reader should notice that in classical topology, the proof of Theorem 1.5 is also technical.

For $r = 1$, the theorem was proven in [59].

For $n = 2$, using the results of [60] one may also establish the theorem when $r = 2$. □

Remark 1.31. It is well known that the result can't hold in general if X is not affine: the functor $X \mapsto \Phi_r(X)$ is not $\mathbb{A}^1$-invariant on the whole category of smooth k-schemes in contrast to the right hand side $X \mapsto Hom_{\mathcal{H}(k)}(X, BGL_r)$ which is $\mathbb{A}^1$-invariant by construction.

For instance one may construct a rank 2 vector bundle over $\mathbb{A}^1 \times \mathbb{P}^1$ whose pull-back through the 0-section and 1-section are non isomorphic vector bundle over $\mathbb{P}^1$. □

To explain our approach, recall that $\Delta^\bullet$ denotes the algebraic cosimplicial simplex ([75] for instance) with $\Delta^n := Spec(k[T_0, \ldots, T_n]/\Sigma_i T_i = 1)$.

Let F be a sheaf of sets on Sm_k. We denote by $Sing^{\mathbb{A}^1}_\bullet(F)$ the space (or simplicial sheaf of sets) $U \mapsto F(\Delta^\bullet \times U)$, and we call it the Suslin–Voevodsky construction on F [75,76]. There is a canonical morphism of spaces $F \to Sing^{\mathbb{A}^1}_\bullet(F)$ which is an $\mathbb{A}^1$-weak equivalence.

The proof of Theorem 1.29 consists, roughly speaking, to observe first that if $\mathbb{G}r_r$ denotes the infinite algebraic Grassmanian of r-plans, then for any k-smooth affine X, there is a canonical bijection between $\Phi_r(X)$ and the set of naive $\mathbb{A}^1$-homotopy classes of morphisms $X \to \mathbb{G}r_r$ (this idea was already used in [49]). The second step is to check that for such an X, the set of naive $\mathbb{A}^1$-homotopy classes of morphisms $X \to \mathbb{G}r_r$ is equal to $Hom_{\mathcal{H}(k)}(X; \mathbb{G}r_r)$. It is obtained by introducing the *affine B.G.-property* (a variant of the B.G.-property of [59]) and by establishing that property for $\underline{Sing}^{\mathbb{A}^1}_\bullet(\mathbb{G}r_r)$, using the known works cited above. This is the very technical step. On the way we proved that for $n \neq 2$, $\underline{Sing}^{\mathbb{A}^1}_\bullet(SL_n)$, $\underline{Sing}^{\mathbb{A}^1}_\bullet(GL_n)$ have also the affine B.G.-property and is are $\mathbb{A}^1$-local. This was our starting point in our approach to the Friedlander–Milnor conjecture. The affine B.G.-property was established for any split semi-simple group G not containing the type SL_2 by Wendt [87], and to SL_2 as well by Moser [60].

For a given smooth affine k-scheme X and an integer $r \geq 4$ we may use the previous results to study the map of "adding the trivial line bundle":

$$\Phi_{r-1}(X) \to \Phi_r(X)$$

following the classical method in homotopy theory. Using classical obstruction theory (which is recalled in the appendix B) this amounts to study the connectivity of the simplicial homotopy fiber $\underline{Sing}^{\mathbb{A}^1}_\bullet(GL_r/GL_{r-1}) \cong \underline{Sing}^{\mathbb{A}^1}_\bullet(\mathbb{A}^r - \{0\})$ of the map:

$$B\underline{Sing}^{\mathbb{A}^1}_\bullet(GL_{r-1}) \to B\underline{Sing}^{\mathbb{A}^1}_\bullet(GL_r)$$

As the space $\mathbb{A}^r - \{0\}$ is $\mathbb{A}^1$-local and $(r-2)$-connected, the Hurewicz theorem gives an isomorphism of sheaves:

$$\pi_{r-1}(\underline{Sing}^{\mathbb{A}^1}_\bullet(\mathbb{A}^r - \{0\})) \cong \underline{\mathbf{K}}^{MW}_r$$

Combining all the previous results, obstruction theory yields the following:

Theorem 1.32 (Theory of Euler class). *Assume $r \geq 4$. Let $X = Spec(A)$ be a smooth affine k-scheme of dimension $\leq r$, and let ξ be an oriented algebraic vector bundle of rank r (that is to say with a trivialization of $\Lambda^r(\xi)$). Define its Euler class*

$$e(\xi) \in H^r_{Nis}(X; \underline{\mathbf{K}}^{MW}_r)$$

to be the obstruction class in

$$H^r_{Nis}(X; \pi^{\mathbb{A}^1}_{r-1}(\underline{Sing}^{\mathbb{A}^1}_\bullet(\mathbb{A}^r - \{0\})) \cong H^r_{Nis}(X; \underline{\mathbf{K}}^{MW}_r)$$

obtained from the above homotopy fibration sequence

$$\underline{Sing}^{\mathbb{A}^1}_\bullet(\mathbb{A}^r - \{0\}) \to B\underline{Sing}^{\mathbb{A}^1}_\bullet(GL_{r-1}) \to B\underline{Sing}^{\mathbb{A}^1}_\bullet(GL_r)$$

Then:

$$\xi \text{ splits off a trivial line bundle } \Leftrightarrow e(\xi) = 0 \in H^r_{Nis}(X; \underline{\mathbf{K}}^{MW}_r)$$

Remark 1.33. 1) The group $H^r(X; \underline{\mathbf{K}}^{MW}_r)$ will be shown in Theorem 5.47 below to coincide with the oriented Chow group $\tilde{CH}^r(X)$ defined in [7]. Our Euler class coincides with the one defined in *loc. cit.*. This proves then the main conjecture of *loc. cit.*, which was proven there if $char(k) \neq 2$ and in the case of rank $r = 2$, see also [23].

2) In [11] an "Euler class group" $E(A)$ introduced by Nori, and a corresponding Euler class, is used to prove an analogous result on a noetherian ring A. However, there is no cohomological interpretation of that group, defined by explicit generators and relations, and its computation seems rather delicate. For A smooth over k, there is an epimorphism of groups $E(A) \twoheadrightarrow \tilde{CH}^r(A)$; that could be an isomorphism.

3) There is an obvious epimorphism

$$H^r(X; \underline{\mathbf{K}}^{MW}_r) \to CH^r(X)$$

The Euler class $e(\xi)$ is mapped to the r-th Chern class $c_r(\xi)$ as defined by Grothendieck. When the multiplicative group of k and of each of its finite extensions $L|k$ is 2-divisible and $dim(X) \leq r$ the above epimorphism is an isomorphism and the theorem contains as a particular case the result of Murthy [61] where k is algebraically closed. $\square$

Chapter 2
Unramified Sheaves and Strongly $\mathbb{A}^1$-Invariant Sheaves

2.1 Unramified Sheaves of Sets

We let $\tilde{Sm}_k$ denote the category of smooth k-schemes and whose morphisms are the smooth morphisms. We start with the following standard definition.

Definition 2.1. An unramified presheaf of sets $\mathcal{S}$ on Sm_k (resp. on $\tilde{Sm}_k$) is a presheaf of sets $\mathcal{S}$ such that the following holds:

(0) For any $X \in Sm_k$ with irreducible components X_α's, $\alpha \in X^{(0)}$, the obvious map $\mathcal{S}(X) \to \Pi_{\alpha \in X^{(0)}} \mathcal{S}(X_\alpha)$ is a bijection.

(1) For any $X \in Sm_k$ and any open subscheme $U \subset X$ the restriction map $\mathcal{S}(X) \to \mathcal{S}(U)$ is injective if U is everywhere dense in X;

(2) For any $X \in Sm_k$, irreducible with function field F, the injective map $\mathcal{S}(X) \hookrightarrow \bigcap_{x \in X^{(1)}} \mathcal{S}(\mathcal{O}_{X,x})$ is a bijection (the intersection being computed in $\mathcal{S}(F)$).

Remark 2.2. An unramified presheaf $\mathcal{S}$ (either on Sm_k or on $\tilde{Sm}_k$) is automatically a sheaf of sets in the Zariski topology. This follows from **(2)**. We also observe that with our convention, for $\mathcal{S}$ an unramified presheaf, the formula in **(2)** also holds for X essentially smooth over k and irreducible with function field F. We will use these facts freely in the sequel. □

Example 2.3. It was observed in [52] that any strictly $\mathbb{A}^1$-invariant sheaf on Sm_k is unramified in this sense. The $\mathbb{A}^1$-invariant sheaves with transfers of [79] as well as the cycle modules[1] of Rost [68] give such unramified sheaves. In characteristic $\neq 2$ the sheaf associated to the presheaf of Witt groups $X \mapsto W(X)$ is unramified by [63] (the sheaf associated in the Zariski topology is in fact already a sheaf in the Nisnevich topology). □

---

[1]These two notions are indeed closely related by [21].

F. Morel, $\mathbb{A}^1$-*Algebraic Topology over a Field*, Lecture Notes in Mathematics 2052, 15
DOI 10.1007/978-3-642-29514-0_2, © Springer-Verlag Berlin Heidelberg 2012

Remark 2.4. Let S be a sheaf of sets in the Zariski topology on Sm_k (resp. on $\widetilde{Sm}_k$) satisfying properties **(0)** and **(1)** of the previous definition. Then it is unramified if and only if, for any $X \in Sm_k$ and any open subscheme $U \subset X$ the restriction map $S(X) \to S(U)$ is bijective if $X - U$ is everywhere of codimension ≥ 2 in X. We left the details to the reader. $\qquad\square$

Remark 2.5. Base change. Let S be a sheaf of sets on $\widetilde{Sm}_k$ or Sm_k, let $K \in \mathcal{F}_k$ be fixed and denote by $\pi : Spec(K) \to Spec(k)$ the structural morphism. One may pull-back S to the sheaf $S|_K := \pi^* S$ on $\widetilde{Sm}_K$ (or Sm_K accordingly). One easily checks that the sections on a separable (finite type) field extension F of K is nothing but $S(F)$ when F is viewed in $\mathcal{F}_k$. If S is unramified so is $S|_K$: indeed $\pi^* S$ is a sheaf and satisfies properties **(0)** and **(1)**. We prove **(3)** using the previous remark. $\qquad\square$

Our aim in this subsection is to give an explicit description of unramified sheaves of sets both on $\widetilde{Sm}_k$ and on Sm_k in terms of their sections on fields $F \in \mathcal{F}_k$ and some extra structure. As usual we will says that a functor $S : \mathcal{F}_k \to Set$ is *continuous* if $S(F)$ is the filtering colimit of the $S(F_\alpha)$'s, where the F_α run over the set of subfields of F of finite type over k.

We start with the simplest case, that is to say unramified sheaves of sets on $\widetilde{Sm}_k$.

Definition 2.6. An unramified $\tilde{\mathcal{F}}_k$-datum consists of:

(D1) A continuous functor $S : \mathcal{F}_k \to Set$;

(D2) For any $F \in \mathcal{F}_k$ and any discrete valuation v on F, a subset

$$S(\mathcal{O}_v) \subset S(F)$$

The previous data should satisfy the following axioms:

(A1) If $i : E \subset F$ is a separable extension in $\mathcal{F}_k$, and v is a discrete valuation on F which restricts to a discrete valuation w on E with ramification index 1 then $S(i)$ maps $S(\mathcal{O}_w)$ into $S(\mathcal{O}_v)$ and moreover if the induced extension $\bar{i} : \kappa(w) \to \kappa(v)$ is an isomorphism, then the following square of sets is cartesian:

$$
\begin{array}{ccc}
S(\mathcal{O}_w) & \to & S(\mathcal{O}_v) \\
\cap & & \cap \\
S(E) & \to & S(F)
\end{array}
$$

(A2) Let $X \in Sm_k$ be irreducible with function field F. If $x \in S(F)$, then x lies in all but a finite number of $S(\mathcal{O}_x)$'s, where x runs over the set $X^{(1)}$ of points of codimension one.

Remark 2.7. The Axiom (A1) is equivalent to the fact that for any discrete valuation v on $F \in \mathcal{F}_k$ with discrete valuation ring $\mathcal{O}_v$, then the following square in which $\mathcal{O}_v^h$ is the henselization and F^h the fraction field of $\mathcal{O}_v^h$ should be cartesian:

$$\begin{array}{ccc} \mathcal{S}(\mathcal{O}_v) & \rightarrow & \mathcal{S}(\mathcal{O}_v^h) \\ \cap & & \cap \\ \mathcal{S}(F) & \rightarrow & \mathcal{S}(F^h) \end{array}$$

□

We observe that an unramified sheaf of sets $\mathcal{S}$ on $\widetilde{Sm}_k$ defines in an obvious way an unramified $\tilde{\mathcal{F}}_k$-datum. First, evaluation on the field extensions (of finite transcendence degree) of k yields a functor:

$$\mathcal{S} : \mathcal{F}_k \rightarrow Set \ , \quad F \mapsto \mathcal{S}(F)$$

For any discrete valuation v on $F \in \mathcal{F}_k$, then $\mathcal{S}(\mathcal{O}_v)$ is a subset of $\mathcal{S}(F)$. We now claim that these data satisfy the axioms **(A1)** and **(A2)** of unramified $\tilde{\mathcal{F}}_k$-datum.

Axiom **(A1)** is easily checked by choosing convenient smooth models over k for the essentially smooth k-schemes $Spec(F)$, $Spec(\mathcal{O}_v)$. To prove axiom **(A2)** one observes that any $x \in \mathcal{S}(F)$ comes, by definition, from an element $x \in \mathcal{S}(U)$ for $U \in Sm_k$ an open subscheme of X. Thus any $\alpha \in \mathcal{S}(F)$ lies in all the $\mathcal{S}(\mathcal{O}_x)$ for $x \in X^{(1)}$ lying in U. But there are only finitely many $x \in X^{(1)}$ not lying in U.

This construction defines a "restriction" functor from the category of unramified sheaves of sets on $\widetilde{Sm}_k$ to that of unramified $\tilde{\mathcal{F}}_k$-data.

Proposition 2.8. *The restriction functor from unramified sheaves on $\widetilde{Sm}_k$ to unramified $\tilde{\mathcal{F}}_k$-data is an equivalence of categories.*

Proof. Given an unramified $\tilde{\mathcal{F}}_k$-datum $\mathcal{S}$, and $X \in Sm_k$ irreducible with function field F, we define the subset $\mathcal{S}(X) \subset \mathcal{S}(F)$ as the intersection $\bigcap_{x \in X^{(1)}} \mathcal{S}(\mathcal{O}_x) \subset \mathcal{S}(F)$. We extend it in the obvious way for X not irreducible so that property **(0)** is satisfied. Given a smooth morphism $f : Y \rightarrow X$ in Sm_k we define a map: $\mathcal{S}(f) : \mathcal{S}(X) \rightarrow \mathcal{S}(Y)$ as follows. By property **(0)** we may assume X and Y are irreducible with field of fractions E and F respectively and f is dominant. The map $\mathcal{S}(f)$ is induced by the map $\mathcal{S}(E) \rightarrow \mathcal{S}(F)$ corresponding to the fields extension $E \subset F$ and the observation that if $x \in X^{(1)}$ then $f^{-1}(x)$ is a finite set of points of codimension 1 in Y. We check that it is a sheaf in the Nisnevich topology using Axiom **(A1)** and the characterization of Nisnevich sheaves from [59]. It is unramified. Finally to check that one has constructed the inverse to the restriction functor, one uses axiom **(A2)**.   □

Definition 2.9. An unramified $\mathcal{F}_k$-datum $\mathcal{S}$ is an unramified $\tilde{\mathcal{F}}_k$-data together with the following additional data:

(D3) For any $F \in \mathcal{F}_k$ and any discrete valuation v on F, a map $s_v : \mathcal{S}(\mathcal{O}_v) \rightarrow \mathcal{S}(\kappa(v))$, called the specialization map associated to v.

These data should satisfy furthermore the following axioms:

(A3) **(i)** If $i : E \subset F$ is an extension in $\mathcal{F}_k$, and v is a discrete valuation on F which restricts to a discrete valuation w on E, then $\mathcal{S}(i)$ maps $\mathcal{S}(\mathcal{O}_w)$ to $\mathcal{S}(\mathcal{O}_v)$ and the following diagram is commutative:

$$\begin{array}{ccc} \mathcal{S}(\mathcal{O}_w) & \to & \mathcal{S}(\mathcal{O}_v) \\ \downarrow & & \downarrow \\ \mathcal{S}(\kappa(w)) & \to & \mathcal{S}(\kappa(v)) \end{array}$$

  (ii) If $i : E \subset F$ is an extension in $\mathcal{F}_k$, and v a discrete valuation on F which restricts to 0 on E then the map $\mathcal{S}(i) : \mathcal{S}(E) \to \mathcal{S}(F)$ has its image contained in $\mathcal{S}(\mathcal{O}_v)$ and if we let $j : E \subset \kappa(v)$ denotes the induced fields extension, the composition $\mathcal{S}(E) \to \mathcal{S}(\mathcal{O}_v) \overset{s_v}{\to} \mathcal{S}(\kappa(v))$ is equal to $\mathcal{S}(j)$.

(A4) **(i)** For any $X \in Sm'_k$ local of dimension 2 with closed point $z \in X^{(2)}$, and for any point $y_0 \in X^{(1)}$ with $\overline{y}_0 \in Sm'_k$, then $s_{y_0} : \mathcal{S}(\mathcal{O}_{y_0}) \to \mathcal{S}(\kappa(y_0))$ maps $\bigcap_{y \in X^{(1)}} \mathcal{S}(\mathcal{O}_y)$ into $\mathcal{S}(\mathcal{O}_{\overline{y}_0,z}) \subset \mathcal{S}(\kappa(y_0))$.
  (ii) The composition

$$\bigcap_{y \in X^{(1)}} \mathcal{S}(\mathcal{O}_y) \to \mathcal{S}(\mathcal{O}_{\overline{y}_0,z}) \to \mathcal{S}(\kappa(z))$$

doesn't depend on the choice of y_0 such that $\overline{y}_0 \in Sm'_k$.

Remark 2.10. When we will construct unramified Milnor-Witt K-theory in Sect. 3.2 below, the axiom (**A4**) will appear to be the most difficult to check. In fact the Sect. 2.3 is devoted to develop some technic to check this axiom in special cases. In Rost's approach [68] this axiom follows from the construction of the Rost's complex for two-dimensional local smooth k-scheme. However the construction of this complex (even for dimension 2 schemes) requires transfers, which we don't want to use at this point.   □

Now we claim that an unramified sheaf of sets $\mathcal{S}$ on Sm_k defines an unramified $\mathcal{F}_k$-datum. From what we have done before, we already have in hand an unramified $\tilde{\mathcal{F}}_k$-datum $\mathcal{S}$. Now, for any discrete valuation v on $F \in \mathcal{F}_k$ with residue field $\kappa(v)$, there is an obvious map $s_v : \mathcal{S}(\mathcal{O}_v) \to \mathcal{S}(\kappa(v))$, obtained by choosing smooth models over k for the closed immersion $Spec(\kappa(v)) \to Spec(\mathcal{O}_v)$. This defines the datum (**D3**). We now claim that these data satisfy the previous axioms for unramified $\mathcal{F}_k$-datum. Axiom (**A3**) is checked by choosing convenient smooth models for $Spec(F)$, $Spec(\mathcal{O}_v)$ and/or $Spec(\kappa(v))$.

To check the axiom (**A4**) we use property (**2**) and the commutative square:

$$\begin{array}{ccc} \mathcal{S}(X) & \subset & \mathcal{S}(\mathcal{O}_{y_0}) \\ \downarrow & & \downarrow \\ \mathcal{S}(\overline{y}_0) = \mathcal{S}(\mathcal{O}_z) & \subset & \mathcal{S}(\kappa(y_0)) \end{array}$$

The following now is the main result of this section:

Theorem 2.11. *The functor just constructed from unramified sheaves of sets on Sm_k to unramified $\mathcal{F}_k$-data is an equivalence of categories.*

The Theorem follows from the following more precise statement:

Lemma 2.12. *Given an unramified $\mathcal{F}_k$-datum $\mathcal{S}$, there is a unique way to extend the unramified sheaf of sets $S : (\widetilde{Sm_k})^{op} \to Set$ to a sheaf $S : (Sm_k)^{op} \to Set$, such that for any discrete valuation v on $F \in \mathcal{F}_k$ with separable residue field, the map $S(\mathcal{O}_v) \to S(\kappa(v))$ induced by the sheaf structure is the specialization map $s_v : S(\mathcal{O}_v) \to S(\kappa(v))$. This sheaf is automatically unramified.*

Proof. We first define a restriction map $s(i) : \mathcal{S}(X) \to \mathcal{S}(Y)$ for a closed immersion $i : Y \hookrightarrow X$ in Sm_k of codimension 1. If $Y = \amalg_\alpha Y_\alpha$ is the decomposition of Y into irreducible components then $\mathcal{S}(Y) = \Pi_\alpha \mathcal{S}(Y_\alpha)$ and $s(i)$ has to be the product of the $s(i_\alpha) : \mathcal{S}(X) \to \mathcal{S}(Y_\alpha)$. We thus may assume Y (and X) irreducible. We then claim there exits a (unique) map $s(i) : \mathcal{S}(X) \to \mathcal{S}(Y)$ which makes the following diagram commute

$$
\begin{array}{ccc}
\mathcal{S}(X) & \overset{s(i)}{\to} & \mathcal{S}(Y) \\
\cap & & \cap \\
\mathcal{S}(\mathcal{O}_{X,y}) & \overset{s_y}{\to} & \mathcal{S}(\kappa(y))
\end{array}
$$

where y is the generic point of Y. To check this it is sufficient to prove that for any $z \in Y^{(1)}$, the image of $\mathcal{S}(X)$ through s_y is contained in $\mathcal{S}(\mathcal{O}_{Y,z})$. But z has codimension 2 in X and this follows from the first part of axiom **(A4)**.

Now we have the following:

Lemma 2.13. *Let $i : Z \to X$ be a closed immersion in Sm_k of codimension $d > 0$. Assume there exists a factorization $Z \overset{j_1}{\to} Y_1 \overset{j_2}{\to} Y_2 \to \cdots \overset{j_d}{\to} Y_d = X$ of i into a composition of codimension 1 closed immersions, with the Y_i closed subschemes of X each of which is smooth over k. Then the composition*

$$
\mathcal{S}(X) \overset{s(j_d)}{\to} \cdots \to \mathcal{S}(Y_2) \overset{s(j_2)}{\to} \mathcal{S}(Y_1) \overset{s(j_1)}{\to} \mathcal{S}(Z)
$$

doesn't depend on the choice of the above factorization of i. We denote this composition by $\mathcal{S}(i)$.

Proof. We proceed by induction on d. For $d = 1$ there is nothing to prove. Assume $d \geq 2$. We may easily reduce to the case Z is irreducible with generic point z. We have to show that the composition

$$
\mathcal{S}(X) \overset{s(j_d)}{\to} \cdots \to \mathcal{S}(Y_2) \overset{s(j_2)}{\to} \mathcal{S}(Y_1) \overset{s(j_1)}{\to} \mathcal{S}(Z) \subset \mathcal{S}(\kappa(z))
$$

doesn't depend on the choice of the flag $Z \to Y_1 \to \cdots \to \ldots \to X$. We may thus replace X by any open neighborhood Ω of z if we want or even by $Spec(A)$ with $A := \mathcal{O}_{X,z}$, which we do.

We first observe that the case $d = 2$ follows directly from the Axiom **(A4)**.

In general as A is regular of dimension d there exists a sequence of elements $(x_1, \ldots, x_d) \in A$ which generates the maximal ideal $\mathcal{M}$ of A and such that the flag

$$Spec(A/(x_1, \ldots, x_d)) \to Spec(A/(x_2, \ldots, x_d)) \to \cdots \to Spec(A/(x_d))$$
$$\to Spec(A)$$

is the induced flag $Z = Spec(\kappa(z)) \subset Y_1 \subset Y_2 \subset \cdots \subset Spec(A)$.

We have thus reduced to proving that under the above assumptions the composition

$$\mathcal{S}(A) \to \mathcal{S}(Spec(A/(x_d))) \to \cdots \to \mathcal{S}(Spec(A/(x_2, \ldots, x_d)) \to \mathcal{S}(\kappa(z))$$

doesn't depend on the choice of $(x_1, \ldots, x_d)$.

By [30, Corollary **(17.12.2)**] the conditions on smoothness on the members of the associated flag to the sequence $(x_1, \ldots, x_d)$ is equivalent to the fact the family $(x_1, \ldots, x_d)$ reduces to a basis of the $\kappa(z)$-vector space $\mathcal{M}/\mathcal{M}^2$.

If $M \in GL_d(A)$, the sequence $M.(x_i)$ also satisfies this assumption. For instance any permutation on the $(x_1, \ldots, x_d)$ yields an other such sequence. By the case $d = 2$ which was observed above, we see that if we permute x_i and x_{i+1} the compositions $\mathcal{S}(A) \to \mathcal{S}(\kappa(v))$ are the same before or after permutation. We thus get by induction that we may permute as we wish the x_i's.

Now assume that $(x'_1, \ldots, x'_d)$ is an other sequence in A satisfying the same assumption. Write the x'_i as linear combination in the x_j. There is a matrix $M \in M_d(A)$ with $(x'_i) = M.(x_j)$. This matrix reduces in $M_d(\kappa)$ to an invertible matrix by what we just observed above; thus M itself is invertible. One may multiply in a sequence $(x_1, \ldots, x_d)$ by a unit of A an element x_i of the sequence without changing the flag (and thus the composition). Thus we may assume $det(M) = 1$. Now for a local ring A we know that the group $SL_d(A)$ is the group $E_d(A)$ of elementary matrices in A (see [39, Chap. VI Corollary 1.5.3] for instance). Thus M can be written as a product of elementary matrices in $M_d(A)$.

As we already know that our statement doesn't depend on the ordering of a sequence, we have reduced to the following claim: given a sequence $(x_1, \ldots, x_d)$ as above and $a \in A$, the sequence $(x_1 + ax_2, x_2, \ldots, x_d)$ induces the same composition $\mathcal{S}(A) \to \mathcal{S}(\kappa(v))$ as $(x_1, \ldots, x_d)$. But in fact the flags are the same. This proves our claim. $\square$

Now we come back to the proof of the Lemma 2.12. Let $i : Z \to X$ be a closed immersion in Sm_k. By what has been recalled above, X can be covered by open subsets U such that for every U the induced closed immersion $Z \cap U \to U$ admits a factorization as in the statement of the previous Lemma 2.13. Thus for each such U we get a canonical map $s_U : S(U) \to S(Z \cap U)$. But applying the same Lemma to the intersections $U \cap U'$, with U' an other such open subset, we see that the s_U are compatible and define a canonical map: $s(i) : S(X) \to S(Z)$.

Let $f : Y \to X$ be any morphism between smooth (quasi-projective) k-schemes. Then f is the composition $Y \hookrightarrow Y \times_k X \to X$ of the closed immersion (given by the graph of f) $\Gamma_f : Y \hookrightarrow Y \times_k X$ and the smooth projection $p_X : Y \times_k X \to X$. We set

$$s(f) := S(X) \xrightarrow{s(p_X)} S(Y \times_k X) \xrightarrow{s(\Gamma_f)} S(Y)$$

To check that this defines a functor on $(Sm_k)^{op}$ is not hard. First given a smooth morphism $\pi : X' \to X$ and a closed immersion $i : Z \to X$ in Sm_k, denote by $i'' : Z' \to X'$ the inverse image of i through π and by $\pi' : Z' \to Z$ the obvious smooth morphism. Then the following diagram is commutative

$$\begin{array}{ccc} S(X) & \xrightarrow{s(\pi)} & S(X') \\ \downarrow s(i) & & \downarrow s(i') \\ S(Z) & \xrightarrow{s(\pi')} & S(Z') \end{array}$$

Then, to prove the functoriality, one takes two composable morphism $Z \xrightarrow{g} Y \xrightarrow{f} X$ and contemplates the diagram

$$\begin{array}{ccc} Z \hookrightarrow Z \times_k Y & \hookrightarrow & Z \times_k Y \times_k X \\ \| \qquad\quad \downarrow & & \downarrow \\ Z \to \quad Y & \hookrightarrow & Y \times_k X \\ \| \qquad\quad \| & & \downarrow \\ Z \to \quad Y & \to & X \end{array}$$

Then one realizes that applying S and s yields a commutative diagram, proving the claim. Now the presheaf S on Sm_k is obviously an unramified sheaf on Sm_k as these properties only depend on its restriction to $\widetilde{Sm}_k$. □

Remark 2.14. From now on in this paper, we will not distinguish between the notion of unramified $\mathcal{F}_k$-datum and that of unramified sheaf of sets on Sm_k. If S is an unramified $\mathcal{F}_k$-datum we still denote by S the associated unramified sheaf of sets on Sm_k and vice versa.

Also, one may in an obvious fashion describe unramified sheaves of groups, abelian groups, etc. on Sm_k in terms of corresponding $\mathcal{F}_k$-group data, $\mathcal{F}_k$-abelian group data, etc., where in the given $\mathcal{F}_k$-datum, everything is endowed with the corresponding structure and each map is a morphism for that structure. □

Remark 2.15. The proof of Lemma 2.12 also shows the following. Let $\mathcal{S}$ and $\mathcal{E}$ be sheaves of sets on Sm_k, with $\mathcal{S}$ unramified and $\mathcal{E}$ satisfying conditions **(0)** and **(1)** of unramified presheaves. Then to give a morphism of sheaves $\Phi : \mathcal{E} \to \mathcal{S}$ is equivalent to give a natural transformation $\phi : \mathcal{E}|_{\mathcal{F}_k} \to \mathcal{S}|_{\mathcal{F}_k}$ such that:

1) For any discrete valuation v on $F \in \mathcal{F}_k$, the image of $\mathcal{E}(\mathcal{O}_v) \subset \mathcal{E}(F)$ through ϕ is contained in $\mathcal{S}(\mathcal{O}_v) \subset \mathcal{S}(F)$;
2) The induced square commutes:

$$\begin{array}{ccc} \mathcal{E}(\mathcal{O}_v) & \xrightarrow{s_v} & \mathcal{E}(\kappa(v)) \\ \downarrow \phi & & \downarrow \phi \\ \mathcal{S}(\mathcal{O}_v) & \to & \mathcal{S}(\kappa(v)) \end{array}$$

We left the details to the reader.   □

$\mathbb{A}^1$-Invariant Unramified Sheaves

Lemma 2.16. *1) Let $\mathcal{S}$ be an unramified sheaf of sets on $\widetilde{Sm}_k$. Then $\mathcal{S}$ is $\mathbb{A}^1$-invariant if and only if it satisfies the following:*

For any k-smooth local ring A of dimension ≤ 1 the canonical map $\mathcal{S}(A) \to \mathcal{S}(\mathbb{A}^1_A)$ is bijective.

2) Let $\mathcal{S}$ be an unramified sheaf of sets on Sm_k. Then $\mathcal{S}$ is $\mathbb{A}^1$-invariant if and only if it satisfies the following:

For any $F \in \mathcal{F}_k$ the canonical map $\mathcal{S}(F) \to \mathcal{S}(\mathbb{A}^1_F)$ is bijective.

Proof. 1) One implication is clear. Let's prove the other one. Let $X \in Sm_k$ be irreducible with function field F. In the following commutative square

$$\begin{array}{ccc} \mathcal{S}(X) & \to & \mathcal{S}(\mathbb{A}^1_X) \\ \downarrow & & \downarrow \\ \mathcal{S}(F) & \to & \mathcal{S}(F(T)) \end{array}$$

each map is injective. We observe that $\mathcal{S}(\mathbb{A}^1_X) \to \mathcal{S}(F(T)$ factors as $\mathcal{S}(\mathbb{A}^1_X) \to \mathcal{S}(\mathbb{A}^1_F) \to \mathcal{S}(F(T))$. By our assumption $\mathcal{S}(F) = \mathcal{S}(\mathbb{A}^1_F)$; this proves that $\mathcal{S}(\mathbb{A}^1_X)$ is contained inside $\mathcal{S}(F)$. Now it is sufficient to prove that for any $x \in X^{(1)}$ one has the inclusion $\mathcal{S}(\mathbb{A}^1_X) \subset \mathcal{S}(\mathcal{O}_{X,x}) \subset \mathcal{S}(F)$. But $\mathcal{S}(\mathbb{A}^1_X) \subset \mathcal{S}(\mathbb{A}^1_{\mathcal{O}_{X,x}}) \subset \mathcal{S}(F(T)$, and our assumption gives $\mathcal{S}(\mathcal{O}_{X,x}) = \mathcal{S}(\mathbb{A}^1_{\mathcal{O}_{X,x}})$. This proves the claim.

2) One implication is clear. Let's prove the other one. Let $X \in Sm_k$ be irreducible with function field F. In the following commutative square

$$\begin{array}{ccc} \mathcal{S}(\mathbb{A}^1_X) & \subset & \mathcal{S}(\mathbb{A}^1_F) \\ \downarrow & & \| \\ \mathcal{S}(X) & \subset & \mathcal{S}(F) \end{array}$$

each map is injective but maybe the left vertical one. The latter is thus also injective which implies the statement.   □

Remark 2.17. Given an unramified sheaf $\mathcal{S}$ of sets on $\tilde{Sm}_k$ with Data **(D3)**, and satisfying the property that for any $F \in \mathcal{F}_k$, the map $\mathcal{S}(F) \to \mathcal{S}(F(T))$ is injective, then $\mathcal{S}$ is an unramified $\mathcal{F}_k$-datum if and only if its extension to $k(T)$ is an unramified $\mathcal{F}_{k(T)}$-datum.

Indeed, given a smooth irreducible k-scheme X, a point $x \in X$ of codimension d, then $X|_{k(T)}$ is still irreducible $k(T)$-smooth and $\overline{x}|_{k(T)}$ is irreducible and has codimension d in $X|_{k(T)}$. Moreover the maps $M(X) \to M(X|_{k(T)})$, $M(X_x) \to M((X|_{k(T)})_{\overline{x}|_{k(T)}})$, etc. are injective. So to check the Axioms involving equality between morphisms, etc., it suffices to check them over $k(T)$ for $M|_{k(T)}$. This allows us to reduce the checking of several Axioms like **(A4)** to the case k is infinite.   □

2.2 Strongly $\mathbb{A}^1$-Invariant Sheaves of Groups

Our aim in this section is to study unramified sheaves of groups $\mathcal{G}$ on Sm_k, their potential strong $\mathbb{A}^1$-invariance property, as well as the comparison between their cohomology in Zariski and Nisnevich topology.

In the sequel, by an unramified sheaf of groups we mean a sheaf of groups on Sm_k whose underlying sheaf of sets is unramified in the sense of the previous section.

Let $\mathcal{G}$ be such an unramified sheaf of groups on Sm_k. For any discrete valuation v on $F \in \mathcal{F}_k$ we introduce the pointed set

$$H_v^1(\mathcal{O}_v; \mathcal{G}) := \mathcal{G}(F)/\mathcal{G}(\mathcal{O}_v)$$

and we observe this is a left $\mathcal{G}(F)$-set.

More generally for y a point of codimension 1 in $X \in Sm_k'$, we set $H_y^1(X; \mathcal{G}) = H_y^1(\mathcal{O}_{X,y}; \mathcal{G})$. By axiom **(A2)**, is X is irreducible with function field F the induced left action of $\mathcal{G}(F)$ on $\Pi_{y \in X^{(1)}} H_y^1(X; \mathcal{G})$ preserves the weak-product

$$\Pi'_{y \in X^{(1)}} H_y^1(X; \mathcal{G}) \subset \Pi_{y \in X^{(1)}} H_y^1(X; \mathcal{G})$$

where the weak-product $\Pi'_{y \in X^{(1)}} H_y^1(X; \mathcal{G})$ means the set of families for which all but a finite number of terms are the base point of $H_y^1(X; \mathcal{G})$. By definition, the isotropy subgroup of this action of $\mathcal{G}(F)$ on the base point of $\Pi'_{y \in X^{(1)}} H_y^1(X; \mathcal{G})$ is exactly $\mathcal{G}(X) = \cap_{y \in X^{(1)}} \mathcal{G}(\mathcal{O}_{X,y})$. We will summarize this property by saying that the diagram (of groups, action and pointed set)

$$1 \to \mathcal{G}(X) \to \mathcal{G}(F) \Rightarrow \Pi'_{y \in X^{(1)}} H_y^1(X; \mathcal{G})$$

is "exact" (the double arrow refereing to a left action).

Definition 2.18. For any point z of codimension 2 in a smooth k-scheme X, we denote by $H_z^2(X;\mathcal{G})$ the orbit set of $\Pi'_{y \in X_z^{(1)}} H_y^1(X;\mathcal{G})$ under the left action of $\mathcal{G}(F)$, where $F \in \mathcal{F}_k$ denotes the field of functions of X_z.

Now for an irreducible essentially smooth k-scheme X with function field F we may define an obvious "boundary" $\mathcal{G}(F)$-equivariant map

$$\Pi'_{y \in X^{(1)}} H_y^1(X;\mathcal{G}) \to \Pi_{z \in X^{(2)}} H_z^2(X;\mathcal{G}) \qquad (2.1)$$

by collecting together the compositions, for each $z \in X^{(2)}$:

$$\Pi'_{y \in X^{(1)}} H_y^1(X;\mathcal{G}) \to \Pi'_{y \in X_z^{(1)}} H_y^1(X;\mathcal{G}) \to H_z^2(X;\mathcal{G})$$

It is not clear in general whether or not the image of the boundary map is always contained in the weak product $\Pi'_{z \in X^{(2)}} H_z^2(X;\mathcal{G})$. For this reason we will introduce the following Axiom depending on $\mathcal{G}$ which completes (**A2**):

(**A2'**) For any irreducible essentially smooth k-scheme X the image of the boundary map (2.1) is contained in the weak product $\Pi'_{z \in X^{(2)}} H_z^2(X;\mathcal{G})$. $\square$

Remark 2.19. Given an unramified sheaf of groups $\mathcal{G}$, and satisfying the property that for any $F \in \mathcal{F}_k$, the map $\mathcal{G}(F) \to \mathcal{G}(F(T))$ is injective, then $\mathcal{G}$ satisfies (**A2'**) if and only if its extension to $k(T)$ does. This is done along the same lines as in Remark 2.17. $\square$

We assume from now on that $\mathcal{G}$ satisfies (**A2'**). Altogether we get for X smooth over k, irreducible with function field F, a "complex" $C^*(X;\mathcal{G})$ of groups, action, and pointed sets of the form:

$$1 \to \mathcal{G}(X) \subset \mathcal{G}(F) \Rightarrow \Pi'_{y \in X^{(1)}} H_y^1(X;\mathcal{G}) \to \Pi_{z \in X^{(2)}} H_z^2(X;\mathcal{G})$$

We will also set for $X \in Sm_k$: $\mathcal{G}^{(0)}(X) := \Pi'_{x \in X^{(0)}} \mathcal{G}(\kappa(x))$, $\mathcal{G}^{(1)}(X) := \Pi'_{y \in X^{(1)}} H_y^1(X;\mathcal{G})$ and $\mathcal{G}^{(2)}(X) := \Pi'_{z \in X^{(2)}} H_z^2(X;\mathcal{G})$. The correspondence $X \mapsto \mathcal{G}^{(i)}(X)$, $i \leq 2$, can be extended to an unramified presheaf of groups on $\tilde{Sm}_k$, which we still denote by $\mathcal{G}^{(i)}$. Note that $\mathcal{G}^{(0)}$ is a sheaf in the Nisnevich topology. However for $\mathcal{G}^{(i)}$, $i \in \{1,2\}$ it is not the case in general, these are only sheaves in the Zariski topology, as any unramified presheaf.

The complex $C^*(X;\mathcal{G}) : 1 \to \mathcal{G}(X) \to \mathcal{G}^{(0)}(X) \Rightarrow \mathcal{G}^{(1)}(X) \to \mathcal{G}^{(2)}(X)$ of sheaves on $\tilde{Sm}_k$ will play in the sequel the role of the (truncated) analogue for $\mathcal{G}$ of the Cousin complex of [19] or of the complex of Rost considered in [68].

Definition 2.20. Let $1 \to H \subset G \Rightarrow E \to F$ be a sequence with G a group acting on the set E which is pointed (as a set not as a G-set), with $H \subset G$ a subgroup and $E \to F$ a G-equivariant map of sets, with F endowed with the trivial action. We shall say this sequence is exact if the isotropy subgroup of the base point of E is H and if the "kernel" of the pointed map $E \to F$ is equal to the orbit under G of the base point of E.

We shall say that it is exact in the strong sense if moreover the map $E \to F$ induces an injection into F of the (left) quotient set $_{G} \backslash E \subset F$.

By construction $C^*(X; \mathcal{G})$ is exact in the strong sense, for X (essentially) smooth local of dimension ≤ 2.

Let us denote by $\mathcal{Z}^1(-; \mathcal{G}) \subset \mathcal{G}^{(1)}$ the sheaf theoretic orbit of the base point under the action of $\mathcal{G}^{(0)}$ in the Zariski topology on $\tilde{Sm}_k$. We thus have an exact sequence of sheaves on $\tilde{Sm}_k$ in the Zariski topology

$$1 \to \mathcal{G} \subset \mathcal{G}^{(0)} \Rightarrow \mathcal{Z}^1(-; \mathcal{G}) \to *$$

As it is clear that $H^1_{Zar}(X; \mathcal{G}^{(0)})$ is trivial (the sheaf $\mathcal{G}^{(0)}$ being flasque), this yields for any $X \in Sm_k$ an exact sequence (of groups and pointed sets)

$$1 \to \mathcal{G}(X) \subset \mathcal{G}^{(0)}(X) \Rightarrow \mathcal{Z}^1(X; \mathcal{G}) \to H^1_{Zar}(X; \mathcal{G}) \to *$$

in the strong sense.

Of course we may extend by passing to the filtering colimit the previous definitions for $X \in Sm'_k$. To be correct, we should introduce a name for the category of essentially smooth k-schemes and smooth morphisms. The previous diagram is then also a diagram of sheaves in the Zariski topology and yields for any $X \in Sm'_k$ an exact sequence as above, which could have been also obtained by passing to the colimit.

Remark 2.21. If X is an essentially smooth k-scheme of dimension ≤ 1, we thus get a bijection $H^1_{Zar}(X; \mathcal{G}) = {}_{\mathcal{G}^{(0)}(X)} \backslash \mathcal{G}^{(1)}(X)$. For instance, when X is a smooth local k-scheme of dimension 2, and if $V \subset X$ is the complement of the closed point, a smooth k-scheme of dimension 1, we thus get a bijection

$$H^2_z(X; \mathcal{G}) = H^1_{Zar}(V; \mathcal{G})$$

Beware that here the Zariski topology is used. This gives a "concrete" interpretation of the "strange" extra cohomology set $H^2_z(X; \mathcal{G})$.   □

For $X \in Sm_k$ (or Sm'_k) as above, let us denote by $\mathcal{K}^1(X; \mathcal{G}) \subset \Pi'_{y \in X^{(1)}} H^1_y(X; \mathcal{G})$ the kernel of the boundary map $\Pi'_{y \in X^{(1)}} H^1_y(X; \mathcal{G}) \to \Pi'_{z \in X^{(2)}} H^2_z(X; \mathcal{G})$. The correspondence $X \mapsto \mathcal{K}^1(X; \mathcal{G})$ is a sheaf in the Zariski topology on $\tilde{Sm}_k$. There is an obvious injective morphism of sheaves in the Zariski topology on $\tilde{Sm}_k$: $\mathcal{Z}^1(-; \mathcal{G}) \to \mathcal{K}^1(-; \mathcal{G})$. As $C^*(X; \mathcal{G})$ is exact for any

k-smooth local X of dimension ≤ 2, $\mathcal{Z}^1(-;\mathcal{G}) \to \mathcal{K}^1(-;\mathcal{G})$ induces a bijection for any (essentially) smooth k-scheme of dimension ≤ 2.

Remark 2.22. In particular if X is an (essentially) smooth k-scheme of dimension ≤ 2, the H^1 of the complex $C^*(X;\mathcal{G})$ is $H^1_{Zar}(X;\mathcal{G})$.   □

Now we introduce the following axiom on $\mathcal{G}$:

(A5) **(i)** For any separable finite extension $E \subset F$ in $\mathcal{F}_k$, any discrete valuation v on F which restricts to a discrete valuation w on E with ramification index 1, and such that the induced extension $\bar{i} : \kappa(w) \to \kappa(v)$ is an isomorphism, the commutative square of groups

$$\begin{array}{ccc} \mathcal{G}(\mathcal{O}_w) & \subset & \mathcal{G}(E) \\ \downarrow & & \downarrow \\ \mathcal{G}(\mathcal{O}_v) & \subset & \mathcal{G}(F) \end{array}$$

induces a bijection $H^1_v(\mathcal{O}_v;\mathcal{G}) \cong H^1_w(\mathcal{O}_w;\mathcal{G})$.

(ii) For any étale morphism $X' \to X$ between smooth local k-schemes of dimension 2, with closed point respectively z' and z, inducing an isomorphism on the residue fields $\kappa(z) \cong \kappa(z')$, the pointed map

$$H^2_z(X;\mathcal{G}) \to H^2_{z'}(X';\mathcal{G})$$

has trivial kernel.   □

Remark 2.23. The Axiom **(A5)(i)** implies that if we denote by $\mathcal{G}_{-1}$ the sheaf of groups

$$X \mapsto Ker(G(\mathbb{G}_m \times X) \overset{ev_1}{\to} G(X))$$

then for any discrete valuation v on $F \in \mathcal{F}_k$ one has a (non canonical) bijection

$$H^1_v(\mathcal{O}_v;\mathcal{G}) \cong \mathcal{G}_{-1}(\kappa(v))$$

Indeed one may reduce to the case where $\mathcal{O}_v$ is henselian, and assume that $\kappa(v) \subset \mathcal{O}_v$. Choosing a uniformizing element then yields a distinguished square

$$\begin{array}{ccc} Spec(F) & \subset & Spec(\mathcal{O}_v) \\ \downarrow & & \downarrow \\ (\mathbb{G}_m)_{\kappa(v)} & \subset & \mathbb{A}^1_{\kappa(v)} \end{array}$$

which in view of Axiom **(A5) (i)** gives the bijection $\mathcal{G}((\mathbb{G}_m)_{\kappa(v)})/\mathcal{G}(\kappa(v)) \cong H^1_v(\mathcal{O}_v;\mathcal{G})$.   □

Lemma 2.24. *Let $\mathcal{G}$ be as above. The following conditions are equivalent:*

(i) *The Zariski sheaf $X \mapsto \mathcal{K}^1(X;\mathcal{G})$ is a sheaf in the Nisnevich topology on $\tilde{Sm}_k$;*

(ii) *For any essentially smooth k-scheme X of dimension ≤ 2 the compari-son map $H^1_{Zar}(X;\mathcal{G}) \to H^1_{Nis}(X;\mathcal{G})$ is a bijection;*

(iii) *$\mathcal{G}$ satisfies Axiom (A5)*

Proof. (i) $\Rightarrow$ (ii). Under (i) we know that $X \mapsto \mathcal{Z}^1(X;\mathcal{G})$ is a sheaf in the Nisnevich topology on essentially smooth k-schemes of dimension ≤ 2 (as $\mathcal{Z}^1(X;\mathcal{G}) \to \mathcal{K}^1(X;\mathcal{G})$ is an isomorphism on essentially smooth k-schemes of dimension ≤ 2). The exact sequence in the Zariski topology $1 \to \mathcal{G} \subset \mathcal{G}^{(0)} \Rightarrow \mathcal{Z}^1(-;\mathcal{G}) \to *$ considered above is then also an exact sequence of sheaves in the Nisnevich topology. The same reasoning as above easily implies (ii), taking into account that $H^1_{Nis}(X;\mathcal{G}^{(0)})$ is also trivial (left to the reader).

(ii) $\Rightarrow$ (iii). Assume (ii). Let's prove (A5) (i). With the assumptions given the square

$$
\begin{array}{ccc}
Spec(F) & \to & Spec(\mathcal{O}_v) \\
\downarrow & & \downarrow \\
Spec(E) & \to & Spec(\mathcal{O}_w)
\end{array}
$$

is a distinguished square in the sense of [59]. Using the corresponding Mayer–Vietoris type exact sequence and the fact by (ii) that $H^1(X;\mathcal{G}) = *$ for any smooth local scheme X yields immediately that $\mathcal{G}(E)/\mathcal{G}(\mathcal{O}_w) \to \mathcal{G}(F)/\mathcal{G}(\mathcal{O}_v)$ is a bijection.

Now let's prove (A5) (ii). Set $V = X - \{z\}$ and $V' = X' - \{z'\}$. The square

$$
\begin{array}{ccc}
V' & \subset & X' \\
\downarrow & & \downarrow \\
V & \subset & X
\end{array}
$$

is distinguished. From the discussion preceding the Lemma and the inter-pretation of $H^2_z(X;\mathcal{G})$ as $H^1_{Zar}(V;\mathcal{G})$, the kernel in question is thus the set of (isomorphism classes) of $\mathcal{G}$-torsors over V (indifferently in Zariski and Nisnevich topology as $H^1_{Zar}(V;\mathcal{G}) \cong H^1_{Nis}(V;\mathcal{G})$ by (ii)) which become trivial over V'; but such a torsor can thus be extended to X' and by a descent argument in the Nisnevich topology, we may extend the torsor on V to X. Thus it is trivial because X is local.

(iii) $\Rightarrow$ (i). Now assume Axiom (A5). We claim that Axiom (A5) (i) gives exactly that $X \mapsto \mathcal{G}^{(1)}(X)$ is a sheaf in the Nisnevich topology. (A5) (ii) is seen to be exactly what is needed to imply that $\mathcal{K}^1(-;\mathcal{G})$ is a sheaf in the Nisnevich topology. $\square$

Now we observe that the monomorphism of Zariski sheaves $\mathcal{Z}^1(-;\mathcal{G}) \to \mathcal{K}^1(-;\mathcal{G})$ is $\mathcal{G}^{(0)}$-equivariant.

Lemma 2.25. *Assume $\mathcal{G}$ satisfies (A5). Let X be an essentially smooth k-scheme. The following conditions are equivalent:*

(i) *For any open subscheme $\Omega \subset X$ the map $\mathcal{Z}^1(\Omega;\mathcal{G}) \to \mathcal{K}^1(\Omega;\mathcal{G})$ is bijective;*

(ii) *For any localization U of X at some point, the map $\mathcal{Z}^1(U;\mathcal{G}) \to \mathcal{K}^1(U;\mathcal{G})$ is bijective;*

(iii) *For any localization U of X at some point, the complex $C^*(U;\mathcal{G}) : 1 \to \mathcal{G}(U) \to \mathcal{G}(F) \Rightarrow \mathcal{G}^{(1)}(U) \to \mathcal{G}^{(2)}(U)$ is exact.*

When moreover these conditions are satisfied for any Y étale over X, then the comparison map $H^1_{Zar}(X;\mathcal{G}) \to H^1_{Nis}(X;\mathcal{G})$ is a bijection.

Proof. (i) $\Leftrightarrow$ (ii) is clear as both are Zariski sheaves. (ii) $\Rightarrow$ (iii) is proven exactly as in the proof of (ii) in Lemma 2.24. (iii) $\Rightarrow$ (i) is also clear using the given expressions of the two sides.

If we assume these conditions are satisfied, then

$$\mathcal{G}^{(0)}(X) \backslash \mathcal{Z}^1(X;\mathcal{G}) = H^1_{Zar}(X;\mathcal{G}) \to H^1_{Nis}(X;\mathcal{G}) = \mathcal{G}^{(0)}(X) \backslash \mathcal{K}^1(X;\mathcal{G})$$

is a bijection. The last equality follows from the fact that $\mathcal{K}^1(;\mathcal{G})$ is a Nisnevich sheaf and the (easy) fact that $H^1_{Nis}(X;\mathcal{G}^{(0)})$ is also trivial. $\square$

Lemma 2.26. *Assume $\mathcal{G}$ is $\mathbb{A}^1$-invariant. Let X be an essentially smooth k-scheme. The following conditions are equivalent:*

(i) *For any open subscheme $\Omega \subset X$ the map*

$$\mathcal{G}^{(0)}(\Omega) \backslash \mathcal{Z}^1(\Omega;\mathcal{G}) = H^1_{Zar}(\Omega;\mathcal{G}) \to H^1_{Zar}(\mathbb{A}^1_\Omega;\mathcal{G}) = \mathcal{G}^{(0)}(\mathbb{A}^1_\Omega) \backslash \mathcal{Z}^1(\mathbb{A}^1_\Omega;\mathcal{G})$$

is bijective;

(ii) *For any localization U of X, $\mathcal{G}^{(0)}(\mathbb{A}^1_U) \backslash \mathcal{Z}^1(\mathbb{A}^1_U;\mathcal{G}) = *$.*

Proof. The implication (i) $\Rightarrow$ (ii) follows from the fact that for U a smooth local k-scheme $H^1_{Zar}(U;\mathcal{G}) = \mathcal{G}^{(0)}(U) \backslash \mathcal{Z}^1(U;\mathcal{G})$ is trivial. Assume (ii). Thus $H^1_{Zar}(\mathbb{A}^1_U;\mathcal{G}) = *$ for any local smooth k-scheme U. Fix $\Omega \subset X$ an open subscheme and denote by $\pi : \mathbb{A}^1_\Omega \to \Omega$ the projection. To prove (i) it is sufficient to prove that the pointed simplicial sheaf of sets $R\pi_*(B(\mathcal{G}|_{\mathbb{A}^1_\Omega}))$ has trivial π_0. Indeed, its π_1 sheaf is $\pi_*(\mathcal{G}|_{\mathbb{A}^1_\Omega}) = \mathcal{G}|_\Omega$ because $\mathcal{G}$ is $\mathbb{A}^1$-invariant. If the π_0 is trivial, $B(\mathcal{G}|_\Omega) \to R\pi_*(B(\mathcal{G}|_{\mathbb{A}^1_\Omega}))$ is a simplicial weak equivalence which implies the result. Now to prove that $\pi_0 R\pi_*((B(\mathcal{G}|_{\mathbb{A}^1_\Omega})))$ is trivial, we just observe that its stalk at a point $x \in \Omega$ is $H^1_{Zar}(\mathbb{A}^1_{X_x};\mathcal{G})$ which is trivial by assumption. $\square$

Now we will use one more Axiom concerning $\mathcal{G}$ and related to $\mathbb{A}^1$-invariance properties:

(A6) For any localization U of a smooth k-scheme at some point u of codimension ≤ 1, the "complex":

$$1 \to \mathcal{G}(\mathbb{A}^1_U) \subset \mathcal{G}^{(0)}(\mathbb{A}^1_U) \Rightarrow \mathcal{G}^{(1)}(\mathbb{A}^1_U) \to \mathcal{G}^{(2)}(\mathbb{A}^1_U)$$

is exact, and moreover, the morphism $\mathcal{G}(U) \to \mathcal{G}(\mathbb{A}^1_U)$ is an isomorphism. $\square$

Observe that if $\mathcal{G}$ satisfies (**A6**) it is $\mathbb{A}^1$-invariant by Lemma 2.16 (as $\mathcal{G}$ is assumed to be unramified). Observe also that if $\mathcal{G}$ satisfies Axioms (**A2'**) and (**A5**), then we know by Lemma 2.24 that $H^1_{Nis}(\mathbb{A}^1_X; \mathcal{G}) = H^1_{Zar}(\mathbb{A}^1_X; \mathcal{G}) = H^1(\mathbb{A}^1_X; \mathcal{G})$ for X smooth of dimension ≤ 1.

Our main result in this section is the following.

Theorem 2.27. *Let $\mathcal{G}$ be an unramified sheaf of groups on Sm_k that satisfies Axioms (**A2'**), (**A5**) and (**A6**). Then it is strongly $\mathbb{A}^1$-invariant. Moreover, for any smooth k-scheme X, the comparison map*

$$H^1_{Zar}(X; \mathcal{G}) \to H^1_{Nis}(X; \mathcal{G})$$

is a bijection.

Remark 2.28. From Corollary 6.9 in Sect. 6.1 below applied to the $\mathbb{A}^1$-local space BG itself, it follows that a strongly $\mathbb{A}^1$-invariant sheaf of groups G on Sm_k is always unramified.

We thus obtain in this way an equivalence between the category of strongly $\mathbb{A}^1$-invariant sheaves of groups on Sm_k and that of unramified sheaves of groups on Sm_k satisfying axioms (**A2'**), (**A5**) and (**A6**). □

To prove theorem 2.27 we fix an unramified sheaf of groups $\mathcal{G}$ on Sm_k which satisfies the Axioms (**A2'**), (**A5**) and (**A6**).

We introduce two properties depending on $\mathcal{G}$, an integer $d \geq 0$:

(**H1**)(**d**) For any localization U of a smooth k-scheme at some point u of codimension $\leq d$ with infinite residue field, the complex $1 \to \mathcal{G}(U) \subset \mathcal{G}^{(0)}(U) \Rightarrow \mathcal{G}^{(1)}(U) \to \mathcal{G}^{(2)}(U)$ is exact. □

(**H2**)(**d**) For any localization U of a smooth k-scheme at some point u of codimension $\leq d$ with infinite residue field, the "complex":

$$1 \to \mathcal{G}(\mathbb{A}^1_U) \subset \mathcal{G}^{(0)}(\mathbb{A}^1_U) \Rightarrow \mathcal{G}^{(1)}(\mathbb{A}^1_U) \to \mathcal{G}^{(2)}(\mathbb{A}^1_U)$$

is exact. □

(**H1**)(**d**) is a reformulation of (ii) of Lemma 2.25. It is a tautology in case $d \leq 2$. (**H2**)(**1**) holds by Axiom (**A6**) and (**H2**)(**d**) implies (ii) of the Lemma 2.26.

Lemma 2.29. *Let $d \geq 0$ be an integer.*

1) (**H1**)(**d**) $\Rightarrow$ (**H2**)(**d**).
2) (**H2**)(**d**) $\Rightarrow$ (**H1**)(**d+1**).

Proof of Theorem 2.27 assuming Lemma 2.29. Lemma 2.29 implies by induction on d that properties (**H1**)(**d**) and (**H2**)(**d**) hold for any d. It follows from Lemmas 2.25 and 2.26 above that for any essentially smooth k-scheme X with infinite residue fields, then $H^1_{Zar}(X; \mathcal{G}) \cong H^1_{Nis}(X; \mathcal{G})$ and $H^1(X; \mathcal{G}) \cong H^1_{Zar}(\mathbb{A}^1_X; \mathcal{G})$.

This implies Theorem 2.27 if k is infinite. Assume now k is finite. Let $\mathcal{G}'$ be the sheaf $\pi_1^{\mathbb{A}^1}(B\mathcal{G}) = \pi_1(L_{\mathbb{A}^1}(B\mathcal{G}))$. By Corollary 6.9 of Sect. 6.1 below applied to the $\mathbb{A}^1$-local space $L_{\mathbb{A}^1}(B\mathcal{G})$, $\mathcal{G}'$ is unramified and by the first part of Theorem 6.11 it satisfies, as $\mathcal{G}$ the Axioms **(A2')**, **(A5)** and **(A6)**.

By general properties of base change through a smooth morphism (see [52]) we see that for any henselian k-smooth local ring A, with infinite residue field, the morphism $G(A) \to G'(A)$ is an isomorphism. Let A be a k-smooth local ring of dimension ≥ 1. By functoriality we see that $G(A) \subset G'(A)$ is injective, as the fraction field of A is infinite. If κ is a finite field (extension of k), $\mathcal{G}(\kappa) = \mathcal{G}(\kappa[T]) \subset \mathcal{G}'(\kappa[T]) = \mathcal{G}'(\kappa)$. We deduce that $\mathcal{G} \to G'$ is always a monomorphism of sheaves, because if κ is a finite extension of k, $\mathcal{G}(\kappa) \subset \mathcal{G}(\kappa(T))$.

Thus we have the monomorphism $\mathcal{G} \subset \mathcal{G}'$ between unramified sheaves satisfying Axioms **(A2')**, **(A5)** and **(A6)** and which is an isomorphism on smooth local ring with infinite residue field. Now using Remark 2.23 and proceeding as below in the proof of Lemma 2.34, we see that, given a discrete valuation ring $A \subset F$, and a uniformizing element π, $H_v^1(A; \mathcal{G}) \to H_v^1(A; \mathcal{G}')$ can be identified to the morphism $G_{-1}(\kappa(v)) \subset G'_{-1}(\kappa(v))$; but this is an injection as $\mathcal{G}(X \times \mathbb{G}_m) \subset \mathcal{G}'(X \times \mathbb{G}_m)$. This implies that $\mathcal{G}(A) = \mathcal{G}'(A)$ $H_v^1(A; \mathcal{G}) \cong H_v^1(A; \mathcal{G}')$ for any discrete valuation ring $A \subset F \in \mathcal{F}_k$. If we prove that $\mathcal{G}(\kappa) \subset \mathcal{G}'(\kappa)$ is an isomorphism for any finite extension κ of k then we conclude that $\mathcal{G} = \mathcal{G}'$, as both are unramified and coincide over each stalk (included the finite fields). To show $\mathcal{G}(\kappa) \subset \mathcal{G}'(\kappa)$ is an isomorphism, we observe that $\mathcal{G}(\kappa[T]) = \mathcal{G}'(\kappa[T])$ by what precedes.

Now that we know $\mathcal{G} = \mathcal{G}'$, we conclude from the fact that the composition $B\mathcal{G} \to L_{\mathbb{A}^1}(B\mathcal{G}) \to B\mathcal{G}'$ is a (simplicial weak-equivalence) that $B\mathcal{G}$ is $\mathbb{A}^1$-local, and $\mathcal{G}$ is thus strongly $\mathbb{A}^1$-invariant, finishing the proof. □

Remark 2.30. The only reason we have to separate the case of a finite residue field and infinite residue field is due to the point (ii) of Lemma 2.31 below. If one could prove this also with finite residue field, we could get rid of the last part of the previous proof. □

Proof of Lemma 2.29 Let $d \geq 2$ be an integer (if $d < 2$ there is nothing to prove).

Let us prove (1). Assume that **(H1)(d)** holds. Let U be an irreducible smooth k-scheme with function field F. Let us study the following diagram whose middle row is $C^*(\mathbb{A}_U^1; \mathcal{G})$, whose bottom row is $C^*(U; \mathcal{G})$ and whose top row is $C^*(\mathbb{A}_F^1; \mathcal{G})$:

$$
\begin{array}{ccccccc}
\mathcal{G}(F) & \subset & \mathcal{G}(F(T)) & \twoheadrightarrow & \Pi'_{y \in (\mathbb{A}_F^1)^{(1)}} H_y^1(\mathbb{A}_F^1; \mathcal{G}) & & \\
\cup & & \| & & \uparrow & & \\
\mathcal{G}(\mathbb{A}_U^1) & \subset & \mathcal{G}(F(T)) & \Rightarrow & \Pi'_{y \in (\mathbb{A}_U^1)^{(1)}} H_y^1(\mathbb{A}_U^1; \mathcal{G}) & \to & \Pi'_{z \in (\mathbb{A}_U^1)^{(2)}} H_z^2(\mathbb{A}_U^1; \mathcal{G}) \\
\| & & \cup & & \uparrow & & \uparrow \\
\mathcal{G}(U) & \subset & \mathcal{G}(F) & \Rightarrow & \Pi'_{y \in U^{(1)}} H_y^1(U; \mathcal{G}) & \to & \Pi'_{z \in U^{(2)}} H_z^2(U; \mathcal{G})
\end{array}
\tag{2.2}
$$

The top horizontal row is exact by Axiom **(A6)**. Assume U is local of dimension $\leq d$. The bottom horizontal row is exact by **(H1) (d)**. The middle vertical column can be explicited as follows. The points y of codimension 1 in $\mathbb{A}^1_U$ are of two types: either the image of y is the generic point of U or it is a point of codimension 1 in U; the first set is in bijection with $(\mathbb{A}^1_F)^{(1)}$ and the second one with $U^{(1)}$ through the map $y \in U^{(1)} \mapsto y[T] := \mathbb{A}^1_{\bar{y}} \subset \mathbb{A}^1_U$. For y of the first type, it is clear that the set $H^1_y(\mathbb{A}^1_U; \mathcal{G})$ is the same as $H^1_y(\mathbb{A}^1_F; \mathcal{G})$. As a consequence, $\Pi'_{y \in (\mathbb{A}^1_U)^{(1)}} H^1_y(\mathbb{A}^1_U; \mathcal{G})$ is exactly the product of $\Pi'_{y \in (\mathbb{A}^1_F)^{(1)}} H^1_y(\mathbb{A}^1_F; \mathcal{G})$ and of $\Pi'_{y \in U^{(1)}} H^1_{y[T]}(\mathbb{A}^1_U; \mathcal{G})$.

To prove **(H2)(d)** we have exactly to prove the exactness of the middle horizontal row in (2.2) and more precisely that the action of $\mathcal{G}(F(T))$ on $\mathcal{K}^1(\mathbb{A}^1_U; \mathcal{G})$ is transitive.

Take $\alpha \in \mathcal{K}^1(\mathbb{A}^1_U; \mathcal{G})$. As the top horizontal row is exact, there is a $g \in \mathcal{G}$ $(F(T))$ such that $g.\alpha$ lies in $\Pi'_{y \in U^{(1)}} H^1_{v[T]}(\mathbb{A}^1_U; \mathcal{G}) \subset \Pi'_{y \in (\mathbb{A}^1_U)^{(1)}} H^1_y(\mathbb{A}^1_U; \mathcal{G})$, which is the kernel of the vertical $\mathcal{G}(F(T))$-equivariant map $\Pi'_{y \in (\mathbb{A}^1_U)^{(1)}} H^1_y(\mathbb{A}^1_U; \mathcal{G}) \to \Pi'_{y \in (\mathbb{A}^1_F)^{(1)}} H^1_y(\mathbb{A}^1_F; \mathcal{G})$.

Thus $g.\alpha$ lies in $\mathcal{K}^1(\mathbb{A}^1_U; \mathcal{G}) \cap \Pi'_{y \in U^{(1)}} H^1_{y[T]}(\mathbb{A}^1_U; \mathcal{G}) \subset \Pi'_{y \in (\mathbb{A}^1_U)^{(1)}} H^1_y(\mathbb{A}^1_U; \mathcal{G})$. Now the obvious inclusion $\mathcal{K}^1(U; \mathcal{G}) \subset \mathcal{K}^1(\mathbb{A}^1_U; \mathcal{G}) \cap \Pi'_{y \in U^{(1)}} H^1_{y[T]}(\mathbb{A}^1_U; \mathcal{G})$ is a bijection. Indeed, from part (1) of Lemma 2.31 below, $\Pi'_{y \in U^{(1)}} H^1_y(U; \mathcal{G}) \subset \Pi'_{y \in U^{(1)}} H^1_{y[T]}(\mathbb{A}^1_U; \mathcal{G})$ is injective and is exactly the kernel of the composition of the boundary map $\Pi'_{y \in U^{(1)}} H^1_{y[T]}(\mathbb{A}^1_U; \mathcal{G}) \to \Pi_{z \in (\mathbb{A}^1_U)^{(2)}} H^2_z(\mathbb{A}^1_U; \mathcal{G})$ and the projection

$$\Pi_{z \in (\mathbb{A}^1_U)^{(2)}} H^2_z(\mathbb{A}^1_U; \mathcal{G}) \to \Pi_{y \in U^{(1)}, z \in (\mathbb{A}^1_{\bar{y}})^{(1)}} H^2_z(\mathbb{A}^1_U; \mathcal{G})$$

This shows that $\mathcal{K}^1(\mathbb{A}^1_U; \mathcal{G}) \cap \Pi'_{y \in U^{(1)}} H^1_{y[T]}(\mathbb{A}^1_U; \mathcal{G})$ is contained in $\Pi'_{y \in U^{(1)}} H^1_y(U; \mathcal{G})$.

But the right vertical map in (2.2), $\Pi_{z \in U^{(2)}} H^2_z(U; \mathcal{G}) \to \Pi_{z \in (\mathbb{A}^1_U)^{(2)}} H^2_z(\mathbb{A}^1_U; \mathcal{G})$, is induced by the correspondence $z \in U^{(2)} \mapsto \mathbb{A}^1_{\bar{z}} \subset \mathbb{A}^1_U$ and the corresponding maps on $H^2_z(-; \mathcal{G})$. By part (2) of Lemma 2.31 below, this map has trivial kernel. This easily implies that $\mathcal{K}^1(\mathbb{A}^1_U; \mathcal{G}) \cap \Pi'_{y \in U^{(1)}} H^1_{y[T]}(\mathbb{A}^1_U; \mathcal{G})$ is contained in $\mathcal{K}^1(U; \mathcal{G})$, proving our claim.

Thus $g.\alpha$ lies in $\mathcal{K}^1(U; \mathcal{G})$. Now by **(H1) (d)** we know there is an $h \in \mathcal{G}(F)$ with $hg.\alpha = *$ as required.

Let us now prove (2). Assume **(H2) (d)** holds. Let's prove **(H1) (d+1)**. Let X be an irreducible smooth k-scheme (of finite type) of dimension $\leq d+1$ with function field F, let $u \in X \in Sm_k$ be a point of codimension $d+1$ and denote by U its associated local scheme, F its function field. We have to check the exactness at the middle of $\mathcal{G}(F) \Rightarrow \Pi'_{y \in U^{(1)}} H^1_y(U; \mathcal{G}) \to \Pi'_{z \in U^{(2)}} H^2_z(U; \mathcal{G})$.

Let $\alpha \in \mathcal{K}^1(U; \mathcal{G}) \subset \Pi'_{y \in U^{(1)}} H^1_y(U; \mathcal{G})$. We want to show that there exists $g \in \mathcal{G}(F)$ such that $\alpha = g.*$. Let us denote by $y_i \in U$ the points of codimension one in U where α is non trivial. Recall that for each $y \in U^{(1)}$, $H^1_y(U; \mathcal{G}) = H^1_y(X; \mathcal{G})$ where we still denote by $y \in X^{(1)}$ the image of y in X. Denote by $\alpha_X \in \Pi'_{y \in X^{(1)}} H^1_y(X; \mathcal{G})$ the canonical element with same support y_i's and same components as α. α_X may not be in $\mathcal{K}^1(X; \mathcal{G})$, but, by Axiom **(A2')**, its boundary its trivial except on finitely many points z_j of codimension 2 in X. Clearly these points are not in $U^{(2)}$, thus we may, up to removing the closure of these z_j's, find an open subscheme Ω' in X which contains u and the y_i's and such that the element $\alpha_{\Omega'} \in \Pi'_{y \in \Omega'^{(1)}} H^1_y(X; \mathcal{G})$, induced by α, is in $\mathcal{K}^1(\Omega'; \mathcal{G})$.

By Gabber's presentation Lemma 1.15, there exists an étale morphism $U \to \mathbb{A}^1_V$, with V the localization of a k-smooth of dimension d, such that if $Y \subset U$ denotes the reduced closed subscheme whose generic points are the y_i, the composition $Y \to U \to \mathbb{A}^1_V$ is still a closed immersion and such that the composition $Y \to U \to \mathbb{A}^1_V \to V$ is a finite morphism.

The étale morphism $U \to \mathbb{A}^1_V$ induces a morphism of complexes of the form:

$$
\begin{array}{ccccc}
\mathcal{G}(F) & - & \Pi'_{y \in U^{(1)}} H^1_y(U; \mathcal{G}) & \to & \Pi'_{z \in U^{(2)}} H^2_z(U; \mathcal{G}) \\
\uparrow & & \uparrow & & \uparrow \\
\mathcal{G}(E(T)) & - & \Pi'_{y \in (\mathbb{A}^1_V)^{(1)}} H^1_y(\mathbb{A}^1_V; \mathcal{G}) & \to & \Pi'_{z \in (\mathbb{A}^1_V)^{(2)}} H^2_z(\mathbb{A}^1_V; \mathcal{G})
\end{array}
$$

where E is the function field of V. Let y'_i be the images of the y_i in $\mathbb{A}^1_V$; these are points of codimension 1 and have the same residue field (because $Y \to \mathbb{A}^1_V$ is a closed immersion). By the axiom **(A5)(i)**, we see that for each i, the map $H^1_{y'_i}(\mathbb{A}^1_V; \mathcal{G}) \to H^1_{y_i}(U; \mathcal{G})$ is a bijection so that there exists in the bottom complex an element $\alpha' \in \Pi'_{y \in (\mathbb{A}^1_V)^{(1)}} H^1_y(\mathcal{G})$ whose image is α.

The boundary of this α' is trivial. To show this, observe that if $z \in (\mathbb{A}^1_V)^{(2)}$ is not contained in Y, then the boundary of α' has a trivial component in $H^2_z(\mathbb{A}^1_V; \mathcal{G})$. Moreover, if $z \in (\mathbb{A}^1_V)^{(2)}$ lies in the image of Y in $\mathbb{A}^1_V$, there is, by construction, a unique point z' of codimension 2 in Ω, lying in Y and mapping to z. It has moreover the same residue field as z. The claim now follows from **(A5)(ii)**.

By the inductive assumption **(H2) (d)** we see that α' is of the form $h.*$ in $\Pi'_{y \in (\mathbb{A}^1_V)^{(1)}} H^1_y(\mathbb{A}^1_V; \mathcal{G})$ with $h \in \mathcal{G}(E(T))$. But if g denotes the image of h in $\mathcal{G}(F)$ we have $\alpha = g.*$, proving our claim. $\square$

Lemma 2.31. *Let $\mathcal{G}$ be an unramified sheaf of groups on Sm_k satisfying* **(A2')**, **(A5)** *and* **(A6)**.

1) Let v be a discrete valuation on $F \in \mathcal{F}_k$. Denote by $v[T]$ the discrete valuation in $F(T)$ corresponding to the kernel of $\mathcal{O}_v[T] \to \kappa(v)(T)$. Then the map

$$
H^1_v(\mathcal{O}_v; \mathcal{G}) \to H^1_{v[T]}(\mathbb{A}^1_{\mathcal{O}_v}; \mathcal{G})
$$

is injective and its image is exactly the kernel of

$$H^1_{v[T]}(\mathbb{A}^1_{\mathcal{O}_v}; \mathcal{G}) \to \Pi'_{z \in (\mathbb{A}^1_\kappa(v))^{(1)}} H^2_z(\mathbb{A}^1_{\mathcal{O}_v}; \mathcal{G})$$

where we see $z \in (\mathbb{A}^1_\kappa(v))^{(1)}$ as a point of codimension 2 in $\mathbb{A}^1_{\mathcal{O}_v}$.
2) *For any k-smooth local scheme U of dimension 2 with closed point u, and infinite residue field, the "kernel" of the map*

$$H^2_u(\mathcal{G}) \to H^2_{u[T]}(\mathcal{G})$$

is trivial.

Proof. Part (1) follows immediately from the fact that we know from our axioms the exactness of each row of the Diagram (2.2) is exact for U smooth local of dimension 1.

To prove (2) we shall use the interpretation of $H^2_z(U; \mathcal{G})$, for U smooth local of dimension 2 with closed point z, as $H^1_{Zar}(V; \mathcal{G})$, with V the complement of the closed point u. By Lemma 2.24, we know that $H^1_{Zar}(V; \mathcal{G}) \cong H^1_{Nis}(V; \mathcal{G})$.

Pick up an element α of $H^2_u(U; \mathcal{G}) = H^1_{Nis}(V; \mathcal{G})$ which becomes trivial in $H^2_{u[T]}(\mathbb{A}^1_U; \mathcal{G}) = H^1_{Nis}(V_T; \mathcal{G})$, where $V_T = (\mathbb{A}^1_U)_{u[T]} - u'$, u' denoting the generic point of $\mathbb{A}^1_u \subset \mathbb{A}^1_U$. This means that the $\mathcal{G}$-torsor over V become trivial over V_T. As V_T is the inverse limit of the schemes of the form $\Omega - \Omega \cap \overline{u'}$, where Ω runs over the open subschemes of $\mathbb{A}^1_U$ which contains u', we see that there exists such an Ω for which the pull-back of α to $\Omega - \Omega \cap \overline{u'}$ is already trivial. As Ω contains u', $\Omega \cap \overline{u'} \subset \mathbb{A}^1_{\kappa(u)}$ is a non empty dense subset; in case $\kappa(u)$ is infinite, we thus know that there exists a $\kappa(u)$-rational point z in $\Omega \cap \overline{u'}$ lying over u. As $\Omega \to U$ is smooth, it follows from [30, Corollary 17.16.3 p. 106] that there exists an immersion $U' \to \Omega$ whose image contains z and such that $U' \to U$ is étale. This immersion is then a closed immersion, and up to shrinking a bit U' we may assume that $\Omega \cap \overline{u'} \cap U' = \{z\}$. Thus the cartesian square

$$\begin{array}{ccc} U' - z & \to & U' \\ \downarrow & & \downarrow \\ V & \to & U \end{array}$$

is a distinguished square [59]. And the pull-back of α to $U' - z$ is trivial. Extending it to U' defines a descent data which defines an extension of α to U; thus as any element of $H^1_{Zar}(U; \mathcal{G}) = H^1_{Nis}(U; \mathcal{G})$ α is trivial we get our claim. $\square$

$\mathbb{G}_m$-*loop spaces.* Recall the following construction, used by Voevodsky in [79]. Given a presheaf of groups G on Sm_k, we let G_{-1} denote the presheaf of groups given by

$$X \mapsto Ker(G(\mathbb{G}_m \times X) \overset{ev_1}{\twoheadrightarrow} G(X))$$

Observe that if G is a sheaf of groups, so is G_{-1}, and that if G is unramified, so is G_{-1}.

Lemma 2.32. *If G is a strongly $\mathbb{A}^1$-invariant sheaf of groups, so is G_{-1}.*

Proof. One might prove this using our description of those strongly $\mathbb{A}^1$-invariant sheaf of groups given in the previous section. We give here another argument. Let BG be the simplicial classifying space of G (see [59] for instance). The assumption that G is strongly $\mathbb{A}^1$-invariant means that it is an $\mathbb{A}^1$-local space. Choose a fibrant resolution $\mathcal{B}G$ of BG. We use the pointed function space

$$R\mathbf{Hom}_\bullet(\mathbb{G}_m, BG) := \mathbf{Hom}_\bullet(\mathbb{G}_m, \mathcal{B}G)$$

It is fibrant and automatically $\mathbb{A}^1$-local, as $\mathcal{B}G$ is. Moreover its π_1 sheaf is G_{-1} and its higher homotopy sheaves vanish. Thus the connected component of $R\mathbf{Hom}_\bullet(\mathbb{G}_m, B(G))$ is BG_{-1}. This suffices for our purpose because, the connected component of the base point in an $\mathbb{A}^1$-local space is $\mathbb{A}^1$-local. This follows formally from the fact (see [59]) that the $\mathbb{A}^1$-localization functor takes a 0-connected space to a 0-connected space.

In fact we may also prove directly that the space $R\mathbf{Hom}_\bullet(\mathbb{G}_m, B(G))$ is 0-connected. Its π_0 is the associated sheaf to the presheaf $X \mapsto H^1_{Nis}(X \times \mathbb{G}_m; G)$, and this amounts to checking that for X the henselization of point in a smooth k-scheme, then $H^1_{Nis}(\mathbb{G}_m \times X; G)$. This follows from the fact $H^1_{Nis}(\mathbb{A}^1 \times X; G)$ is trivial and the description of $H^1(-; G)$ in terms of our complex. $\qquad\qquad\qquad\qquad\qquad\qquad\qquad\qquad\qquad\qquad\qquad\qquad\qquad\quad\square$

Remark 2.33. In fact given any pointed smooth k-scheme Z, and any strongly $\mathbb{A}^1$-invariant sheaf G we may consider the pointed function object $G^{(Z)}$ which is the sheaf $X \mapsto Ker(G(Z \times X) \to G(X))$. The same argument as in the previous proof shows that the connected component of $R\mathbf{Hom}_\bullet(Z, B(G))$ is indeed $B(M^{(Z)})$. Consequently, the sheaf $G^{(Z)}$ is also strongly $\mathbb{A}^1$-invariant. $\qquad\qquad\square$

Let F be in $\mathcal{F}_k$ and let v be a discrete valuation on F, with valuation ring $\mathcal{O}_v \subset F$. We may choose an irreducible smooth k-scheme X with function field F and a closed irreducible subscheme $i : Y \subset X$ of codimension 1 which induces v on F. In particular the function field of Y is $\kappa(v)$. Assume furthermore that $\kappa(v)$ is separable over k. Then we may also assume up to shrinking X that Y is also smooth over k. Consider the pointed sheaf $X/(X - Y)$ which is called the Thom space of i. By the $\mathbb{A}^1$-purity theorem of [59] there is a canonical pointed $\mathbb{A}^1$-weak equivalence (in the pointed $\mathbb{A}^1$-homotopy category)

$$X/(X - Y) \cong Th(\nu_i)$$

where $Th(\nu_i)$ is the Thom space of the normal bundle ν_i of i, that is to say the pointed sheaf $E(\nu_i)/E(\nu_i)^\times$. Let π be a uniformizing element for v; one may see the class of π modulo $(\mathcal{M}_v)^2$ as a (non zero) basis element of

$(\nu_i)_y = \mathcal{M}_v/(\mathcal{M}_v)^2$, the fiber of the normal bundle at the generic point y of Y. Consequently, π (or its class in $\mathcal{M}_v/(\mathcal{M}_v)^2$) induces a trivialization of ν_i at least in a Zariski neighborhood of y. In case ν_i is trivialized, it follows from [59] that the pointed sheaf $Th(\nu_i)$ is canonically isomorphic to $T \wedge (Y_+)$, with $T := \mathbb{A}^1/\mathbb{G}_m$.

Lemma 2.34. *Let $\mathcal{G}$ be a strongly $\mathbb{A}^1$-invariant sheaf. Let Y be a smooth k-scheme. Then there is a canonical bijection*

$$\mathcal{G}_{-1}(Y) \cong H^1(T \wedge (Y_+); \mathcal{G})$$

which is a group isomorphism if $\mathcal{G}$ is abelian.

Proof. We use the cofibration sequence

$$\mathbb{G}_m \times Y \subset \mathbb{A}^1 \times Y \to T \wedge (Y_+)$$

to get a long exact sequence in the usual sense

$$0 \to H^0(\mathbb{A}^1 \times Y; \mathcal{G}) \to H^0(\mathbb{G}_m \times Y; \mathcal{G}) \Rightarrow H^1(T \wedge (Y_+); \mathcal{G})$$
$$\to H^1(\mathbb{A}^1 \times Y; \mathcal{G}) \to H^1(\mathbb{G}_m \times Y; \mathcal{G}) \to \cdots$$

The pointed map $H^1(Y; \mathcal{G}) = H^1(\mathbb{A}^1 \times Y; \mathcal{G}) \to H^1(\mathbb{G}_m \times Y; \mathcal{G})$ being split injective (use the evaluation at 1), we get an exact sequence

$$0 \to \mathcal{G}(Y) \subset \mathcal{G}(\mathbb{G}_m \times Y) \Rightarrow H^1(T \wedge (Y_+); \mathcal{G}) \to *$$

As $\mathcal{G}_{-1}(Y)$ is the kernel of $ev_1 : \mathcal{G}(\mathbb{G}_m \times Y) \to \mathcal{G}(Y)$, this exact sequence implies that the action of $\mathcal{G}_{-1}(Y)$ on the base point $*$ of $H^1(T \wedge (Y_+); \mathcal{G})$ induces the claimed bijection $\mathcal{G}_{-1}(Y) \cong H^1(T \wedge (Y_+); \mathcal{G})$. The statement concerning the abelian case is easy. $\square$

From what we did before, it follows at once by passing to the filtering colimit over the set of open neighborhoods of y the following:

Corollary 2.35. *Let F be in $\mathcal{F}_k$ and let v be a discrete valuation on F, with valuation ring $\mathcal{O}_v \subset F$. For any strongly $\mathbb{A}^1$-invariant sheaf of groups $\mathcal{G}$, a choice of a non-zero element μ in $\mathcal{M}_v/(\mathcal{M}_v)^2$ (that is to say the class a uniformizing element π of $\mathcal{O}_v$) induces a canonical bijection*

$$\theta_\mu : \mathcal{G}_{-1}(\kappa(v)) \cong H_v^1(\mathcal{O}_v; \mathcal{G})$$

which is an isomorphism of abelian groups in case $\mathcal{G}$ is a sheaf of abelian groups.

Using the previous bijection, we may define in the situation of the corollary a map

$$\partial_v^\pi : \mathcal{G}(F) \to \mathcal{G}_{-1}(\kappa(v))$$

as the composition $\mathcal{G}(F) \twoheadrightarrow H_v^1(\mathcal{O}_v; \mathcal{G}) \cong \mathcal{G}_{-1}(\kappa(v))$ which we call *the residue map* associated to π. If $\mathcal{G}$ is abelian, the residue map is a morphism of abelian groups.

2.3 $\mathbb{Z}$-Graded Strongly $\mathbb{A}^1$-Invariant Sheaves of Abelian Groups

In this section we want to give some criteria which imply the Axioms **(A4)** in some particular cases of $\tilde{\mathcal{F}}_k$-data. Our method is inspired by Rost [68] but avoids the use of transfers. The results of this section will be used in Sect. 3.2 below to construct the sheaves of unramified Milnor-Witt K-theory and unramified Milnor K-theory, etc., without using any transfers as it is usually done. As a consequence, our construction of transfers in Chap. 4 gives indeed a new construction of the transfers on the previous sheaves.

Let M_* be a functor $\mathcal{F}_k \to Ab_*$ to the category of $\mathbb{Z}$-graded abelian groups. We assume throughout this section that M_* is endowed with the following extra structures.

(D4) (i) For any $F \in \mathcal{F}_k$ a structure of $\mathbb{Z}[F^\times/(F^{\times 2})]$-module on $M_*(F)$, which we denote by $(u, \alpha) \mapsto\, <u> \alpha \in M_n(F)$ for $u \in F^\times$ and for $\alpha \in M_n(F)$. This structure should be functorial in the obvious sense in $\mathcal{F}_k$.   □

(D4) (ii) For any $F \in \mathcal{F}_k$ and any $n \in \mathbb{Z}$, a map $F^\times \times M_{n-1}(F) \to M_n(F)$, $(u, \alpha) \mapsto [u].\alpha$, functorial (in the obvious sense) in $\mathcal{F}_k$.   □

(D4) (iii) For any discrete valuation v on $F \in \mathcal{F}_k$ and uniformizing element π a graded epimorphism of degree -1

$$\partial_v^\pi : M_*(F) \to M_{*-1}(\kappa(v))$$

which is functorial, in the obvious sense, with respect to extensions $E \to F$ such that v restricts to a discrete valuation on E, with ramification index 1, if we choose as uniformizing element an element π in E.   □

We assume furthermore that the following axioms hold:

(B0) For $(u, v) \in (F^\times)^2$ and $\alpha \in M_n(F)$, one has

$$[uv]\alpha = [u]\alpha + <u> [v]\alpha$$

and moreover $[u][v]\alpha = -<-1> [v][u]\alpha$.

(B1) For a k-smooth integral domain A with field of fractions F, for any $\alpha \in M_n(F)$, then for all but only finitely many point $x \in Spec(A)^{(1)}$, one has that for any uniformizing element π for x, $\partial_x^\pi(\alpha) \neq 0$. $\qquad \square$

(B2) For any discrete valuation v on $F \in \mathcal{F}_k$ with uniformizing element π one has $\partial_v^\pi([u]\alpha) = [\overline{u}]\partial_v^\pi(\alpha) \in M_n(\kappa(v))$ and $\partial_v^\pi(<u>\alpha) =<\overline{u}>\partial_v^\pi(\alpha) \in M_{(n-1)}(\kappa(v))$, for any unit u in $(\mathcal{O}_v)^\times$ and any $\alpha \in M_n(F)$. $\qquad \square$

(B3) For any field extension $E \subset F \in \mathcal{F}_k$ and for any discrete valuation v on $F \in \mathcal{F}_k$ which restricts to a discrete valuation w on E, with ramification index e, let $\pi \in \mathcal{O}_v$ be a uniformizing element for v and $\rho \in \mathcal{O}_w$ be a uniformizing element for w. Write $\rho = u\pi^e$, with u a unit in $\mathcal{O}_v$. Then one has for $\alpha \in M_*(E)$, $\partial_v^\pi(\alpha|_F) = e_\epsilon < \overline{u} >$ $(\partial_w^\rho(\alpha))|_{\kappa(v)} \in M_*(\kappa(v))$. $\qquad \square$

Here we set for any integer n,

$$n_\epsilon = \sum_{i=1}^{n} < (-1)^{(i-1)} >$$

We observe that as a particular case of **(B3)** we may choose $E = F$ so that $e = 1$ and we get that for any discrete valuation v on $F \in \mathcal{F}_k$, any uniformizing element π, and any unit $u \in \mathcal{O}_v^\times$, then one has $\partial_v^{u\pi}(\alpha) =< \overline{u} >$ $\partial_v^\pi(\alpha) \in M_{(n-1)}(\kappa(v))$ for any $\alpha \in M_n(F)$.

Thus in case Axiom **(B3)** holds, the kernel of the surjective homomorphism ∂_v^π only depends on the valuation v, not on any choice of π. In that case we then simply denote by

$$M_*(\mathcal{O}_v) \subset M_*(F)$$

this kernel. Axiom **(B1)** is then exactly equivalent to Axiom **(A2)** for unramified $\tilde{\mathcal{F}}_k$-sets. The following is easy:

Lemma 2.36. *Assume M_* satisfies Axioms* **(B1)**, **(B2)** *and* **(B3)**. *Then it satisfies (in each degree) the axioms for a unramified $\tilde{\mathcal{F}}_k$-abelian group datum. Moreover, it satisfies Axiom* **(A5) (i)**.

We assume from now on (in this section) that M_* satisfies Axioms **(B0)**, **(B1)**, **(B2)** and **(B3)**. Thus we may (and will) consider each M_n as a sheaf of abelian groups on $\tilde{Sm}_k$.

We recall that we denote, for any discrete valuation v on $F \in \mathcal{F}_k$, by $H_v^1(\mathcal{O}_v, M_n)$ the quotient group $M_n(F)/M_n(\mathcal{O}_v)$ and by $\partial_v : M_n(F) \to H_v^1(\mathcal{O}_v, M_n)$ the projection. Of course, if one chooses a uniformizing element π, one gets an isomorphism $\theta_\pi : M_{(n-1)}(\kappa(v)) \cong H_v^1(\mathcal{O}_v, M_n)$ with $\partial_v = \theta_\pi \circ \partial_v^\pi$.

For each discrete valuation v on $F \in \mathcal{F}_k$, and any uniformizing element π set

$$s_v^\pi : M_*(F) \to M_*(\kappa(v)) , \quad \alpha \mapsto \partial_v^\pi([\pi]\alpha)$$

Lemma 2.37. *Assume M_* satisfies Axioms* **(B0)**, **(B1)**, **(B2)** *and* **(B3)**. *Then for each discrete valuation v the homomorphism $s_v^\pi : M_*(\mathcal{O}_v) \subset M_*(F)$ doesn't depend on the choice of a uniformizing element π.*

Proof. From Axiom **(B0)** we get for any unit $u \in \mathcal{O}^\times$, any uniformizing element π and any $\alpha \in M_n(F)$: $[u\pi]\alpha = [u]\alpha + <u>[\pi]\alpha$. Thus if moreover $\alpha \in M(\mathcal{O}_v)$, one has $s_v^{u\pi}(\alpha) = \partial_v^{u\pi}([u\pi]\alpha) = \partial_v^{u\pi}([u]\alpha) + \partial_v^{u\pi}(<u>[\pi]\alpha) = \partial_v^{u\pi}(<u>[\pi]\alpha)$, as by Axiom **(B2)** $\partial_v^{u\pi}([u]\alpha) = [\bar{u}]\partial_v^{u\pi}(\alpha) = [\bar{u}]0 = 0$. But by the same Axiom **(B2)**, $\partial_v^{u\pi}(<u>[\pi]\alpha) = <\bar{u}> \partial_v^{u\pi}([\pi]\alpha)$, which by Axiom **(B3)** is equal to $<\bar{u}><\bar{u}>\partial_v^\pi([\pi]\alpha) = \partial_v^\pi([\pi]\alpha)$. This proves the claim. $\square$

We will denote by

$$s_v : M_*(\mathcal{O}_v) \to M_n(\kappa(v))$$

the common value of all the s_v^π's. In this way M_* is endowed with a datum **(D3)**.

We introduce the following Axiom:

(HA) (i) For any $F \in \mathcal{F}_k$, the following diagram

$$0 \to M_*(F) \to M_*(F(T)) \xrightarrow{\Sigma \partial_{(P)}^P} \oplus_{P \in \mathbb{A}_F^1} M_{*-1}(F[T]/P) \to 0$$

is a short exact sequence. Here P runs over the set of irreducible monic polynomials, and (P) means the associated discrete valuation. $\square$

(HA) (ii) For any $\alpha \in M(F)$, one has $\partial_{(T)}^T([T]\alpha|_{F(T)}) = \alpha$. $\square$

This axiom is obviously related to the Axiom **(A6)**, as it immediately implies that for any $F \in \mathcal{F}_k$, $M(F) \to M(\mathbb{A}_F^1)$ is an isomorphism and $H_{Zar}^1(\mathbb{A}_F^1; M) = 0$.

We next claim:

Lemma 2.38. *Let M_* be as in Lemma 2.37, and suppose it additionally satisfies Axioms* **(HA) (i)** *and* **(HA) (ii)**. *Then Axioms* **(A1) (ii)**, **(A3)** **(i)** *and* **(A3) (ii)** *hold.*

Proof. The first part of Axiom **(A1) (ii)** follows from Axiom **(B4)**. For the second part we choose a uniformizing element π in $\mathcal{O}_w$, which is still a uniformizing element for $\mathcal{O}_v$ and the square

$$
\begin{array}{ccc}
M_*(F) & \xrightarrow{\partial_v^\pi} & M_{(*-1)}(\kappa(v)) \\
\uparrow & & \uparrow \\
M_*(E) & \xrightarrow{\partial_w^\pi} & M_{(*-1)}(\kappa(w))
\end{array}
$$

is commutative by our definition **(D4) (iii)**. Moreover the morphism $M_*(E) \to M_*(F)$ preserve the product by π by **(D4) (i)**.

To prove Axiom **(A3)** we proceed as follows. By assumption we have $E \subset \mathcal{O}_v \subset F$. Choose a uniformizing element π of v. We consider the extension

$E(T) \subset F$ induced by $T \mapsto \pi$. The restriction of v is the valuation defined by T on $E[T]$. The ramification index is 1. Using the previous point, we see that we can reduce to the case $E \subset F$ is $E \subset E(T)$ and $v = (T)$. In that case, the claim follows from our Axioms **(HA) (i)** and **(HA) (ii)**.   □

From now on, we assume that M_* satisfies all the Axioms previously met in this subsection. We observe that by construction the Axiom **(A5) (i)** is clear.

Fix a discrete valuation v on $F \in \mathcal{F}_k$. We denote by $v[T]$ the discrete valuation on $F(T)$ defined by the divisor $\mathbb{G}_m|_{\kappa(v)} \subset \mathbb{G}_m|_{\mathcal{O}_v}$ whose open complement is $\mathbb{G}_m|_F$. Choose a uniformizing element π for v. Observe that $\pi \in F(T)$ is still a uniformizing element for $v[T]$.

We want to analyze the following commutative diagram in which the horizontal rows are short exact sequences (given by Axiom **(HA)**):

$$
\begin{array}{ccccccccc}
0 \to & M_*(F) & \to & M_*(F(T)) & \xrightarrow{\Sigma_P \partial^P_{(P)}} & \oplus_{P \in (\mathbb{A}^1_F)^{(1)}} M_{*-1}(F[T]/P) & \to 0 \\
& \downarrow \partial^\pi_v & & \downarrow \partial^\pi_{v[T]} & & \downarrow \Sigma_{P,Q} \partial^{\pi,P}_Q \\
0 \to & M_{*-1}(\kappa(v)) & \to & M_{*-1}(\kappa(v)(T)) & \xrightarrow{\Sigma_Q \partial^Q_{(Q)}} & \oplus_{Q \in (\mathbb{A}^1_{\kappa(v)})^{(1)}} M_{*-2}(\kappa(v)[T]/Q) & \to 0
\end{array}
$$
(2.3)

and where the morphisms $\partial^{\pi,P}_Q : M_*(F[T]/P) \to M_{*-1}(\kappa(v)[T]/Q)$ are defined by the diagram.

For this we need the following Axiom:

(B4) Let v be discrete valuation on $F \in \mathcal{F}_k$ and let π be a uniformizing element. Let $P \in (\mathbb{A}^1_F)^{(1)}$ and $Q \in (\mathbb{A}^1_{\kappa(v)})^{(1)}$ be fixed.

(i) If the closed point $Q \in \mathbb{A}^1_{\kappa(v)} \subset \mathbb{A}^1_{\mathcal{O}_v}$ is not in the divisor $D_P \subset \mathbb{A}^1_{\mathcal{O}_v}$ with generic point $P \in \mathbb{A}^1_F \subset \mathbb{A}^1_{\mathcal{O}_v}$ then the morphism $\partial^{\pi,P}_Q$ is zero.

(ii) If Q is in $D_P \subset \mathbb{A}^1_{\mathcal{O}_v}$ and if the local ring $\mathcal{O}_{D_P,Q}$ is a discrete valuation ring with π as uniformizing element then

$$
\partial^{\pi,P}_Q = - < -\frac{\overline{P'}}{\overline{Q'}} > \partial^Q_Q \quad : M_*(F[T]/P) \to M_{*-1}(\kappa(v)[T]/Q) \qquad □
$$

We will set $U = Spec(\mathcal{O}_v)$ in the sequel. We first observe that $(\mathbb{A}^1_U)^{(1)} = (\mathbb{A}^1_F)^{(1)} \amalg \{v[T]\}$, where as usual $v[T]$ means the generic point of $\mathbb{A}^1_{\kappa(v)} \subset \mathbb{A}^1_U$. For each $P \in (\mathbb{A}^1_F)^{(1)}$, there is a canonical isomorphism $M_{*-1}(F[T]/P) \cong H^1_P(\mathbb{A}^1_U; M_*)$, as P itself is a uniformizing element for the discrete valuation (P) on $F(T)$. For $v[T]$, there is also a canonical isomorphism $M_{*-1}(\kappa(v)[T]) \cong H^1_{v[T]}(\mathbb{A}^1_U; M_*)$ as π is also a uniformizing element for the discrete valuation $v[T]$ on $F(T)$.

Using the previous isomorphisms, we see that the beginning of the complex $C^*(\mathbb{A}^1_U; M_*)$ (see Sect. 2.2) is isomorphic to

$$0 \to M_*(\mathbb{A}^1_U) \to M_*(F(T)) \xrightarrow{\partial^\pi_{v[T]} + \sum_P \partial^P_{(P)}} M_{*-1}(\kappa(v)(T)) \oplus$$
$$\left(\oplus_{P \in (\mathbb{A}^1_F)^{(1)}} M_{*-1}(F[T]/P) \right)$$

The diagram (2.3) can be used to compute the cokernel of the previous morphism $\partial : M_*(F(T)) \to M_{*-1}(\kappa(v)(T)) \oplus \left(\oplus_{P \in (\mathbb{A}^1_F)^{(1)}} M_{*-1}(F[T]/P) \right)$. Indeed the epimorphism ∂'

$$M_{*-1}(\kappa(v)(T)) \oplus (\oplus_P M_{*-1}(F[T]/P)) \xrightarrow{\sum_Q \partial^Q_{(Q)} - \sum_{P,Q} \partial^{\pi,P}_Q} \oplus_{Q \in (\mathbb{A}^1_{\kappa(v)})^{(1)}}$$
$$M_{*-2}(\kappa(v)[T]/Q)$$

composed with ∂ is trivial, and the diagram

$$M_*(F(T)) \xrightarrow{\partial} M_{*-1}(\kappa(v)(T)) \oplus (\oplus_P M_{*-1}(F[T]/P))$$
$$\xrightarrow{\partial'} \oplus_Q M_{*-2}(\kappa(v)[T]/Q) \to 0 \tag{2.4}$$

is an exact sequence: this is just an obvious reformulation of the properties of (2.3).

Now fix $Q_0 \in (\mathbb{A}^1_{\kappa(v)})^{(1)}$. Let $(\mathbb{A}^1_F)^{(1)}_0$ be the set of P's such that Q_0 lies in the divisor D_P of $\mathbb{A}^1_U$ defined by P.

Lemma 2.39. *Assume M_* satisfies all the previous Axioms (including* **(B4)***). The obvious quotient*

$$M_*(F(T)) \xrightarrow{\partial} M_{*-1}(\kappa(v)(T)) \oplus \left(\oplus_{P \in (\mathbb{A}^1_F)^{(1)}_0} M_{*-1}(F[T]/P) \right)$$
$$\xrightarrow{\partial'_Q} M_{*-2}(\kappa(v)[T]/Q_0) \to 0$$

of the previous diagram is also an exact sequence.

Proof. Using the snake Lemma, it is sufficient to prove that the image of the composition $\oplus_{P \notin (\mathbb{A}^1_U)^{(1)}_0} M_{*-1}(F[T]/P) \subset \oplus_{P \in (\mathbb{A}^1_U)^{(1)}} M_{*-1}(F[T]/P) \to \oplus_{Q \in (\mathbb{A}^1_{\kappa(v)})^{(1)}} M_{*-2}(\kappa(v)[T]/Q$ is exactly $\oplus_{Q \in (\mathbb{A}^1_{\kappa(v)})^{(1)} - \{Q_0\}} M_{*-2}(\kappa(v)[T]/Q$. Axiom **(B4)(i)** readily implies that this image is contained in

$$\oplus_{Q \in (\mathbb{A}^1_{\kappa(v)})^{(1)} - \{Q_0\}} M_{*-2}(\kappa(v)[T]/Q).$$

Now we want to show that the image entirely reaches each $M_{*-2}(\kappa(v)[T]/Q$, $Q \neq Q_0$. For any such Q, there is a P, irreducible, such that Q is $\alpha\overline{P}$, for some unit $\alpha \in \kappa(v)^\times$. Thus Q lies over D_P, but not Q_0. Moreover, (π, P) is a system of generators of the maximal ideal of the local dimension 2 regular ring $(\mathcal{O}_v[T])_{(Q)}$, thus $(\mathcal{O}_v[T]/P)_{(Q)}$ is a discrete valuation ring with uniformizing element the image of π. By Axiom **(B4)(ii)** now, we conclude that $\partial_Q^{\pi,P}$ is onto, proving the claim. $\qquad\qquad\qquad\qquad\qquad\qquad\square$

Now let X be a local smooth k-scheme of dimension 2 with closed point z and function field E. Recall from the beginning of Sect. 2.2 that we denote by $H_z^2(X; M)$ the cokernel of the sum of the residues $M_*(E) \overset{\Sigma_{y \in X^{(1)}} \partial_y}{\longrightarrow}$ $\oplus_{y \in X^{(1)}} H_y^1(X; M_*)$. We thus have a canonical exact sequence of the form:

$$0 \to M_*(X) \to M_*(E) \overset{\Sigma_{y \in X^{(1)}} \partial_y}{\longrightarrow} \oplus_{y \in X^{(1)}} H_y^1(X; M_*)$$

$$\overset{\Sigma_{y \in X^{(1)}} \partial_z^y}{\longrightarrow} H_z^2(X; M_*) \to 0 \qquad\qquad (2.5)$$

where the homomorphisms denoted ∂_z^y are defined by the diagram. This diagram is the complex $C^*((\mathbb{A}_U^1)_0; M_*)$.

For X the localization $(\mathbb{A}_U^1)_0$ of $\mathbb{A}_U^1$ at some closed point $Q_0 \in \mathbb{A}_{\kappa(v)}^1$, with $U = Spec(\mathcal{O}_v)$ where v is a discrete valuation on some $F \in \mathcal{F}_k$, we thus get immediately:

Corollary 2.40. *Assume M_* satisfies all the previous Axioms. The complex $C^*((\mathbb{A}_U^1)_0; M_*)$ is canonically isomorphic to exact sequence:*

$$0 \to M_*((\mathbb{A}_U^1)_Q) \to M_*(F(T)) \to M_{*-1}(\kappa(v)(T))$$

$$\oplus \left(\oplus_{P \in (\mathbb{A}_F^1)_0^{(1)}} M_{*-1}(F[T]/P) \right)$$

$$\to M_{*-2}(\kappa(v)[T]/Q) \to 0$$

This isomorphism provides in particular a canonical isomorphism

$$M_{*-2}(\kappa(v)[T]/Q_0) \cong H_{Q_0}^2(\mathbb{A}_U^1; M_*)$$

Corollary 2.41. *Assume M_* satisfies all the previous Axioms. For each n, the unramified sheaves of abelian groups (on $\widetilde{Sm}_k$) M_n satisfies Axiom* **(A2')**.

Proof. From Remark 2.19, it suffices to check this when k is infinite.

Now assume X is a smooth k-scheme. Let $y \in X^{(1)}$ be a point of codimension 1. We wish to prove that given $\alpha \in H_y^1(X; M_*)$, there are only finitely many $z \in X^{(2)}$ such that $\partial_z^y(\alpha)$ is non trivial. By Gabber's Lemma, there is an open neighborhood $\Omega \subset X$ of y and an étale morphism $\Omega \to \mathbb{A}_V^1$, for V some open subset of an affine space over k, such that the morphism $\overline{y} \cap \Omega \to \mathbb{A}_V^1$ is a closed immersion.

The complement $\overline{y} - \overline{y} \cap \Omega$ is a closed subset everywhere of > 0-dimension and thus contains only finitely many points of codimension 1 in $\overline{y}$.

For any $z \in (\overline{y} \cap \Omega)^{(1)}$, the étale morphism $\Omega \to \mathbb{A}_V^1$ obviously induces a commutative square

$$H^1_y(X; M_*) \overset{\partial^y_z}{\to} H^2_z(X; M_*)$$
$$\uparrow \wr \qquad\qquad \uparrow \wr$$
$$H^1_y(\mathbb{A}^1_V; M_*) \overset{\partial^y_z}{\to} H^2_z(\mathbb{A}^1_V; M_*)$$

(because $\overline{y} \cap \Omega \to \mathbb{A}^1_V$ is a closed immersion), we reduce to proving the claim for the image of y in $\mathbb{A}^1_V$, which follows from our previous results.   □

Now that we know that M_* satisfies Axiom **(A2')**, for X a smooth k-scheme with function field E we may define as in Sect. 2.2 a (whole) complex $C^*(X; M_*)$ of the form

$$0 \to M_*(X) \to M_*(E) \overset{\Sigma_{y \in X^{(1)}} \partial_y}{\longrightarrow} \oplus_{y \in X^{(1)}} H^1_y(X; M_*)$$

$$\overset{\Sigma_{y,z} \partial^y_z}{\longrightarrow} \oplus_{z \in X^{(2)}} H^2_z(X; M_*) \tag{2.6}$$

We thus get as an immediate consequence:

Corollary 2.42. *Assume M_* satisfies all the previous Axioms. For any discrete valuation v on $F \in \mathcal{F}_k$, setting $U = Spec(\mathcal{O}_v)$, the complex $C^*(\mathbb{A}^1_U; M_*)$ is canonically isomorphic to the exact sequence (2.4):*

$$0 \to M_*(\mathbb{A}^1_U) \to M_*(F(T)) \to M_{*-1}(\kappa(v)(T)) \oplus \left(\oplus_{P \in (\mathbb{A}^1_F)^{(1)}} M_{*-1}(F[T]/P) \right)$$

$$\to \oplus_{Q \in (\mathbb{A}^1_U)^{(1)}} M_{*-2}(\kappa(v)[T]/Q) \to 0$$

Consequently, the complex $C^*(\mathbb{A}^1_U; M_*)$ is an exact complex, and in particular, for each n, the unramified sheaves of abelian groups (on $\tilde{Sm}_k$) M_n satisfies Axiom **(A6)**.

Proof. Only the statement concerning Axiom **(A6)** is not completely clear: we need to prove that $M_n(U) \to M_n(\mathbb{A}^1_U)$ is an isomorphism for U a smooth local k-scheme of dimension ≤ 1. The rest of the Axiom is clear. This claim is clear by Axiom **(HA)** for U of dimension 0. We need to prove it for U of the form $Spec(\mathcal{O}_v)$ for some discrete valuation v on some $F \in \mathcal{F}_k$ (observe that for the moment M_* only defines an unramified sheaf on $\tilde{Sm}_k$, and we can only apply point (1) of Lemma 2.16. But this statement follows rather easily by contemplating the diagram (2.3).   □

We next prepare the statement of our last Axiom. Let X be a local smooth k-scheme of dimension 2, with field of functions F and closed point z. Consider the complex $C^*(X; M_*)$ associated to X in (2.5). By definition we have a short exact sequence:

$$0 \to M_*(F)/M_*(X) \to \oplus_{y \in X^{(1)}} H^1_y(X; M_*) \to H^2_z(X; M_*) \to 0$$

Let $y_0 \in X^{(1)}$ be such that $\overline{y_0}$ is smooth over k.

The properties of the induced morphism

$$M_*(F)/M_*(X) \to \oplus_{y \in X^{(1)} - \{y_0\}} H^1_y(X; M_*) \tag{2.7}$$

will play a very important role. We first observe:

Lemma 2.43. *Assume M_* satisfies all the previous Axioms (including* **(B4)***). Let X be a local smooth k-scheme of dimension 2, with field of functions F and closed point z, let $y_0 \in X^{(1)}$ be such that $\overline{y_0}$ is smooth over k. Then the homomorphism (2.7) is onto.*

Proof. We first observe that this property is true for any localization of a scheme of the form $\mathbb{A}^1_U$ at a point z of codimension 2, with $U = Spec(\mathcal{O}_v)$, for some discrete valuation v on F. If $\overline{y_0}$ is $\mathbb{A}^1_{\kappa(v)}$ this is just Axiom **(HA)**. If $\overline{y_0}$ is not $\mathbb{A}^1_{\kappa(v)}$ we observe that the complex $C^*((\mathbb{A}^1_U)_z; M_*)$:

$$M(F(T)) \xrightarrow{\Sigma_{y \in ((\mathbb{A}^1_U)_z)^1} \partial_y} \oplus_{y \in ((\mathbb{A}^1_U)_z)^{(1)}} H^1_y(X; M) \to H^2_z(\mathbb{A}^1_{\kappa(v)}; M_*) \to 0$$

is isomorphic to the one of Corollary 2.40. By Axiom **(B4)(ii)** we deduce that the map $\partial^y_z : H^1_{y_0}(X; M) \to H^2_z(\mathbb{A}^1_{\kappa(v)}; M_*)$ is surjective. This implies the statement.

To prove the general case we use Gabber's Lemma. Let α be an element in $\oplus_{y \in X^{(1)} - \{y_0\}} H^1_y(X; M)$. Let $y_1, .., y_r$ be the points in the support of α. There exists an étale morphism $X \to \mathbb{A}^1_U$, for some local smooth scheme U of dimension 1, and with function field K, such that $\overline{y_i} \to \mathbb{A}^1_U$ is a closed immersion for each i. But then use the commutative square

$$
\begin{array}{ccc}
M_*(F) & \xrightarrow{\Sigma_{y \in X^1 - \{y_0\}} \partial_y} & \oplus_{y \in X^{(1)} - \{y_0\}} H^1_y(X; M_*) \\
\uparrow & & \uparrow \\
M_*(K(T)) & \xrightarrow{\Sigma_{y \in ((\mathbb{A}^1_U)_z)^1 - \{y_0\}} \partial_y} & \oplus_{y \in ((\mathbb{A}^1_U)_z)^{(1)} - \{y_0\}} H^1_y(\mathbb{A}^1_U; M_*)
\end{array}
$$

We now conclude that $\alpha = \Sigma_i \alpha_i$, with $\alpha_i \in H^1_{y_i}(X; M_*) \cong H^1_{y_i}(\mathbb{A}^1_U; M_*)$, $i \in \{1, \ldots, r\}$ comes from an element from the bottom right corner. The isomorphism $H^1_{y_i}(X; M_*) \cong H^1_{y_i}(\mathbb{A}^1_U; M_*)$ is a consequence of our definition of $H^1_y(-; M_*)$ and **(D4)(iii)**. The bottom horizontal morphism is onto by the first case we treated. Thus α lies in the image of our morphism. □

Now for our X local smooth k-scheme of dimension 2, with field of functions F and closed point z, with $y_0 \in X^{(1)}$ such that $\overline{y_0}$ is smooth over k, choose a uniformizing element π of y_0 (in $\mathcal{O}_{X,y_0}$). This produces by definition an isomorphism $M_{*-1}(\kappa(y_0)) \cong H^1_{y_0}(X; M_*)$. Now the kernel of the morphism (2.7) is contained in $M_{*-1}(\kappa(y_0)) \cong H^1_{y_0}(X; M_*)$. We may now state our last Axiom:

(B5) Let X be a local smooth k-scheme of dimension 2, with field of functions F and closed point z, let $y_0 \in X^{(1)}$ be such that $\overline{y_0}$ is smooth over k. Choose a uniformizing element π of y_0 (in $\mathcal{O}_{X,y_0}$). Then the kernel of the morphism (2.7) is (identified to a subgroup of $M_{*-1}(\kappa(y_0))$) equal to $M_{*-1}(\mathcal{O}_{y_0,z}) \subset M_{*-1}(\kappa(y_0))$.   □

Remark 2.44. Thus if M_* satisfies Axiom **(B5)** one gets an exact sequence

$$0 \to M_{*-1}(\mathcal{O}_{y_0,z}) \to M_*(F)/M_*(X) \to \oplus_{y \in X^{(1)} - \{y_0\}} H^1_y(X; M_*)$$

Lemma 2.43 shows that it is in fact a short exact sequence.   □

Lemma 2.45. *Assume that M_* satisfies all the previous Axioms of this section, including* **(B4)**, **(B5)**.

1) Let X be a local smooth k-scheme of dimension 2, with field of functions F and closed point z, let $y_0 \in X^{(1)}$ be such that $\overline{y_0}$ is smooth over k. Choose a uniformizing element π of $\mathcal{O}_{X,y_0}$. Then the homomorphism

$$M_{*-1}(\kappa(y_0)) \cong H^1_{y_0}(X; M) \overset{\partial^{y_0}_z}{\to} H^2_z(X; M) \text{ induces an isomorphism}$$

$$\Theta_{y_0,\pi} : M_{*-1}(\kappa(y_0))/M_{*-1}(\mathcal{O}_{y_0,z}) = H^1_z(\overline{y_0}; M_{*-1}) \cong H^2_z(X; M)$$

2) Assume $f : X' \to X$ is an étale morphisms between smooth local k-schemes of dimension 2, with closed points respectively z' and z and with the same residue field $\kappa(z) = \kappa(z')$. Then the induced morphism $H^2_z(X; M_*) \to H^2_{z'}(X'; M_*)$ is an isomorphism. In particular, M_* satisfies Axiom **(A5) (ii)**.

Proof. 1) We know from the previous Remark that the sequence $0 \to M_{*-1}(\mathcal{O}_{y_0}) \to M_*(F)/M_*(X) \to \oplus_{y \in X^{(1)} - \{y_0\}} H^1_y(X; M_*) \to 0$ is a short exact sequence. By the definition of $H^2_z(X; M)$ given by the short exact sequence (2.5), this provides a short exact sequence of the form

$$0 \to M_{*-1}(\mathcal{O}_{y_0,z}) \to M_{*-1}(\kappa(y_0)) \to H^2_z(X; M) \to 0$$

and produces the required isomorphism $\Theta_{y_0,\pi}$.

2) Choose $y_0 \in X^{(1)}$ such that $\overline{y_0}$ is smooth over k and a uniformizing element $\pi \in \mathcal{O}_{X,y_0}$. Clearly the pull back of y_0 to X' is still a smooth divisor denoted by y'_0, and the image of π is a uniformizing element for $\mathcal{O}_{y'_0}$. Then the following diagram commutes

$$
\begin{array}{ccc}
H^1_{z'}(\overline{y'_0}; M_{*-1}) & \overset{\Theta_{y'_0,\pi'}}{\to} & H^2_{z'}(X'; M) \\
\uparrow & & \uparrow \\
H^1_z(\overline{y_0}; M_{*-1}) & \overset{\Theta_{y,\pi}}{\to} & H^2_z(X; M_*)
\end{array}
$$

Thus all the morphisms in this diagram are isomorphisms.   □

Theorem 2.46. *Let M_* be a functor $\mathcal{F}_k \to \mathcal{A}b_*$ endowed with data (**D4**) (i), (**D4**) (ii) and (**D4**) (iii) and satisfying the Axioms (**B0**), (**B1**), (**B2**), (**B3**), (**HA**), (**B4**) and (**B5**).*

Then for each n, endowed with the s_v's constructed in Lemma 2.37, M_n is an unramified $\mathcal{F}_k$-abelian group datum in the sense of Definition 2.9. By Lemma 2.12 it thus defines an unramified sheaf of abelian groups on Sm_k that we still denote by M_n.

Moreover M_n is strongly $\mathbb{A}^1$-invariant.

Proof of Theorem 2.46. The previous results (Lemmas 2.36 and 2.38) have already established that M_n is an unramified sheaf of abelian groups on $\tilde{Sm}_k$, satisfying all the Axioms for unramified sheaves on Sm_k except Axiom (**A4**) that we establish below.

Axiom (**A2'**) is proven in Corollary 2.41. Axiom (**A5**)(i) is clear and Axiom (**A5**)(ii) holds by Lemma 2.45. Axiom (**A6**) holds by Corollary 2.42. Theorem 2.27 then establishes that each M_n is strongly $\mathbb{A}^1$-invariant.

The only remaining point is thus to check Axiom (**A4**). By Remark 2.17 to prove (**A4**) in general it is sufficient to treat the case where the residue fields are infinite. We will freely use this remark in the proof below.

We start by checking the first part of Axiom (**A4**). Let $X = Spec(A)$ be a local smooth k-scheme of dimension 2 with closed point z and function field F. Let $y_0 \in X^{(1)}$ be such that $\overline{y_0}$ is smooth over k. Choose a pair (π_0, π_1) of generators for the maximal ideal of A, such that π_0 defines y_0. Clearly $\overline{\pi_1} \in \mathcal{O}(\overline{y_0})$ is a uniformizing element for $z \in \mathcal{O}(\overline{y_0})$.

We consider the complex (2.5) of X with coefficients in M_* and the induced commutative square:

$$
\begin{array}{ccc}
M_*(F) & \xrightarrow{\Sigma_{y\in X^{(1)}-\{y_0\}}\partial_y} & \oplus_{y\in X^{(1)}-\{y_0\}}H_y^1(X;M_*) \\
\downarrow \partial_{y_0} & & \downarrow -\Sigma_{y\in X^{(1)}-\{y_0\}}\partial_z^y \\
H_{y_0}^1(X;M_*) & \xrightarrow{\partial_z^{y_0}} & H_z^2(X;M_*)
\end{array}
$$

We put this square at the top of the commutative square

$$
\begin{array}{ccc}
H_{y_0}^1(X;M_*) & \xrightarrow{\partial_z^{y_0}} & H_z^2(X;M_*) \\
\downarrow \wr & & \downarrow \wr \\
M_{*-1}(\kappa(y_0)) & \xrightarrow{\partial_z^{\pi_1}} & M_{*-2}(\kappa(z))
\end{array}
$$

where $H_{y_0}^1(X;M_*) \xrightarrow{\sim} M_{*-1}(\kappa(y_0))$ is the inverse to the canonical isomorphism θ_{π_0} induced by π_0, and where $H_z^2(X;M_*) \xrightarrow{\sim} M_{*-2}(\kappa(z))$ is obtained by composing the inverse to the isomorphism Θ_{y_0,π_0} obtained by the previous lemma and $\theta_{\overline{\pi_1}}$.

Now we add on the left top corner the morphism $M_{*-1}(\mathcal{O}_{X,y_0}) \to M_*(F)$, $\alpha \mapsto [\pi_0]\,\alpha$. We thus get a commutative square of the form:

$$
\begin{array}{ccccc}
M_{*-1}(\mathcal{O}_{\pi_0}) & \overset{[\pi_0]\cdot -}{\to} & M_*(F) & \overset{\Sigma_{y \in X^{(1)} - \{y_0\}} \partial_y}{\longrightarrow} & \oplus_{y \in X^{(1)} - \{y_0\}} H_y^1(X; M_*) \\
\downarrow \partial_{y_0}^{\pi_0} & & & & \downarrow \\
M_{*-1}(\kappa(y_0)) & \overset{\partial_z^{\overline{\pi_1}}}{\longrightarrow} & & & M_{*-2}(\kappa(z))
\end{array}
$$

$$(2.8)$$

As for $y \neq y_0$, π_0 is unit in $\mathcal{O}_{X,y}$ we see that if $\alpha \in \cap_{y \in X^{(1)}} M_*(\mathcal{O}_y)$ the image of α through the composition $M_{*-1}(\mathcal{O}_{y_0}) \overset{[\pi_0]\cdot -}{\to} M_*(F) \overset{\Sigma_{y \in X^{(1)} - \{y_0\}} \partial_y}{\longrightarrow}$ $\oplus_{y \in X^{(1)} - \{y_0\}} H_y^1(X; M_*)$ is zero. By the commutativity of the above diagram this shows that the image of such an α through $s_{y_0} = \partial_{y_0}^{\pi_0}([y_0].-)$ lies in the kernel of $\partial_z^{\overline{\pi_1}}$. But this kernel is $M_{*-1}(\mathcal{O}_{\overline{y_0},z})$ and this proves the first part of Axiom (**A4**) (for M_{*-1} thus) for M_*.

Now we prove the second part of Axiom (**A4**). Let $y_1 \in X^{(1)}$ be such that $\overline{y_1}$ is smooth over k and different from $\overline{y_0}$. The intersection $\overline{y_0} \cap \overline{y_1}$ is the point z as a closed subset. If $\overline{y_0}$ and $\overline{y_1}$ do not intersect transversally, we may choose (at least when $\kappa(z)$ is infinite which we may assume by Remark 2.17) a $y_2 \in X^{(1)}$ which intersects transversally both $\overline{y_0}$ and $\overline{y_1}$. Thus we may reduce to the case, that $\overline{y_0}$ and $\overline{y_1}$ do intersect transversally.

Choose $\pi_1 \in A$ which defines $\overline{y_1}$; (π_0, π_1) generate the maximal ideal of A. Now we want to prove that the two morphisms $\cap_{y \in X^{(1)}} M_*(\mathcal{O}_y) \to M_{*-2}(\kappa(z))$ obtained by using y_0 is the same as the one obtained by using y_1.

We contemplate the complex (2.5) for X and expand the equation $\partial \circ \partial = 0$ for the elements of the form $[\pi_0][\pi_1]\alpha$ with $\alpha \in \cap_{y \in X^{(1)}} M_*(\mathcal{O}_y)$. From our axioms it follows that if $y \neq y_0$ and $y \neq y_1$ then $\partial_y([\pi_0][\pi_1]\alpha) = 0$. Now $\partial_{y_1}^{\pi_1}([\pi_0][\pi_1]\alpha)$ is $[\overline{\pi_0}]s_{y_1}(\alpha) \in M_{*-1}(\kappa(y_1)) \overset{\theta_{y_1}}{\cong} H_{y_1}^1(X; M_*)$ and $\partial_{y_0}^{\pi_0}([\pi_0][\pi_1]\alpha)$ is (using Axiom (**B0**)) $- <-1> [\overline{\pi_1}]s_{y_0}(\alpha) \in M_{*-1}(\kappa(y_0)) \overset{\theta_{y_0}}{\cong} H_{y_0}^1(X; M_*)$. Now we compute the last boundary morphism and find that the sum

$$
\Theta_{y_1, \pi_1} \circ \theta_{\overline{\pi_0}}(s_z^{\overline{\pi_0}} \circ s_{y_1}(\alpha)) + \Theta_{y_0, \pi_0} \circ \theta_{\overline{\pi_1}}(- <-1> s_z^{\overline{\pi_1}} \circ s_{y_0}(\alpha)) = 0
$$

vanishes in $H_z^2(X; M)$ (as $\partial \circ \partial = 0$). Lemma 2.47 below exactly yields, from this, the required equality $s_z \circ s_{y_1}(\alpha) = s_z \circ s_{y_0}(\alpha)$. $\qquad\square$

Lemma 2.47. *Assume that M_* is as above. Let $X = Spec(A)$ be a local smooth k-scheme of dimension 2, with field of functions F and closed point z. Let (π_0, π_1) be elements of A generating the maximal ideal of A and let $y_0 \in X^{(1)}$ the divisor of X corresponding to π_0 and $y_1 \in X^{(1)}$ that corresponding to π_0. Assume both are smooth over k. Then the composed isomorphism*

$$M_{*-2}(\kappa(v)) \overset{\theta_{\pi_1}}{\cong} H^1_z(\overline{y_0}; M_{*-1}) \overset{\Theta_{y_0,\pi_0}}{\cong} H^2_z(X;M)$$

is equal to $< -1 >$ times the isomorphism

$$M_{*-2}(\kappa(v)) \overset{\theta_{\pi_0}}{\cong} H^1_z(\overline{y_1}; M_{*-1}) \overset{\Theta_{y_1,\pi_1}}{\cong} H^2_z(X;M)$$

Proof. We first observe that if $f : X' \to X$ is an étale morphism, with X' smooth local of dimension two, with closed point z' having the same residue field as z, and if y'_0 and y'_1 denote respectively the pull-back of y_0 and y_1, then the elements (π_0, π_1) of $A' = \mathcal{O}(X')$ satisfy the same conditions. Clearly, by the previous Lemma, the assertion is true for X if and only if it is true for X', because the θ_π's and $\Theta_{y,\pi}$'s are compatible. Now there is a Nisnevich neighborhood of z: $\Omega \to X$ and an étale morphism $\Omega \to (\mathbb{A}^2_{\kappa(z)})(0,0)$ which is also an étale neighborhood and such that (π_0, π_1) corresponds to the coordinates (T_0, T_1). In this way we reduce to the case $X = (\mathbb{A}^2_{\kappa(z)})(0,0)$ and $(\pi_0, \pi_1) = (T_0, T_1)$.

Now one reapplies exactly the same computation as in the proof of the Theorem to elements of the form $[T_0][T_1](\alpha|_{F(T_0,T_1)}) \in M_*(F(T_0, T_1))$ with $\alpha \in M_{*-2}(F)$. Now the point is that using our axioms $s^{\overline{T_0}}_{(0,0)} \circ s_{Y_1}(\alpha|_{F(T_0,T_1)}) = s^{\overline{T_0}}_{(0,0)}(\alpha|_{F(T_0)}) = \alpha$ and the same holds for the other term. We thus get from the proof the equality, for each $\alpha \in M_{*-2}(F)$

$$\Theta_{Y_1,T_1} \circ \theta_{\overline{T_0}}(\alpha) = \Theta_{Y_0,T_0} \circ \theta_{\overline{T_1}}(< -1 > \alpha)$$

which proves our claim. □

Let M_* be as above. For any discrete valuation v on $F \in \mathcal{F}_k$ the image of $(\mathcal{O}_v)^\times \times M_{(*-1)}(\mathcal{O}_v) \to M_*(F)$, $(u, \alpha) \mapsto [u]\alpha$ lies in $M_*(\mathcal{O}_v)$. This produces for each $n \in \mathbb{Z}$ a morphism of sheaves on Sm_k: $\mathbb{G}_m \times M_{(*-1)} \to M_*$.

Lemma 2.48. *The previous morphism of sheaves induces for any n, an isomorphism* $(M_n)_{-1} \cong M_{(n-1)}$.

Proof. This follows from the short exact sequence

$$0 \to M_n(F) = M_n(\mathbb{A}^1_F) \to M_n(\mathbb{G}_m|_F) \overset{\partial^T_{D_0}}{\longrightarrow} M_{n-1}(F) \to 0$$

given by Axiom **(HA) (i)**. □

Remark 2.49. 1) Conversely given a $\mathbb{Z}$-graded abelian sheaf M_* on Sm_k, consisting of strongly $\mathbb{A}^1$-invariant sheaves, together with isomorphisms $(M_n)_{-1} \cong M_{(n-1)}$, then one may show that evaluation on fields yields a functor $\mathcal{F}_k \to \mathcal{A}b_*$ to $\mathbb{Z}$-graded abelian groups together with Data **(D4)** (i), (D4) (ii) and (D4) (iii) satisfying Axioms (B0), (B1), (B2), (B3), (HA), (B4) and (B5). This is an equivalence of categories.

2) We will prove in Chap. 5 that any strongly $\mathbb{A}^1$-invariant sheaf is strictly $\mathbb{A}^1$-invariant. Thus the previous category is also equivalent to that of homotopy modules over k consisting of $\mathbb{Z}$-graded strictly $\mathbb{A}^1$-invariant abelian sheaves M_* on Sm_k, together with isomorphisms $(M_n)_{-1} \cong M_{(n-1)}$; see also [21]. This category is known to be the heart of the *homotopy t-structure* on the stable $\mathbb{A}^1$-homotopy category of $\mathbb{P}^1$-spectra over k, see [50–52]. □

Remark 2.50. Our approach can be used also to analyze Rost cycle modules [68] over a perfect field k. Then Rost's Axioms imply the existence of a obvious forgetful functor from his category of cycle modules over k to the category of M_* as above in the Theorem, with trivial $\mathbb{Z}[F^\times]$-module structure, that is to say $< u >= 1$ for each $u \in F^\times$. This can be shown to be an equivalence of categories (using for instance [21] or by direct inspection using our construction of transfers in Sect. 4.2). In particular, in the concept of cycle module, one might forget the transfers but should keep track of some consequences like Axioms (B4) and (B5) to get an equivalent notion. □

Chapter 3
Unramified Milnor–Witt K-Theories

Our aim in this section is to compute (or describe), for any integer $n > 0$, the free strongly $\mathbb{A}^1$-invariant sheaf generated by the n-th smash power of $\mathbb{G}_m$, in other words the "free strongly $\mathbb{A}^1$-invariant sheaf on n units". As we will prove in Chap. 5 that any strongly $\mathbb{A}^1$-invariant sheaf of abelian groups is also strictly $\mathbb{A}^1$-invariant, this is also the free strictly $\mathbb{A}^1$-invariant sheaf on $(\mathbb{G}_m)^{\wedge n}$.

3.1 Milnor–Witt K-Theory of Fields

The following definition was found in collaboration with Mike Hopkins:

Definition 3.1. Let F be a commutative field. The Milnor–Witt K-theory of F is the graded associative ring $K_*^{MW}(F)$ generated by the symbols $[u]$, for each unit $u \in F^\times$, of degree $+1$, and one symbol η of degree -1 subject to the following relations:

1 (Steinberg relation) For each $a \in F^\times - \{1\}$: $[a].[1-a] = 0$
2 For each pair $(a, b) \in (F^\times)^2$: $[ab] = [a] + [b] + \eta.[a].[b]$
3 For each $u \in F^\times$: $[u].\eta = \eta.[u]$
4 Set $h := \eta.[-1] + 2$. Then $\eta \, . \, h = 0$

These Milnor–Witt K-theory groups were introduced by the author in a different (and more complicated) way, until the previous presentation was found with Mike Hopkins. The advantage of this presentation was made clear in our computations of the stable $\pi_0^{\mathbb{A}^1}$ in [50,51] as the relations all have very natural explanations in the stable $\mathbb{A}^1$-homotopical world. To perform these computations in the unstable world and also to produce unramified Milnor–Witt K-theory sheaves in a completely elementary way, over any field (any characteristic) we will need to use an "unstable" variant of that presentation in Lemma 3.4.

F. Morel, $\mathbb{A}^1$-*Algebraic Topology over a Field*, Lecture Notes in Mathematics 2052, 49
DOI 10.1007/978-3-642-29514-0_3, © Springer-Verlag Berlin Heidelberg 2012

Remark 3.2. The quotient ring $K_*^{MW}(F)/\eta$ is the Milnor K-theory $K_*^M(F)$ of F defined in [47]: indeed if η is killed, the symbol $[u]$ becomes additive. Observe precisely that η controls the failure of $u \mapsto [u]$ to be additive in Milnor–Witt K-theory.

With all this in mind, it is natural to introduce the Witt K-theory of F as the quotient $K_*^W(F) := K_*^{MW}(F)/h$. It was studied in [54] and will also be used in our computations below. In *loc. cit.* it was proven that the non-negative part is the quotient of the ring $Tens_{W(F)}(I(F))$ by the Steinberg relation $<< u >> . << 1 - u >>$. This can be shown to still hold in characteristic 2.

Proceeding along the same line, it is easy to prove that the non-negative part $K_{\geq 0}^{MW}(F)$ is isomorphic to the quotient of the ring $Tens_{K_0^{MW}(F)}(K_1^{M\widetilde{W}}(F))$ by the Steinberg relation $[u].[1-u]$. This is related to our old definition of $K_*^{MW}(F)$. $\qquad\square$

We will need at some point a presentation of the group of weight n Milnor–Witt K-theory. The following one will suffice for our purpose. One may give some simpler presentation but we won't use it:

Definition 3.3. Let F be a commutative field. Let n be an integer. We let $\tilde{K}_n^{MW}(F)$ denote the abelian group generated by the symbols of the form $[\eta^m, u_1, \ldots, u_r]$ with $m \in \mathbb{N}$, $r \in \mathbb{N}$, and $n = r - m$, and with the u_i's unit in F, and subject to the following relations:

1_n (Steinberg relation) $[\eta^m, u_1, \ldots, u_r] = 0$ if $u_i + u_{i+1} = 1$, for some i.
2_n For each pair $(a, b) \in (F^\times)^2$ and each i: $[\eta^m, \ldots, u_{i-1}, ab, u_{i+1}, \ldots] = [\eta^m, \ldots, u_{i-1}, a, u_{i+1}, \ldots] + [\eta^m, \ldots, u_{i-1}, b, u_{i+1}, \ldots] + [\eta^{m+1}, \ldots, u_{i-1}, a, b, u_{i+1}, \ldots]$.
4_n For each i, $[\eta^{m+2}, \ldots, u_{i-1}, -1, u_{i+1}, \ldots] + 2[\eta^{m+1}, \ldots, u_{i-1}, u_{i+1}, \ldots] = 0$.

The following lemma is straightforward:

Lemma 3.4. *For any field F, any integer $n \geq 1$, the correspondence*

$$[\eta^m, u_1, \ldots, u_n] \mapsto \eta^m [u_1] \ldots [u_n]$$

induces an isomorphism

$$\tilde{K}_n^{MW}(F) \cong K_n^{MW}(F)$$

Proof. The proof consists in expressing the possible relations between elements of degree n. That is to say the element of degree n in the two-sided ideal generated by the relations of Milnor–Witt K-theory, except the number 3, which is encoded in our choices. We leave the details to the reader. $\qquad\square$

Now we establish some elementary but useful facts. For any unit $a \in F^\times$, we set $< a >= 1 + \eta[a] \in K_0^{MW}(F)$. Observe then that $h = 1 + < -1 >$.

Lemma 3.5. *Let* $(a, b) \in (F^\times)^2$ *be units in* F. *We have the followings formulas:*

1) $[ab] = [a] + < a > . [b] = [a] . < b > + [b]$;
2) $< ab > = < a > . < b >$; $K_0^{MW}(F)$ *is central in* $K_*^{MW}(F)$;
3) $< 1 >= 1$ *in* $K_0^{MW}(F)$ *and* $[1] = 0$ *in* $K_1^{MW}(F)$;
4) $< a >$ *is a unit in* $K_0^{MW}(F)$ *whose inverse is* $< a^{-1} >$;
5) $[\frac{a}{b}] = [a] - < \frac{a}{b} > . [b]$. *In particular one has:* $[a^{-1}] = - < a^{-1} > . [a]$.

Proof. (1) is obvious. One obtains the first relation of (2) by applying η to relation **2** and using relation **3**. By (1) we have for any a and b: $< a > . [b] = [b] . < a >$ thus the elements $< a >$ are central.

Multiplying relation **4** by $[1]$ (on the left) implies that $(< 1 > -1) . (< -1 > +1) = 0$ (observe that $h = 1 + < -1 >$). Using **2** this implies that $< 1 >= 1$. By (1) we have now $[1] = [1] + < 1 > . [1] = [1] + 1 . [1] = [1] + [1]$; thus $[1] = 0$. (4) follows clearly from (2) and (3). (5) is an easy consequence of (1)–(4). $\square$

Lemma 3.6. 1) *For each* $n \geq 1$, *the group* $K_n^{MW}(F)$ *is generated by the products of the form* $[u_1]. \ldots . [u_n]$, *with the* $u_i \in F^\times$.
2) *For each* $n \leq 0$, *the group* $K_n^{MW}(F)$ *is generated by the products of the form* $\eta^n . < u >$, *with* $u \in F^\times$. *In particular the product with* η: $K_n^{MW}(F) \to K_{n-1}^{MW}(F)$ *is always surjective if* $n \leq 0$.

Proof. An obvious observation is that the group $K_n^{MW}(F)$ is generated by the products of the form $\eta^m . [u_1]. \ldots . [u_\ell]$ with $m \geq 0$, $\ell \geq 0$, $\ell - m = n$ and with the u_i's units. The relation **2** can be rewritten $\eta . [a] . [b] = [ab] - [a] - [b]$. This easily implies the result using the fact that $< 1 >= 1$. $\square$

Remember that $h = 1 + < -1 >$. Set $\epsilon := - < -1 > \in K_0^{MW}(F)$. Observe then that relation **4** in Milnor–Witt K-theory can also be rewritten $\epsilon . \eta = \eta$.

Lemma 3.7. 1) *For* $a \in F^\times$ *one has:* $[a] . [-a] = 0$ *and* $< a > + < -a >= h$;
2) *For* $a \in F^\times$ *one has:* $[a] . [a] = [a] . [-1] = \epsilon [a] [-1] = [-1] . [a] = \epsilon [-1] [a]$;
3) *For* $a \in F^\times$ *and* $b \in F^\times$ *one has* $[a] . [b] = \epsilon . [b] . [a]$;
4) *For* $a \in F^\times$ *one has* $< a^2 >= 1$.

Corollary 3.8. *The graded* $K_0^{MW}(F)$-*algebra* $K_*^{MW}(F)$ *is* ϵ-*graded commutative: for any element* $\alpha \in K_n^{MW}(F)$ *and any element* $\beta \in K_m^{MW}(F)$ *one has*

$$\alpha . \beta = (\epsilon)^{n.m} \beta . \alpha$$

Proof. It suffices to check this formula on the set of multiplicative generators $F^\times \amalg \{\eta\}$: for products of the form $[a] . [b]$ this is (3) of the previous Lemma. For products of the form $[a] . \eta$ or $\eta . \eta$, this follows from the relation **3** and relation **4** (reading $\epsilon . \eta = \eta$) in Milnor–Witt K-theory. $\square$

Proof of Lemma 3.7. We adapt [47]. Start from the equality (for $a \neq 1$) $-a = \frac{1-a}{1-a^{-1}}$. Then $[-a] = [1-a] - <-a> .[1-a^{-1}]$. Thus

$$[a].[-a] = [a][1-a] - <-a> .[a].[1-a^{-1}] = 0 - <-a> .[a].[1-a^{-1}]$$

$$= <-a><a> [a^{-1}][1-a^{-1}] = 0$$

by **1** and (1) of lemma 3.5. The second relation follows from this by applying η^2 and expanding.

As $[-a] = [-1] + <-1> [a]$ we get

$$0 = [a].[-1] + <-1> [a][a]$$

so that $[a].[a] = - <-1> [a].[-1] = [a].[-1]$ because $0 = [1] = [-1] + < -1 > [-1]$. Using $[-a][a] = 0$ we find $[a][a] = - <-1> [-1][a] = [-1][a]$.

Finally expanding

$$0 = [ab].[-ab] = ([a] + <a> .[b])([-a] + <-a> [b])$$

gives

$$0 = <a> ([b][-a] + <-1> [a][b]) + <-1> [-1][b]$$

as $[-a] = [a] + <a> [-1]$ we get

$$0 = <a> ([b][a] + <-1> [a][b]) + [b][-1] + <-1> [-1][b]$$

the last term is 0 by (3) so that we get the third claim.

The fourth one is obtained by expanding $[a^2] = 2[a] + \eta[a][a]$; now due to point (2) we have $[a^2] = 2[a] + \eta[-1][a] = (2 + \eta[-1])[a] = h[a]$. Applying η we thus get 0. □

Let us denote (in any characteristic) by $GW(F)$ the Grothendieck–Witt ring of isomorphism classes of non-degenerate symmetric bilinear forms [48]: this is the group completion of the commutative monoid of isomorphism classes of non-degenerate symmetric bilinear forms for the direct sum.

For $u \in F^\times$, we denote by $< u > \in GW(F)$ the form on the vector space of rank one F given by $F^2 \to F$, $(x, y) \mapsto uxy$. By the results of *loc. cit.*, these $< u >$ generate $GW(F)$ as a group. The following Lemma is (essentially) [48, Lemma (1.1) Chap. IV]:

Lemma 3.9. *[48] The group $GW(F)$ is generated by the elements $< u >$, $u \in F^\times$, and the following relations give a presentation of $GW(F)$:*

(i) $< u(v^2) > = < u >$;
(ii) $< u > + < -u > = 1 + < -1 >$;
(iii) $< u > + < v > = < u + v > + < (u + v)uv >$ *if* $(u + v) \neq 0$.

When $char(F) \neq 2$ the first two relations imply the third one and one obtains the standard presentation of the Grothendieck–Witt ring $GW(F)$, see [69]. If $char(F) = 2$ the third relation becomes $2(< u > -1) = 0$.

We observe that the subgroup (h) of $GW(F)$ generated by the hyperbolic plan $h = 1+ < -1 >$ is actually an ideal (use the relation (ii)). We let $W(F)$ be the quotient (both as a group or as a ring) $GW(F)/(h)$ and let $W(F) \to \mathbb{Z}/2$ be the corresponding mod 2 rank homomorphism; $W(F)$ is the Witt ring of F [48], and [69] in characteristic $\neq 2$. Observe that the following commutative square of commutative rings

$$
\begin{array}{ccc}
GW(F) & \to & \mathbb{Z} \\
\downarrow & & \downarrow \\
W(F) & \to & \mathbb{Z}/2
\end{array}
\tag{3.1}
$$

is cartesian. The kernel of the mod 2 rank homomorphism $W(F) \to \mathbb{Z}/2$ is denoted by $I(F)$ and is called the fundamental ideal of $W(F)$.

It follows from our previous results that $u \mapsto < u > \in K_0^{MW}(F)$ satisfies all the relations defining the Grothendieck–Witt ring. Only the last one requires a comment. As the symbol $< u >$ is multiplicative in u, we may reduce to the case $u + v = 1$ by dividing by $< u + v >$ if necessary. In that case, this follows from the Steinberg relation to which one applies η^2. We thus get a ring epimorphism (surjectivity follows from Lemma 3.6)

$$
\phi_0 : GW(F) \twoheadrightarrow K_0^{MW}(F)
$$

For $n > 0$ the multiplication by $\eta^n : K_0^{MW}(F) \to K_{-n}^{MW}(F)$ kills h (because $h.\eta = 0$ and thus we get an epimorphism:

$$
\phi_{-n} : W(F) \twoheadrightarrow K_{-n}^{MW}(F)
$$

Lemma 3.10. *For each field F, each $n \geq 0$ the homomorphism ϕ_{-n} is an isomorphism.*

Proof. Following [7], let us define by $J^n(F)$ the fiber product $I^n(F) \times_{i^n(F)} K_n^M(F)$, where we use the Milnor epimorphism $s_n : K_n^M(F)/2 \twoheadrightarrow i^n(F)$, with $i^n(F) := I^n(F)/I^{(n+1)}(F)$. For $n \leq 0$, $I^n(F)$ is understood to be $W(F)$. Now altogether the $J^*(F)$ form a graded ring and we denote by $\eta \in J^{-1}(F) = W(F)$ the element $1 \in W(F)$. For any $u \in F^\times$, denote by $[u] \in J^1(F) \subset I(F) \times F^\times$ the pair $(< u > -1, u)$. Then the four relations hold in $J^*(F)$ which produces an epimorphism $K_*^{MW}(F) \twoheadrightarrow J^*(F)$. For $n > 0$ the composition of epimorphisms $W(F) \to K_{-n}^{MW}(F) \to J^{-n}(F) = W(F)$ is the identity. For $n = 0$ the composition $GW(F) \to K_0^{MW}(F) \to J^0(F) = GW(F)$ is also the identity. The Lemma is proven. $\qquad\square$

Corollary 3.11. *The canonical morphism of graded rings*

$$K_*^{MW}(F) \to W(F)[\eta, \eta^{-1}]$$

induced by $[u] \mapsto \eta^{-1}(<u> -1)$ *induces an isomorphism*

$$K_*^{MW}(F)[\eta^{-1}] = W(F)[\eta, \eta^{-1}]$$

Remark 3.12. For any F let $I^*(F)$ denote the graded ring consisting of the powers of the fundamental ideal $I(F) \subset W(F)$. We let $\eta \in I^{-1}(F) = W(F)$ be the generator. Then the product with η acts as the inclusions $I^n(F) \subset I^{n-1}(F)$. We let $[u] =<u> -1 \in I(F)$ be the opposite to the Pfister form $<< u >>= 1- < u >$. Then these symbol satisfy the relations of Milnor–Witt K-theory [54] and the image of h is zero. We obtain in this way an epimorphism $K_*^W(F) \twoheadrightarrow I^*(F)$, $[u] \mapsto < u > -1 = - << u >>$. This ring $I^*(F)$ is exactly the image of the morphism $K_*^{MW}(F) \to W(F)[\eta, \eta^{-1}]$ considered in the Corollary above.

We have proven that this is always an isomorphism in degree ≤ 0. In fact this remains true in degree 1, see Corollary 3.47 for a stronger version. In fact it was proven in [54] (using [1] and Voevodsky's proof of the Milnor conjectures) that

$$K_*^W(F) \twoheadrightarrow I^*(F) \tag{3.2}$$

is an isomorphism in characteristic $\neq 2$. Using Kato's proof of the analogues of those conjectures in characteristic 2 [37] we may extend this result for any field F.

From that we may also deduce (as in [54]) that the obvious epimorphism

$$K_*^W(F) \twoheadrightarrow J^*(F) \tag{3.3}$$

is always an isomorphism.   □

Here is a very particular case of the last statement, but completely elementary:

Proposition 3.13. *Let F be a field for which any unit is a square. Then the epimorphism*

$$K_*^{MW}(F) \to K_*^M(F)$$

is an isomorphism in degrees ≥ 0, and the epimorphism

$$K_*^{MW}(F) \to K_*^W(F)$$

is an isomorphism in degrees < 0. In fact $I^n(F) = 0$ for $n > 0$ and $I^n(F) = W(F) = \mathbb{Z}/2$ for $n \leq 0$. In particular the epimorphisms (3.2) and (3.3) are isomorphisms.

Proof. The first observation is that $< -1 >= 1$ and thus $2\eta = 0$ (fourth relation in Milnor–Witt K-theory). Now using Lemma 3.14 below we see that for any unit $a \in F^\times$, $\eta[a^2] = 2\eta[a] = 0$, thus as any unit b is a square, we get that for any $b \in F^\times$, $\eta[b] = 0$. This proves that the second relation of Milnor–Witt K-theory gives for units (a, b) in F: $[ab] = [a] + [b] + \eta[a][b] = [a] + [b]$. The proposition now follows easily from these observations.   □

Lemma 3.14. *Let $a \in F^\times$ and let $n \in \mathbb{Z}$ be an integer. Then the following formula holds in $K_1^{MW}(F)$:*

$$[a^n] = n_\epsilon [a]$$

where for $n \geq 0$, where $n_\epsilon \in K_0^{MW}(F)$ is defined as follows

$$n_\epsilon = \sum_{i=1}^{n} < (-1)^{(i-1)} >$$

(and satisfies for $n > 0$ the relation $n_\epsilon =< -1 > (n-1)_\epsilon + 1)$ and where for $n \leq 0$, $n_\epsilon := - < -1 > (-n)_\epsilon$.   □

Proof. The proof is quite straightforward by induction: one expands $[a^n] = [a^{n-1}] + [a] + \eta[a^{n-1}][a]$ as well as $[a^{-1}] = - < a > [a] = -([a] + \eta[a][a])$.   □

3.2   Unramified Milnor–Witt K-Theories

In this section we will define for each $n \in \mathbb{Z}$ an explicit sheaf $\mathbf{K}_n^{MW}$ on Sm_k called *unramified* Milnor–Witt K-theory in weight n, whose sections on any field $F \in \mathcal{F}_k$ is the group $K_n^{MW}(F)$. In the next section we will prove that for $n > 0$ this sheaf $\mathbf{K}_n^{MW}$ is the free strongly $\mathbb{A}^1$-invariant sheaf generated by $(\mathbb{G}_m)^{\wedge n}$.

Residue homomorphisms. Recall from [47], that for any discrete valuation v on a field F, with valuation ring $\mathcal{O}_v \subset F$, and residue field $\kappa(v)$, one can define a unique homomorphism (of graded groups)

$$\partial_v : K_*^M(F) \to K_{*-1}^M(\kappa(v))$$

called "residue" homomorphism, such that

$$\partial_v(\{\pi\}\{u_2\}\ldots\{u_n\}) = \{\overline{u_2}\}\ldots\{\overline{u_n}\}$$

for any uniformizing element π and units $u_i \in \mathcal{O}_v^\times$, and where $\overline{u}$ denotes the image of $u \in \mathcal{O}_v \cap F^\times$ in $\kappa(v)$.

In the same way, given a uniformizing element π, one has:

Theorem 3.15. *There exists one and only one morphism of graded groups*

$$\partial_v^\pi : K_*^{MW}(F) \to K_{*-1}^{MW}(\kappa(v))$$

which commutes to product by η and satisfying the formulas:

$$\partial_v^\pi([\pi][u_2]\dots[u_n]) = [\overline{u_2}]\dots[\overline{u_n}]$$

and

$$\partial_v^\pi([u_1][u_2]\dots[u_n]) = 0$$

for any units $u_1, \dots, u_n$ of $\mathcal{O}_v$.

Proof. Uniqueness follows from the following Lemma as well as the formulas $[a][a] = [a][-1]$, $[ab] = [a] + [b] + \eta[a][b]$ and $[a^{-1}] = - < a > [a] = -([a] + \eta[a][a])$. The existence follows from Lemma 3.16 below.   □

To define the residue morphism ∂_v^π we use the method of Serre [47]. Let ξ be a variable of degree 1 which we adjoin to $K_*^{MW}(\kappa(v))$ with the relation $\xi^2 = \xi[-1]$; we denote by $K_*^{MW}(\kappa(v))[\xi]$ the graded ring so obtained.

Lemma 3.16. *Let v be a discrete valuation on a field F, with valuation ring $\mathcal{O}_v \subset F$ and let π be a uniformizing element of v. The map*

$$\mathbb{Z} \times \mathcal{O}_v^\times = F^\times \to K_*^{MW}(\kappa(v))[\xi]$$

$$(\pi^n.u) \mapsto \Theta_\pi(\pi^n.u) := [\overline{u}] + (n_\epsilon < \overline{u} >).\xi$$

and $\eta \mapsto \eta$ satisfies the relations of Milnor–Witt K-theory and induce a morphism of graded rings:

$$\Theta_\pi : K_*^{MW}(F) \to K_*^{MW}(\kappa(v))[\xi]$$

Proof. We first prove the first relation of Milnor–Witt K-theory. Let $\pi^n.u \in F^\times$ with u in $\mathcal{O}_v^\times$. We want to prove $\Theta_\pi(\pi^n.u)\Theta_\pi(1 - \pi^n.u) = 0$ in $K_*^{MW}(\kappa(v))[\xi]$. If $n > 0$, then $1 - \pi^n.u$ is in $\mathcal{O}_v^\times$ and by definition $\Theta_\pi(1 - \pi^n.u) = 0$. If $n = 0$, then write $1 - u = \pi^m.v$ with v a unit in $\mathcal{O}_v$. If $m > 0$ the symmetric reasoning allows to conclude. If $m = 0$, then $\Theta_\pi(u) = [\overline{u}]$ and $\Theta_\pi(1 - u) = [1 - \overline{u}]$ in which case the result is also clear.

It remains to consider the case $n < 0$. Then $\Theta_\pi(\pi^n.u) = [\overline{u}] + (n_\epsilon < \overline{u} >)\xi$. Moreover we write $(1 - \pi^n.u)$ as $\pi^n(-u)(1 - \pi^{-n}u^{-1})$ and we observe that $(-u)(1-\pi^{-n}u^{-1})$ is a unit on $\mathcal{O}_v$ so that $\Theta_\pi(1-\pi^n.u) = [-\overline{u}]+n_\epsilon < -\overline{u} > \xi$. Expanding $\Theta_\pi(\pi^n.u)\Theta_\pi(1 - \pi^n.u)$ we find $[\overline{u}][-\overline{u}] + n_\epsilon < \overline{u} > \xi[-\overline{u}] + n_\epsilon < -\overline{u} > [\overline{u}][\xi] + (n_\epsilon)^2 < -1 > \xi^2$. We observe that $[\overline{u}][-\overline{u}] = 0$ and that $(n_\epsilon)^2 < -1 > \xi^2 = (n_\epsilon)^2[-1] < -1 > \xi = n_\epsilon < -1 > \xi[-1]$ because $(n_\epsilon)^2[-1] = n_\epsilon[-1]$ (this follows from Lemma 3.14 : $(n_\epsilon)^2[-1] = n_\epsilon[(-1)^n] = [(-1)^{n^2}] = [(-1)^n]$ as $n^2 - n$ is even). Thus $\Theta_\pi(\pi^n.u)\Theta_\pi(1 - \pi^n.u) = n_\epsilon\{??\}\xi$

where the expression $\{??\}$ is

$$< -\overline{u} > ([\overline{u}] - [-\overline{u}]) + < -1 > [-1]$$

But $[\overline{u}] - [-\overline{u}] = [\overline{u}] - [\overline{u}] - [-1] - \eta[\overline{u}][-1] = - < \overline{u} > [-1]$ thus $< -\overline{u} > ([\overline{u}] - [-\overline{u}]) = - < -1 > [-1]$, proving the result.

We now check relation 2 of Milnor–Witt K-theory. Expanding we find that the coefficient which doesn't involve ξ is 0 and the coefficient of ξ is

$$n_\epsilon < \overline{u} > + m_\epsilon < \overline{v} > - n_\epsilon < -\overline{u} > (< \overline{v} > -1) + m_\epsilon < \overline{v} > (< u > -1)$$

$$+ n_\epsilon m_\epsilon < \overline{uv} > (< -1 > -1)$$

A careful computation (using $< \overline{u} > + < -\overline{u} > = < 1 > + < -1 > = < \overline{uv} > + < -\overline{uv} >$ yields that this term is

$$n_\epsilon + m_\epsilon - n_\epsilon m_\epsilon + < -1 > n_\epsilon m_\epsilon$$

which is shown to be $(n + m)_\epsilon$. The last two relations of the Milnor–Witt K-theory are very easy to check. $\qquad\square$

We now proceed as in [47], we set for any $\alpha \in K_n^{MW}(F)$:

$$\Theta_\pi(\alpha) := s_v^\pi(\alpha) + \partial_v^\pi(\alpha).\xi$$

The homomorphism ∂_v^π so defined is easily checked to have the required properties. Moreover $s_v^\pi : K_*^{MW}(F) \to K_*^{MW}(\kappa(v))$ is a morphism of rings, and as such is the unique one mapping η to η and $\pi^n u$ to $[\overline{u}]$.

Proposition 3.17. *We keep the previous notations and assumptions. For any $\alpha \in K_*^{MW}(F)$:*

1) $\partial_v^\pi([-\pi].\alpha) = < -1 > s_v^\pi(\alpha)$;
2) $\partial_v^\pi([u].\alpha) = - < -1 > [\overline{u}]\partial_v^\pi(\alpha)$ *for any* $u \in \mathcal{O}_v^\times$.
3) $\partial_v^\pi(< u > .\alpha) = < \overline{u} > \partial_v^\pi(\alpha)$ *for any* $u \in \mathcal{O}_v^\times$.

Proof. We observe that, for $n \geq 1$, $K_n^{MW}(F)$ is generated as group by elements of the form $\eta^m[\pi][u_2] \ldots [u_{n+m}]$ or of the form $\eta^m[u_1][u_2] \ldots [u_{n+m}]$, with the u_i's units of $\mathcal{O}_v$ and with $n + m \geq 1$. Thus it suffices to check the formula on these elements, which is straightforward. $\qquad\square$

Remark 3.18. A heuristic but useful explanation of this "trick" of Serre is the following. $Spec(F)$ is the open complement in $Spec(\mathcal{O}_v)$ of the closed point $Spec(\kappa(v))$. If one had a tubular neighborhood for that closed immersion, there should be a morphism $E(\nu_v) - \{0\} \to Spec(F)$ of the complement of the zero section of the normal bundle to $Spec(F)$; the map θ_π is the map induced in cohomology by this "hypothetical" morphism. Observe that choosing π corresponds to trivializing ν_v, in which case $E(\nu_v) - \{0\}$ becomes $(\mathbb{G}_m)_{Spec(\kappa(v))}$. Then the ring $K_*^{MW}(\kappa(v))[\xi]$ is just the ring of sections of K_*^{MW} on $(\mathbb{G}_m)_{Spec(\kappa(v))}$. The "funny" relation $\xi^2 = \xi[-1]$ which is true for any element in $K_*^{MW}(F)$, can also be explained by the fact that the reduced diagonal $(\mathbb{G}_m)_{Spec(\kappa(v))} \to (\mathbb{G}_m)_{Spec(\kappa(v))}^{\wedge 2}$ is equal to the multiplication by $[-1]$.   □

Lemma 3.19. *For any field extension $E \subset F$ and for any discrete valuation on F which restricts to a discrete valuation w on E with ramification index e. Let π be a uniformizing element of v and ρ a uniformizing element of w. Write it $\rho = u\pi^e$ with $u \in \mathcal{O}_v^\times$. Then for each $\alpha \in K_*^{MW}(E)$ one has*

$$\partial_v^\pi(\alpha|_F) = e_\epsilon < \overline{u} > (\partial_w^\rho(\alpha))|_{\kappa(v)}$$

Proof. We just observe that the square (of rings)

$$
\begin{array}{ccc}
K_*^{MW}(F) & \xrightarrow{\Theta_{\overline{\xi}}} & K_*^{MW}(\kappa(v))[\xi] \\
\uparrow & & \uparrow \Psi \\
K_*^{MW}(E) & \xrightarrow{\Theta_\varsigma} & K_*^{MW}(\kappa(w))[\xi]
\end{array}
$$

where Ψ is the ring homomorphism defined by $[a] \mapsto [a|_F]$ for $a \in \kappa(v)$ and $\xi \mapsto [\overline{u}] + e_\epsilon < \overline{u} > \xi$ is commutative. It is sufficient to check the commutativity in degree 1, which is not hard.   □

Using the residue homomorphism and the previous Lemma one may define for any discrete valuation v on F the subgroup $\underline{\mathbf{K}}_n^{MW}(\mathcal{O}_v) \subset K_n^{MW}(F)$ as the kernel of ∂_v^π. From our previous Lemma (applied to $E = F$, $e = 1$), it is clear that the kernel doesn't depend on π, only on v. We define $H_v^1(\mathcal{O}_v; \underline{\mathbf{K}}_n^{MW})$ as the quotient group $K_n^{MW}(F)/K_n^{MW}(\mathcal{O}_v)$. Once we choose a uniformizing element π we get of course a canonical isomorphism $K_n^{MW}(\kappa(v)) = H_v^1(\mathcal{O}_v; \underline{\mathbf{K}}_n^{MW})$.

Remark 3.20. One important feature of residue homomorphisms is that in the case of Milnor K-theory, these residues homomorphisms don't depend on the choice of π, only the valuation, but in the case of Milnor–Witt K-theory, they do depend on the choice of π: for $u \in \mathcal{O}^\times$, as one has $\partial_v^\pi([u.\pi]) = \partial_v^\pi([\pi]) + \eta.[\overline{u}] = 1 + \eta.[\overline{u}]$.

This property of independence of the residue morphisms on the choice of π is a general fact (in fact equivalent) for the $\mathbb{Z}$-graded unramified sheaves M_* considered above for which the $\mathbb{Z}[F^\times/F^{\times 2}]$-structure is trivial, like Milnor K-theory.   □

Remark 3.21. To make the residue homomorphisms "canonical" (see [7,8,70] for instance), one defines for a field κ and a one dimensional κ-vector space L, twisted Milnor–Witt K-theory groups: $K_*^{MW}(\kappa; L) = K_*^{MW}(\kappa) \otimes_{\mathbb{Z}[\kappa^\times]} \mathbb{Z}[L - \{0\}]$, where the group ring $\mathbb{Z}[\kappa^\times]$ acts through $u \mapsto < u >$ on $K_*^{MW}(\kappa)$ and through multiplication on $\mathbb{Z}[L - \{0\}]$. The canonical residue homomorphism is of the following form

$$\partial_v : K_*^{MW}(F) \to K_{*-1}^{MW}(\kappa(v); m_v/(m_v)^2)$$

with $\partial_v([\pi].[u_2]\ldots[u_n]) = [\overline{u_2}]\ldots[\overline{u_n}] \otimes \overline{\pi}$, where $m_v/(m_v)^2$ is the cotangent space at v (a one dimensional $\kappa(v)$-vector space). We will make this precise in Sect. 4.1 below. $\qquad\square$

The following result and its proof follow closely Bass–Tate [9]:

Theorem 3.22. *Let v be a discrete valuation ring on a field F. Then the subring*

$$\mathbf{K}_*^{MW}(\mathcal{O}_v) \subset K_*^{MW}(F)$$

is as a ring generated by the elements η and $[u] \in K_1^{MW}(F)$, with $u \in \mathcal{O}_v^\times$ a unit of $\mathcal{O}_v$.

Consequently, the group $\mathbf{K}_n^{MW}(\mathcal{O}_v)$ is generated by symbols $[u_1]\ldots[u_n]$ with the u_i's in $\mathcal{O}_v^\times$ for $n \geq 1$ and by the symbols $\eta^{-n} < u >$ with the u's in $\mathcal{O}_v^\times$ for $n \leq 0$

Proof. The last statement follows from the first one as in Lemma 3.6.

We consider the quotient graded abelian group Q_* of $K_*^{MW}(F)$ by the subring A_* generated by the elements and $\eta \in K_{-1}^{MW}(F)$ and $[u] \in K_1^{MW}(F)$, with $u \in \mathcal{O}_v^\times$ a unit of $\mathcal{O}_v$. We choose a uniformizing element π. The valuation morphism induces an epimorphism $Q_* \to K_{*-1}^{MW}(\kappa(v))$. It suffices to check that this is an isomorphism. We will produce an epimorphism $K_{*-1}^{MW}(\kappa(v)) \to Q_*$ and show that the composition $K_{*-1}^{MW}(\kappa(v)) \to Q_* \to K_{*-1}^{MW}(\kappa(v))$ is the identity.

We construct a $K_*^{MW}(\kappa(v))$-module structure on $Q_*(F)$. Denote by $\mathcal{E}_*$ the graded ring of endomorphisms of the graded abelian group $Q_*(F)$. First the element η still acts on Q_* and yields an element $\eta \in \mathcal{E}_{-1}$. Let $a \in \kappa(v)^\times$ be a unit in $\kappa(v)$. Choose a lifting $\tilde{\alpha} \in \mathcal{O}_v^\times$. Then multiplication by $\tilde{\alpha}$ induces a morphism of degree $+1$, $Q_* \to Q_{*+1}$. We first claim that it doesn't depend on the choice of $\tilde{\alpha}$. Let $\tilde{\alpha}' = \beta\tilde{\alpha}$ be another lifting so that $u \in \mathcal{O}_v^\times$ is congruent to 1 mod π. Expanding $[\tilde{\alpha}'] = [\tilde{\alpha}] + [\beta] + \eta[\tilde{\alpha}][\beta]$ we see that it is sufficient to check that for any $a \in F^\times$, the product $[\beta][a]$ lies in the subring A_*. Write $a = \pi^n.u$ with $u \in \mathcal{O}_v^\times$. Then expanding $[\pi^n.u]$ we end up to checking the property for the product $[\beta][\pi^n]$, and using Lemma 3.14 we may even assume $n = 1$. Write $\beta = 1 - \pi^n.v$, with $n > 0$ and $v \in \mathcal{O}_v^\times$.

Thus we have to prove that the products of the above form $[1 - \pi^n.v][\pi]$ are in A_*. For $n = 1$, the Steinberg relation yields $[1 - \pi.v][\pi.v] = 0$. Expanding $[\pi.v] = [\pi](1 + \eta[v]) + [v]$, implies $[1 - \pi.v][\pi](1 + \eta[v])$ is in A_*. But by Lemma 3.7, $1 + \eta[v] =< v >$ is a unit of A_*, with inverse itself. Thus $[1 - \pi.v][\pi] \in A_*$. Now if $n \geq 2$, $1 - \pi^n.v = (1 - \pi) + \pi(1 - \pi^{n-1}v) = (1 - \pi)(1 + \pi(\frac{1-\pi^{n-1}}{1-\pi})) = (1 - \pi)(1 - \pi w)$, with $w \in \mathcal{O}_v^\times$. Expending, we get

$[1 - \pi^n.v][\pi] = [1 - \pi][\pi] + [1 - \pi w][\pi] + \eta[1 - \pi][1 - \pi w][\pi] = [1 - \pi w][\pi]$.
Thus the result holds in general.

We thus define this way elements $[u] \in \mathcal{E}_1$. We now claim these elements (together with η) satisfy the four relations in Milnor–Witt K-theory: this is very easy to check, by the very definitions. Thus we get this way a $K_*^{MW}(\kappa(v))$-module structure on Q_*. Pick up the element $[\pi] \in Q_1 = K_1^{MW}(F)/A_1$. Its image through ∂_v^π is the generator of $K_*^{MW}(\kappa(v))$ and the homomorphism $K_{*-1}^{MW}(\kappa(v)) \to Q_*, \alpha \mapsto \alpha.[\pi]$ provides a section of $\partial_v^\pi : Q_* \to K_{*-1}^{MW}(\kappa(v))$. This is clear from our definitions.

It suffices now to check that $K_{*-1}^{MW}(\kappa(v)) \to Q_*$ is onto. Using the fact that any element of F can be written $\pi^n u$ for some unit $u \in \mathcal{O}_v^\times$, we see that $K_*^{MW}(F)$ is generated as a group by elements of the form $\eta^m[\pi][u_2]\dots[u_n]$ or $\eta^m[u_1]\dots[u_n]$, with the u_i's in $\mathcal{O}_v^\times$. But the latter are in A_* and the former are, modulo A_*, in the image of $K_{*-1}^{MW}(\kappa(v)) \to Q_*$. $\square$

Remark 3.23. In fact one may also prove as in *loc. cit.* the fact that the morphism Θ_π defined in the Lemma 3.16 is onto and its kernel is the ideal generated by η and the elements $[u] \in K_1^{MW}(F)$ with $u \in \mathcal{O}_v^\times$ a unit of $\mathcal{O}_v$ congruent to 1 modulo π. We will not give the details here, we do not use these results. $\square$

Theorem 3.24. *For any field F the following diagram is a (split) short exact sequence of $K_*^{MW}(F)$-modules:*

$$0 \to K_n^{MW}(F) \to K_n^{MW}(F(T)) \xrightarrow{\Sigma \partial_{(P)}^P} \oplus_P K_{n-1}^{MW}(F[T]/P) \to 0$$

(where P runs over the set of monic irreducible polynomials of $F[T]$).

Proof. It it is again very much inspired from [47]. We first observe that the morphism $K_*^{MW}(F) \to K_*^{MW}(F(T))$ is a split monomorphism; from our previous computations we see that $K_*^{MW}(F(T)) \xrightarrow{\partial_{(T)}^T([T] \cup -)} K_*^{MW}(F)$ provides a retraction.

Now we define a filtration on $K_*^{MW}(F(T))$ by sub-rings L_d's

$$L_0 = K_*^{MW}(F) \subset L_1 \subset \dots \subset L_d \subset \dots \subset K_*^{MW}(F(T))$$

such that L_d is exactly the sub-ring generated by $\eta \in K_{-1}^{MW}(F(T))$ and all the elements $[P] \in K_1^{MW}(F(T))$ with $P \in F[T] - \{0\}$ of degree less or equal to d. Thus L_0 is indeed $K_*^{MW}(F) \subset K_*^{MW}(F(T))$. Observe that $\bigcup_d L_d = K_*^{MW}(F(T))$. Observe that each L_d is actually a sub $K_*^{MW}(F)$-algebra.

Also observe that using the relation $[a.b] = [a] + [b] + \eta[a][b]$ that if $[a] \in L_d$ and $[b] \in L_d$ then so are $[ab]$ and $[\frac{a}{b}]$. As a consequence, we see that for $n \geq 1$, $L_d(K_n^{MW}(F(T)))$ is the sub-group generated by symbols $[a_1]\dots[a_n]$ such that each a_i itself is a fraction which involves only polynomials of degree

$\leq d$. In degree ≤ 0, we see in the same way that $L_d(K_n^{MW}(F(T)))$ is the subgroup generated by symbols $< a > \eta^n$ with a a fraction which involves only polynomials of degree $\leq d$.

It is also clear that for $n \geq 1$, $L_d(K_n^{MW}(F(T)))$ is generated as a group by elements of the form $\eta^m[a_1]\ldots[a_{n+m}]$ with the a_i of degree $\leq d$. $\qquad \square$

Lemma 3.25. *1) For $n \geq 1$, $L_d(K_n^{MW}(F(T)))$ is generated by the elements of $L_{(d-1)}(K_n^{MW}(F(T)))$ and elements of the form $\eta^m[a_1]\ldots[a_{n+m}]$ with a_1 of degree d and the a_i's, $i \geq 2$ of degree $\leq (d-1)$.*

2) Let $P \in F[T]$ be a monic polynomial of degree $d > 0$. Let $G_1, \ldots, G_i$ be polynomials of degrees $\leq (d-1)$. Finally let G be the rest of the Euclidean division of $\Pi_{j \in \{1,\ldots,i\}} G_j$ by P, so that G has degree $\leq (d-1)$. Then one has in the quotient group $K_2^{MW}(F(T))/L_{d-1}$ the equality

$$[P][G_1 \ldots G_i] = [P][G]$$

Proof. 1) We proceed as in Milnor's paper. Let f_1 and f_2 be polynomials of degree d. We may write $f_2 = -af_1 + g$, with $a \in F^\times$ a unit and g of degree $\leq (d-1)$. If $g = 0$, the we have $[f_1][f_2] = [f_1][a(-f_1)] = [f_1][a]$ (using the relation $[f_1, -f_1] = 0$). If $g \neq 0$ then as in *loc. cit.* we get $1 = \frac{af_1}{g} + \frac{f_2}{g}$ and the Steinberg relation yields $[\frac{af_1}{g}][\frac{f_2}{g}] = 0$. Expanding with η we get: $([f_1] - [\frac{g}{a}] - \eta[\frac{g}{a}][\frac{af_1}{g}])[\frac{f_2}{g}] = 0$, which readily implies (still in $K_2^{MW}(F(T))$):

$$([f_1] - [\frac{g}{a}])[\frac{f_2}{g}] = 0$$

But expanding the right factor now yields

$$([f_1] - [\frac{g}{a}])([f_2] - [g] - \eta[g][\frac{f_2}{g}]) = 0$$

which implies (using again the previous vanishing):

$$([f_1] - [\frac{g}{a}])([f_2] - [g]) = 0$$

We see that $[f_1][f_2]$ can be expressed as a sum of symbols in which at most one of the factor as degree d, the other being of smaller degree. An easy induction proves (1).

2) We first establish the case $i = 2$. We start with the Euclidean division $G_1 G_2 = PQ + G$. We get from this the equality $1 = \frac{G}{G_1.G_2} + \frac{PQ}{G_1.G_2}$ which gives $[\frac{PQ}{G_1.G_2}][\frac{G}{G_1.G_2}] = 0$. We expand the left term as $[\frac{PQ}{G_1.G_2}] = < \frac{Q}{G_1.G_2} > [P] + [\frac{Q}{G_1.G_2}]$. We thus obtain $[P][\frac{G}{G_1.G_2}] = - < \frac{Q}{G_1.G_2} > [\frac{Q}{G_1.G_2}][\frac{G}{G_1.G_2}]$ but the right hand side is in $L_{(d-1)}$ (observe Q has degree

$\leq (d-1))$ thus $[P][\frac{G}{G_1.G_2}] \in L_{(d-1)} \subset K_2^{MW}(F(T))$. Now $[\frac{G}{G_1G_2}] = [G] - [G_1G_2] - \eta[G_1G_2][\frac{G}{G_1G_2}]$. Thus $[P][\frac{G}{G_1.G_2}] = [P][G] - [P][G_1G_2] + < -1 > \eta[G_1G_2][P][\frac{G}{G_1G_2}]$. This shows that modulo $L_{(d-1)}$, $[P][G] - [P][G_1G_2]$ is zero, as required.

For the case $i \geq 3$ we proceed by induction. Let $\Pi_{j\in\{2,...,i\}}G_j = P.Q + G'$ be the Euclidean division of $\Pi_{j\in\{2,...,i\}}G_j$ by P with G' of degree $\leq (d-1)$. Then the rest G of the Euclidean division by P of $G_1 \ldots G_i$ is the same as the rest of the Euclidean division of G_1G' by P. Now $[P][G_1 \ldots G_i] = [P][G_1] + [P][G_2 \ldots G_i] + \eta[P][G_2 \ldots G_i][G_1]$. By the inductive assumption this is equal, in $K_2^{MW}(F(T))/L_{d-1}$, to $[P][G_1] + [P][G'] + \eta[P][G'][G_1] = [P][G'G_1]$. By the case 2 previously proven we thus get in $K_2^{MW}(F(T))/L_{d-1}$,

$$[P][G_1 \ldots G_i] = [P][G_1G'] = [P][G]$$

which proves our claim.　　　□

Now we continue the proof of Theorem 3.24 following Milnor's proof of [47, Theorem 2.3]. Let $d \geq 1$ be an integer and let $P \in F[T]$ be a monic irreducible polynomial of degree d. We denote by $\mathcal{K}_P \subset L_d/L_{(d-1)}$ the sub-graded group generated by elements of the form $\eta^m[P][G_1] \ldots [G_n]$ with the G_i of degree $(d-1)$. For any polynomial G of degree $\leq (d-1)$, the multiplication by $\epsilon[G]$ induces a morphism:

$$\epsilon[G]. : \mathcal{K}_P \to \mathcal{K}_P$$

$$\eta^m[P][G_1] \ldots [G_n] \mapsto \epsilon[G]\eta^m[P][G_1] \ldots [G_n] = \eta^m[P][G][G_1] \ldots [G_n]$$

of degree $+1$. Let $\mathcal{E}_P$ be the graded associative ring of graded endomorphisms of $\mathcal{K}_P$. We claim that the map $(F[T]/P)^\times \to (\mathcal{E}_P)_1, (G) \mapsto \epsilon[G]$. (where G has degree $\leq (d-1)$) and the element $\eta \in (\mathcal{E}_P)_{-1}$ (corresponding to the multiplication by η) satisfy the four relations of the Milnor–Witt K-theory. Let us check the Steinberg relation. Let $G \in F[T]$ be of degree $\leq (d-1)$. Then so is $1 - G$ and the relation $(\epsilon[G].) \circ (\epsilon[1-G].) = 0 \in \mathcal{E}_P$ is clear. Let us check relation 2. We let H_1 and H_2 be polynomials of degree $\leq (d-1)$. Let G be the rest of division of H_1H_2 by P. By definition $\epsilon[(H_1)(H_2)]$. is $\epsilon[(G)]$.. But by the part (2) of the Lemma we have (in $\mathcal{K}_P \subset K_m^{MW}(F(T))/L_{(d-1)}$):

$$\epsilon[(G)].(\eta^m[P][G_1] \ldots [G_n]) = \eta^m[P][G][G_1] \ldots [G_n]$$
$$= \eta^m[P][H_1H_2][G_1] \ldots [G_n]$$

which easily implies the claim. The last two relations are easy to check.

We thus obtain a morphism of graded ring $K_*^{MW}(F[T]/P) \to \mathcal{E}_P$. By letting $K_*^{MW}(F[T]/P)$ act on $[P] \in L_d/L_{(d-1)} \subset K_1^{MW}(F(T))/L_{(d-1)}$ we obtain a graded homomorphism

$$K_*^{MW}(F[T]/P) \to \mathcal{K}_P \subset L_d/L_{(d-1)}$$

which is an epimorphism. By the first part of the Lemma, we see that the induced homomorphism

$$\oplus_P K_*^{MW}(F[T]/P) \to L_d/L_{(d-1)} \tag{3.4}$$

is an epimorphism. Now using our definitions, one checks as in [47] that for P of degree d, the residue morphism ∂^P vanishes on $L_{(d-1)}$ and that moreover the composition

$$\oplus_P K_*^{MW}(F[T]/P) \to L_d(K_n^{MW}(F(T)))/L_{(d-1)}(K_n^{MW}(F(T)))$$

$$\xrightarrow{\Sigma_P \partial^P} \oplus_P K_*^{MW}(F[T]/P)$$

is the identity. As in *loc. cit.* this implies the theorem, with the observation that the quotients L_d/L_{d-1} are $K_*^{MW}(F)$-modules and the residues maps are morphisms of $K_*^{MW}(F)$-modules. $\qquad\square$

Remark 3.26. We observe that the previous theorem in negative degrees is exactly [53, Theorem 5.3].

Now we come back to our fixed base field k and work in the category $\mathcal{F}_k$. We will make constant use of the results of Sect. 2.3. We endow the functor $F \mapsto K_*^{MW}(F)$, $\mathcal{F}_k \to \mathcal{A}b_*$ with Data **(D4) (i)**, **(D4) (ii)** and **(D4) (iii)**. The datum **(D4) (i)** comes from the $K_0^{MW}(F) = GW(F)$-module structure on each $K_n^{MW}(F)$ and the datum **(D4) (ii)** comes from the product $F^\times \times K_n^{MW}(F) \to K_{(n+1)}^{MW}(F)$. The residue homomorphisms ∂_v^π gives the Data **(D4) (iii)**. We observe of course that these Data are extended from the prime field of k.

Axioms **(B0)**, **(B1)** and **(B2)** are clear from our previous results. The Axiom **(B3)** follows at once from Lemma 3.19.

Axiom **(HA) (ii)** is clear, Theorem 3.24 establishes Axiom **(HA) (i)**.

For any discrete valuation v on $F \in \mathcal{F}_k$, and any uniformizing element π, define morphisms of the form $\partial_z^y : K_n^{MW}(\kappa(y)) \to K_{n-1}^{MW}(\kappa(z))$ for any $y \in (\mathbb{A}_F^1)^{(1)}$ and $z \in (\mathbb{A}_{\kappa(v)}^1)^{(1)}$ fitting in the following diagram:

$$\begin{array}{ccccccccc} 0 \to & K_*^{MW}(F) & \to & K_*^{MW}(F(T)) & \to & \oplus_{y \in (\mathbb{A}_F^1)^{(1)}} K_{*-1}^{MW}(\kappa(y)) \to 0 \\ & \downarrow \partial_v^\pi & & \downarrow \partial_{v[T]}^\pi & & \downarrow \Sigma_{y,z} \partial_z^{\pi,y} & \\ 0 \to & K_{*-1}^{MW}(\kappa(v)) \to & K_{n-1}^{MW}(\kappa(v)(T)) & \to & \oplus_{z \in \mathbb{A}_{\kappa(v)}^1} K_{*-2}^{MW}(\kappa(v)) & \to 0 \end{array} \tag{3.5}$$

The following Theorem establishes Axiom **(B4)**.

Theorem 3.27. *Let v be a discrete valuation on $F \in \mathcal{F}_k$, let π be a uniformizing element. Let $P \in \mathcal{O}_v[T]$ be an irreducible primitive polynomial, and $Q \in \kappa(v)[T]$ be an irreducible monic polynomial.*

(i) *If the closed point $Q \in \mathbb{A}^1_{\kappa(v)} \subset \mathbb{A}^1_{\mathcal{O}_v}$ is not in the divisor D_P then the morphism $\partial_Q^{\pi,P}$ is zero.*

(ii) *If Q is in $D_P \subset \mathbb{A}^1_{\mathcal{O}_v}$ and if the local ring $\mathcal{O}_{D_P,Q}$ is a discrete valuation ring with π as uniformizing element then*

$$\partial_Q^{\pi,P} = - < -\frac{\overline{P'}}{Q'} > \partial_Q^Q$$

Proof. Let $d \in \mathbb{N}$ be an integer. We will say that Axiom **(B4)** holds in degree $\leq d$ if for any field $F \in \mathcal{F}_k$, any irreducible primitive polynomial $P \in \mathcal{O}_v[T]$ of degree $\leq d$, any monic irreducible $Q \in \kappa(v)[T]$ then: if Q doesn't lie in the divisor D_P, the homomorphism ∂_Q^P is 0 on $K_*^{MW}(F[T]/P)$ and if Q lies in D_P and that the local ring $\mathcal{O}_{\overline{y},z}$ is a discrete valuation ring with π as uniformizing element , then the homomorphism ∂_Q^P is equal to $-\partial_Q^\pi$.

We now proceed by induction on d to prove that Axiom **(B4)** holds in degree $\leq d$ for any d. For $d = 0$ this is trivial, the case $d = 1$ is also easy.

We may use Remark 2.17 to reduce to the case the residue field $\kappa(v)$ is infinite.   □

We will use:

Lemma 3.28. *Let P be a primitive irreducible polynomial of degree d in $F[T]$. Let Q be a monic irreducible polynomial in $\kappa(v)[T]$.*

Assume either that $\overline{P}$ is prime to Q, or that Q divides $\overline{P}$ and that the local ring $\mathcal{O}_{D_P,Q}$ is a discrete valuation ring with uniformizing element π.

Then the elements of the form $\eta^m[\overline{G_1}]\ldots[\overline{G_n}]$, where all the G_i's are irreducible elements in $\mathcal{O}_v[T]$ of degree $< d$, such that, either G_1 is equal to π or $\overline{G_1}$ is prime to Q, and for any $i \geq 2$, $\overline{G_i}$ is prime to Q, generate $K_*^{MW}(F[T]/P)$ as a group.

Proof. First the symbols of the form $\eta^m[\overline{G_1}]\ldots[\overline{G_n}]$ with the G_i irreducible elements of degree $< d$ of $\mathcal{O}_v[T]$ generate the Milnor–Witt K-theory of $f[T]/P$ as a group.

1) We first assume that $\overline{P}$ is prime to Q. It suffices to check that those element above are expressible in terms of symbols of the form of the Lemma. Pick up one such $\eta^m[\overline{G_1}]\ldots[\overline{G_n}]$. Assume that there exists i such that $\overline{G_i}$ is divisible by Q (otherwise there is nothing to prove), for instance G_1.

If the field $\kappa(v)$ is infinite, which we may assume by Remark 2.17, we may find an $\alpha \in \mathcal{O}_v$ such that $G_1(\alpha)$ is a unit in $\mathcal{O}_v^\times$. Then there exists a unit u in $\mathcal{O}_v^\times$ and an integer v (actually the valuation of $P(\alpha)$ at π) such that $P + u\pi^v G$ is divisible by $T - \alpha$ in $\mathcal{O}_v[T]$. Write $P + u\pi^v G_1 = (T - \alpha)H_1$. Observe that Q which divides $\overline{G_1}$ and is prime to $\overline{P}$ must be prime to both $T - \overline{\alpha}$ and $\overline{H_1}$.

Observe that $\frac{(T-\alpha)}{u\pi^v}H_1 = \frac{P}{u\pi^v} + G_1$ is the Euclidean division of $\frac{(T-\alpha)}{u\pi^v}H_1$ by P. By Lemma 3.25 one has in $K_*^{MW}(F(T))$, modulo L_{d-1}

$$\eta^m[P][G_1][G_2]\ldots[G_n] = \eta^m[P][\frac{(T-\alpha)}{u\pi^v}H_1][G_2]\ldots[G_n]$$

Because $\partial^P_{D_P}$ vanishes on L_{d-1}, applying $\partial^P_{D_P}$ to the previous congruence yields the equality in $K^{MW}_*(F[T]/P)$

$$\eta^m\overline{[G_1]}\ldots\overline{[G_n]} = \eta^m[\frac{(T-\alpha)}{u\pi^v}\overline{H_1}][G_2]\ldots\overline{[G_n]}$$

Expanding $[\frac{(T-\alpha)}{u\pi^v}\overline{H_1}]$ as $[\frac{(T-\alpha)}{u\pi^v}] + [\overline{H_1}] + \eta[\frac{(T-\alpha)}{u\pi^v}][\overline{H_1}]$ shows that we may strictly reduce the number of G_i's whose mod π reduction is divisible by Q. This proves our first claim (using the relation $[\pi][\pi] = [\pi][-1]$ we may indeed assume that only G_1 is maybe equal to π).

2) Now assume that Q divides $\overline{P}$ and that the local ring $\mathcal{O}_{D_P,Q}$ is a discrete valuation ring with uniformizing element π. By our assumption, any non-zero element in the discrete valuation ring $\mathcal{O}_{D_P,Q} = (\mathcal{O}_v[T]/P)_Q$ can be written as

$$\pi^v\frac{\overline{R}}{\overline{S}}$$

with R and S polynomials in $\mathcal{O}_v[T]$ of degree $< d$ whose mod π reduction in $\kappa(v)[T]$ is prime to Q. From this, it follows easily that the symbols of the form $\eta^m\overline{[G_1]}\ldots\overline{[G_n]}$, with the G_i's being either a polynomial in $\mathcal{O}_v[T]$ of degree $< d$ whose mod π reduction in $\kappa(v)[T]$ is prime to Q, either equal to π.

The Lemma is proven.   □

Now let $d > 0$ and assume the claim is proven in degrees $< d$, for all fields. Let P be a primitive irreducible polynomial of degree d in $\mathcal{O}_v[T]$. Let Q be a monic irreducible polynomial in $\kappa(v)[T]$.

Under our inductive assumption, we may compute $\partial^{\pi,P}_Q(\eta^m[G_1]\ldots\overline{[G_n]})$ for any sequence $G_1, .., G_n$ as in the Lemma.

Indeed, the symbol $\eta^m[P][G_1]\ldots\overline{[G_n]} \in K^{MW}_{n-m}$ has residue at P the symbol $\eta^m\overline{[G_1]}\ldots\overline{[G_n]}$. All its other potentially non trivial residues concern irreducible polynomials of degree $< d$. By the (proof of) Theorem 3.24, we know that there exists an $\alpha \in L_{d-1}(K^{MW}_{n-m}(F(T)))$ such that

$$\eta^m[P][G_1]\ldots\overline{[G_n]} + \alpha$$

has only one non vanishing residue, which is at P, and which equals $\eta^m\overline{[G_1]}\ldots\overline{[G_n]}$.

Then the support of α (which means the set of points of codimension one in $\mathbb{A}^1_F$ where α has a non trivial residue) consists of the divisors defined by the G_i's (P doesn't appear). But those don't contain Q.

Using the commutative diagram which defines the ∂_Q^P's, we may compute $\partial_Q^{\pi,P}(\eta^m[\overline{G_1}]\dots[\overline{G_n}])$ as

$$\partial_Q^Q(\partial_v^\pi(\eta^m[P][G_1]\dots[G_n]+\alpha)) = \partial_Q^Q(\partial_v^\pi(\eta^m[P][G_1]\dots[G_n]) + \sum_i \partial_Q^{\pi,G_i}(\partial_{D_{G_i}}^{G_i}(\alpha))$$

By our inductive assumption, $\sum_i \partial_Q^{\pi,G_i}(\partial_{D_{G_i}}^{G_i}(\alpha)) = 0$ because the supports G_i do not contain Q.

We then have two cases:

(1) G_1 is not π. Then
$$\partial_v^\pi(\eta^m[P][G_1]\dots[G_n]) = 0$$
as every element lies in $\mathcal{O}_{v[T]}^\times$. Thus in that case, $\partial_Q^{\pi,P}(\eta^m[\overline{G_1}]\dots[\overline{G_n}]) = 0$ which is compatible with our claim.

(2) $G_1 = \pi$. Then

$$\partial_v^\pi(\eta^m[P][\pi][G_2]\dots[G_n]) = - < -1 > \partial_v^\pi(\eta^m[\pi][P][G_2]\dots[G_n])$$
$$= - < -1 > \eta^m[\overline{P}][\overline{G_2}]\dots[\overline{G_n}]$$

Applying ∂_Q^Q yields 0 if $\overline{P}$ is prime to Q, as all the terms are units. If $\overline{P} = QR$, then R is a unit in $(\mathbb{A}_{\kappa v}^1)_Q$ by our assumptions. Expending $[QR] = [Q]+[R]+\eta[Q][R]$, we get

$$\partial_Q^{\pi,P}(\eta^m[\overline{G_1}]\dots[\overline{G_n}]) = - < -1 > \eta^m([\overline{G_2}]\dots[\overline{G_n}] + \eta[\overline{R}][\overline{G_2}]\dots[\overline{G_n}])$$
$$= - < -\overline{R} > \eta^m[\overline{G_2}]\dots[\overline{G_n}]$$

It remains to observe that $\overline{R} = \frac{\overline{P'}}{Q'}$.

By the previous Lemma the symbols we used generate $K_*^{MW}(F[T]/P)$. Thus the previous computations prove the Theorem. $\square$

Now we want to prove Axiom **(B5)**. Let X be a local smooth k-scheme of dimension 2, with field of functions F and closed point z, let $y_0 \in X^{(1)}$ be such that $\overline{y_0}$ is smooth over k. Choose a uniformizing element π of $\mathcal{O}_{X,y_0}$. Denote by $\mathcal{K}_n(X;y_0)$ the kernel of the map

$$K_n^{MW}(F) \xrightarrow{\Sigma_{y\in X^{(1)}-\{y_0\}}\partial_y} \oplus_{y\in X^{(1)}-\{y_0\}} H_y^1(X;\underline{\mathbf{K}}_n^{MW}) \qquad (3.6)$$

By definition $\underline{\mathbf{K}}_n^{MW}(X) \subset \mathcal{K}_n(X;y_0)$. The morphism $\partial_{y_0}^\pi : K_n^{MW}(F) \to K_{n-1}^{MW}(\kappa(y_0))$ induces an injective homomorphism $\mathcal{K}_n(X;y_0)/\underline{\mathbf{K}}_n^{MW}(X) \subset K_{n-1}^{MW}(\kappa(y_0))$.

We first observe:

Lemma 3.29. *Keep the previous notations and assumptions. Then* $\underline{\mathbf{K}}_{n-1}^{MW}(\mathcal{O}_{y_0}) \subset \mathcal{K}_n(X; y_0)/\underline{\mathbf{K}}_n^{MW}(X) \subset K_{n-1}^{MW}(\kappa(y_0)).$

Proof. We apply Gabber's lemma to y_0, and in this way, we see (by diagram chase) that we can reduce to the case $X = (\mathbb{A}_U^1)_z$ where U is a smooth local k-scheme of dimension 1. As Theorem 3.27 implies Axiom **(B4)**, we know by Lemma 2.43 that the following complex

$$
0 \to \underline{\mathbf{K}}_n^{MW}(X) \to K_n^{MW}(F) \xrightarrow{\Sigma_{y \in X^{(1)}} \partial_y} \oplus_{y \in X^{(1)}} H_y^1(X; \underline{\mathbf{K}}_n^{MW})
$$
$$
\to H_z^2(X; \underline{\mathbf{K}}_n^{MW}) \to 0
$$

is an exact sequence. Moreover, we know also from there that for $\overline{y}_0$ smooth, the morphism $H_y^1(X; \underline{\mathbf{K}}_n^{MW}) \to H_z^2(X; \underline{\mathbf{K}}_n^{MW})$ can be "interpreted" as the residue map. Its kernel is thus $\underline{\mathbf{K}}_{n-1}^{MW}(\mathcal{O}_{y_0}) \subset K_{n-1}^{MW}(\kappa(y_0)) \cong H_y^1(X; \underline{\mathbf{K}}_n^{MW})$. The exactness of the previous complex implies that

$$
\mathcal{K}_n(X; y_0)/\underline{\mathbf{K}}_n^{MW}(X) = \underline{\mathbf{K}}_{n-1}^{MW}(\mathcal{O}_{y_0})
$$

proving the statement. $\square$

Our last objective is now to show that in fact $\underline{\mathbf{K}}_{n-1}^{MW}(\mathcal{O}_{y_0}) = \mathcal{K}_n(X; y_0)/\underline{\mathbf{K}}_n^{MW}(X) \subset K_{n-1}^{MW}(\kappa(y_0))$. To do this we observe that by Lemma 2.43, for k infinite, the morphism (3.6) above is an epimorphism. Thus the previous statement is equivalent to the fact that the diagram

$$
0 \to \underline{\mathbf{K}}_{n-1}^{MW}(\mathcal{O}_{y_0}) \to K_n^{MW}(F)/\underline{\mathbf{K}}_n^{MW}(X) \xrightarrow{\Sigma_{y \in X^{(1)} - \{y_0\}} \partial_y} \oplus_{y \in X^{(1)} - \{y_0\}}
$$

$H_y^1(X; \underline{\mathbf{K}}_n^{MW}) \to 0$

is a short exact sequence or in other words that the epimorphism

$$
\Phi_n(X; y_0): K_n^{MW}(F)/\underline{\mathbf{K}}_n^{MW}(X) + \underline{\mathbf{K}}_{n-1}^{MW}(\mathcal{O}_{y_0}) \xrightarrow{\Sigma_{y \in X^{(1)} - \{y_0\}} \partial_y} \oplus_{y \in X^{(1)} - \{y_0\}}
$$

$H_y^1(X; \underline{\mathbf{K}}_n^{MW}) \tag{3.7}$

is an isomorphism. We also observe that the group $K_n^{MW}(F)/\underline{\mathbf{K}}_n^{MW}(X) + \underline{\mathbf{K}}_{n-1}^{MW}(\mathcal{O}_{y_0})$ doesn't depend actually on the choice of a local parametrization of $\overline{y}_0$.

Theorem 3.30. *Let X be a local smooth k-scheme of dimension 2, with field of functions F and closed point z, let $y_0 \in X^{(1)}$ be such that $\overline{y_0}$ is smooth over k. Then the epimorphism $\Phi_n(X; y_0)$ (3.7) is an isomorphism.*

Proof. We know from Axiom **(B1)** (that is to say Theorem 3.27) and Lemma 2.43 that the assertion is true for X a localization of $\mathbb{A}^1_U$ at some codimension 2 point, where U is a smooth local k-scheme of dimension 1. □

Lemma 3.31. *Given any element $\alpha \in K^{MW}_n(F)$, write it as $\alpha = \sum_i \alpha_i$, where the α_i's are pure symbols. Let $Y \subset X$ be the union of the hypersurfaces defined by each factor of each pure symbol α_i. Let $X \to \mathbb{A}^1_U$ be an étale morphism with U smooth local of dimension 1, with field of functions E, such that $Y \to \mathbb{A}^1_U$ is a closed immersion. Then for each i there exists a pure symbol $\beta_i \in K^{MW}_n(E(T))$ which maps to α_i modulo $\underline{\mathbf{K}}^{MW}_n(X) \subset K^{MW}_n(F)$.*

As a consequence, if $\partial_y(\alpha) \neq 0$ in $H^1_y(X; \underline{\mathbf{K}}^{MW}_n)$ for some $y \in X^{(1)}$ then $y \in Y$ and $\partial_y(\alpha) = \partial_y(\beta) = \in H^1_y(X; \underline{\mathbf{K}}^{MW}_n) = H^1_y(\mathbb{A}^1_U; \underline{\mathbf{K}}^{MW}_n)$.

Proof. Let us denote by π_j the irreducible elements in the factorial ring $\mathcal{O}(U)[T]$ corresponding to the irreducible components of $Y \subset \mathbb{A}^1_U$. Each $\alpha_i = [\alpha^1_i] \dots [\alpha^n_i]$ is a pure symbol in which each term α^s_i decomposes as a product $\alpha^s_i = u^s_i \alpha'^s_i$ of a unit u^s_i in $\mathcal{O}(X)^\times$ and a product α'^s_i of π_j's (this follows from our choices and the factoriality property of $A := \mathcal{O}(X)$. Thus α'_i is in the image of $K^{MW}_n(E(T)) \to K^{MW}_n(F)$. Now by construction, $A/(\Pi\pi_j) = B/(\Pi\pi_j)$, where $B = \mathcal{O}(U)[T]$. Thus one may choose unit v^s_i in $B^\times$ with $w^s_i := \frac{u^s_i}{v^s_i} \equiv 1 [\Pi\pi_j]$.

Now set $\beta^s_i = v^s_i \alpha'^s_i$, $\beta_i := [\beta^1_i] \dots [\beta^n_i]$. Then we claim that β_i maps to α_i modulo $\underline{\mathbf{K}}^{MW}_n(X) \subset K^{MW}_n(F)$. In other words, we claim that $[\alpha^1_i] \dots [\alpha^n_i] - [\beta^1_i] \dots [\beta^n_i]$ lies in $\underline{\mathbf{K}}^{MW}_n(X)$ which means that each of its residue at any point of codimension one in X vanishes. Clearly, by construction the only non-zero residues can only occur at each π_j.

We end up in showing the following: given elements $\beta^s \in A - \{0\}$, $s \in \{1, \dots, n\}$ and $w^s \in A^\times$ which is congruent to 1 modulo each irreducible element π which divides one of the β^s, then for each such π, $\partial^\pi([\beta^1] \dots [\beta^n]) = \partial^\pi([w^1\beta^1] \dots [w^n\beta^n])$. We expand $[w^1\beta^1] \dots [w^n\beta^n]$ as $[w^1][w^2\beta^2] \dots [w^n\beta^n] + [\beta^1][w^2\beta^2] \dots [w^n\beta^n] + \eta[w^1][\beta^1][w^2\beta^2] \dots [w^n\beta^n]$. Now using Proposition 3.17 and the fact that $\overline{w^{i^\pi}} = 1$, we immediately get $\partial^\pi([w^1\beta^1] \dots [w^n\beta^n]) = \partial^\pi([\beta^1][w^2\beta^2] \dots [w^n\beta^n])$ which gives the result. An easy induction gives the result. This proof can obviously be adapted for pure symbols of the form $\eta^n[\alpha]$. □

Now the theorem follows from the Lemma. Let $\overline{\alpha} \in K^{MW}_n(F)/\underline{\mathbf{K}}^{MW}_n(X) + \underline{\mathbf{K}}^{MW}_{n-1}(\mathcal{O}_{y_0})$ be in the kernel of $\Phi_n(X; y_0)$. Assume $\alpha \in K^{MW}_n(F)$ represents $\overline{\alpha}$. By Gabber's Lemma there exists an étale morphism $X \to \mathbb{A}^1_U$ with U smooth local of dimension 1, with field of functions E, such that $Y \cup \overline{y_0} \to \mathbb{A}^1_U$ is a closed immersion, where Y is obtained by writing α as a sum of pure symbols α_i's. By the previous Lemma, we may find β_i in $K^{MW}_n(E(T))$ mapping to α modulo $\underline{\mathbf{K}}^{MW}_n(X)$ yo α_i. Let β be the sum of the β_i's. Then $\overline{\beta} \in K^{MW}_n(E(T))/\underline{\mathbf{K}}^{MW}_n((\mathbb{A}^1_U)_z) + \underline{\mathbf{K}}^{MW}_{n-1}(\mathcal{O}_{y_0})$ is also in the kernel of our morphism $\Phi_n((\mathbb{A}^1_U)_z; y_0)$. Thus $\overline{\beta} = 0$ and so $\overline{\alpha} = 0$. □

Unramified $K^{\mathcal{R}}$-theories. We now slightly generalize our construction by allowing some "admissible" relations in $K^{MW}_*(F)$. An admissible set of relations $\mathcal{R}$ is the datum for each $F \in \mathcal{F}_k$ of a graded ideal $\mathcal{R}_*(F) \subset K^{MW}_*(F)$ with the following properties:

1. For any extension $E \subset F$ in $\mathcal{F}_k$, $\mathcal{R}_*(E)$ is mapped into $\mathcal{R}_*(F)$.
2. For any discrete valuation v on $F \in \mathcal{F}_k$, any uniformizing element π,
$\partial^\pi_v(\mathcal{R}_*(F)) \subset \mathcal{R}_*(\kappa(v))$.
3. For any $F \in \mathcal{F}_k$ the following sequence is a short exact sequence:

$$0 \to \mathcal{R}_*(F) \to \mathcal{R}_*(F(T)) \xrightarrow{\Sigma_P \, \partial^P_{DP}} \oplus_P \mathcal{R}_{*-1}(F[t]/P) \to 0 \qquad \Box$$

The third one is usually more difficult to check.

Given an admissible relation $\mathcal{R}$, for each $F \in \mathcal{F}_k$ we simply denote by $K^{\mathcal{R}}_*(F)$ the quotient graded ring $K^{MW}_*(F)/\mathcal{R}_*(F)$. The property (1) above means that we get this way a functor

$$\mathcal{F}_k \to \mathcal{Ab}_*$$

This functor is moreover endowed with data **(D4) (i)** and **(D4) (ii)** coming from the K^{MW}_*-algebra structure. The property (2) defines the data **(D4) (iii)**. The axioms **(B0)**, **(B1)**, **(B2)**, **(B3)** are immediate consequences from those for K^{MW}_*. Property (3) implies axiom **(HA) (i)**. Axiom **(HA) (ii)** is clear. Axioms **(B4)** and **(B5)** are also consequences from the corresponding axioms just established for K^{MW}_*. We thus get as in Theorem 2.46 a $\mathbb{Z}$-graded strongly $\mathbb{A}^1$-invariant sheaf, denoted by $\underline{\mathbf{K}}^{\mathcal{R}}_*$ with isomorphisms $(\underline{\mathbf{K}}^{\mathcal{R}}_n)_{-1} \cong \underline{\mathbf{K}}^{\mathcal{R}}_{n-1}$. There is obviously a structure of $\mathbb{Z}$ graded sheaf of algebras over $\underline{\mathbf{K}}^{MW}_*$.

Lemma 3.32. *Let $R_* \subset K^{MW}_*(k)$ be a graded ideal. For any $F \in \mathcal{F}_k$, denote by $\mathcal{R}_*(F) := R_*.K^{MW}_*(F)$ the ideal generated by R_*. Then $\mathcal{R}_*(F)$ is an admissible relation on K^{MW}_*. We denote the quotient simply by $K^{MW}_*(F)/R_*$.*

Proof. Properties (1) and (2) are easy to check. We claim that the property (3) also hold: this follows from Theorem 3.24 which states that the morphisms and maps are $K^{MW}_*(F)$-module morphisms. $\qquad \Box$

Of course when $R_* = 0$, we get the $\mathbb{Z}$-graded sheaf of unramified Milnor–Witt K-theory

$$\underline{\mathbf{K}}^{MW}_*$$

itself.

Example 3.33. For instance we may take an integer n and $R_* = (n) \subset K^{MW}_*(k)$; we obtain mod n Milnor–Witt unramified sheaves. For $R_* = (\eta)$ the ideal generated by η, this yields unramified Milnor K-theory $\underline{\mathbf{K}}^M_*$. For

$R_* = (n, \eta)$ this yields mod n Milnor K-theory. For $\mathcal{R} = (h)$, this yields Witt
K-theory $\underline{\mathbf{K}}_*^W$, for $\mathcal{R} = (\eta, \ell)$ this yields mod ℓ Milnor K-theory. $\square$

Example 3.34. Let $\mathcal{R}_*^I(F)$ be the kernel of the epimorphism $K_*^{MW}(F) \to$
$I^*(F)$, $[u] \mapsto < u > -1 = - << u >>$ described in [54], see also
Remark 3.12. Then $\mathcal{R}_*^I(F)$ is admissible. Recall from the Remark 3.12 that
$K_*^{MW}(F)[\eta^{-1}] = W(F)[\eta, \eta^{-1}]$ and that $I^*(F)$ is the image of $K_*^{MW}(F) \to$
$W(F)[\eta, \eta^{-1}]$. Now the morphism $K_*^{MW}(F) \to W(F)[\eta, \eta^{-1}]$ commutes to
every data. We conclude using the Lemma 4.5 below. Thus we get in this way
unramified sheaves of powers of the fundamental ideal $\underline{\mathbf{I}}^*$ (see also [53]). $\square$

Let $\phi : M_* \to N_*$ be a morphism (in the obvious sense) of between functors
$\mathcal{F}_k \to \mathcal{A}b_*$ endowed with data **(D4) (i)**, **(D4) (ii)** and **(D4) (iii)** and
satisfying the Axioms **(B0)**, **(B1)**, **(B2)**, **(B3)**, **(HA)**, **(B4)** and **(B5)** of
Theorem 2.46.

Denote for each $F \in \mathcal{F}_k$ by $Im(\phi)_*(F)$ (resp. $Ker(\phi)_*(F)$) the image
(resp. the kernel) of $\phi(F) : M_*(F) \to N_*(F)$. One may extend both to
functor $\mathcal{F}_k \to \mathcal{A}b_*$ with data **(D4) (i)**, **(D4) (ii)** and **(D4) (iii)** induced
from the one on M_* and N_*.

Lemma 3.35. *Let* $\phi : M_* \to N_*$ *be a morphism of as above. Then* $Im(\phi)_*$
and $Ker(\phi)_*$ *with the induced Data* **(D4) (i)**, **(D4) (ii)** *and* **(D4) (iii)**
satisfy the Axioms **(B0)**, **(B1)**, **(B2)**, **(B3)**, **(HA)**, **(B4)** *and* **(B5)** *of*
Theorem 2.46.

Proof. The only difficulty is to check axiom **(HA) (i)**. It is in fact very easy
to check it using the axioms **(HA) (i)** and **(HA) (ii)** for M_* and N_*. Indeed
(HA) (ii) provides a splitting of the short exact sequences of **(HA) (i)** for
M_* and N_* which are compatible. One gets the axiom **(HA) (i)** for $Im(\phi)_*$
and $Ker(\phi)_*$ using the snake lemma. We leave the details to the reader. $\square$

3.3   Milnor–Witt K-Theory and Strongly
$\mathbb{A}^1$-Invariant Sheaves

Fix a natural number $n \geq 1$. Recall from [59] that $(\mathbb{G}_m)^{\wedge n}$ denotes the
n-th smash power of the pointed space $\mathbb{G}_m$. We first construct a canonical
morphism of pointed spaces

$$\sigma_n : (\mathbb{G}_m)^{\wedge n} \to \underline{\mathbf{K}}_n^{MW}$$

$(\mathbb{G}_m)^{\wedge n}$ is *a priori* the associated sheaf to the naive presheaf $\Theta_n : X \mapsto$
$(\mathcal{O}^\times(X))^{\wedge n}$ but in fact:

Lemma 3.36. *The presheaf* $\Theta_n : X \mapsto (\mathcal{O}(X)^\times)^{\wedge n}$ *is an unramified sheaf*
of pointed sets.

Proof. It is as a presheaf unramified in the sense of our Definition 2.1 thus automatically a sheaf in the Zariski topology. One may check it is a sheaf in the Nisnevich topology by checking Axiom **(A1)**. One has only to use the following observation: let E_α be a family of pointed subsets in a pointed set E. Then $\cap_\alpha (E_\alpha)^{\wedge n} = (\cap_\alpha E_\alpha)^{\wedge n}$, where the intersection is computed inside $E^{\wedge n}$. □

Fix an irreducible $X \in Sm_k$ with function field F. There is a tautological symbol map $(\mathcal{O}(X)^\times)^{\wedge n} \subset (F^\times)^{\wedge n} \to K_n^{MW}(F)$ that takes a symbol $(u_1, \ldots, u_n) \in (\mathcal{O}(X)^\times)^{\wedge n}$ to the corresponding symbol in $[u_1] \ldots [u_n] \in K_n^{MW}(F)$. But this symbol $[u_1] \ldots [u_n] \in K_n^{MW}(F)$ lies in $\mathbf{K}_n^{MW}(X)$, that is to say each of its residues at points of codimension 1 in X is 0. This follows at once from the definitions and elementary formulas for the residues.

This defines a morphism of sheaves on $\tilde{Sm}_k$. Now to show that this extends to a morphism of sheaves on Sm_k, using the equivalence of categories of Theorem 2.11 (and its proof) we end up to show that our symbol maps commutes to restriction maps s_v, which is also clear from the elementary formulas we proved in Milnor–Witt K-theory. In this way we have obtained our canonical symbol map

$$\sigma_n : (\mathbb{G}_m)^{\wedge n} \to \mathbf{K}_n^{MW}$$

From what we have done in Chaps. 2 and 3, we know that $\mathbf{K}_n^{MW}$ is a strongly $\mathbb{A}^1$-invariant sheaf.

Theorem 3.37. *Let $n \geq 1$. The morphism σ_n is the universal morphism from $(\mathbb{G}_m)^{\wedge n}$ to a strongly $\mathbb{A}^1$-invariant sheaf of abelian groups. In other words, given a morphism of pointed sheaves $\phi : (\mathbb{G}_m)^{\wedge n} \to M$, with M a strongly $\mathbb{A}^1$-invariant sheaf of abelian groups, then there exists a unique morphism of sheaves of abelian groups $\Phi : \mathbf{K}_n^{MW} \to M$ such that $\Phi \circ \sigma_n = \phi$.*

Remark 3.38. The statement is wrong if we release the assumption that M is a sheaf of abelian groups. The free strongly $\mathbb{A}^1$-invariant sheaf of groups generated by $\mathbb{G}_m$ will be seen in Sect. 7.3 to be non commutative. For $n = 2$, it is a sheaf of abelian groups. For $n > 2$ it is not known to us.

The statement is also false for $n = 0$: $(\mathbb{G}_m)^{\wedge 0}$ is just $Spec(k)_+$, that is to say $Spec(k)$ with a base point added, and the free strongly $\mathbb{A}^1$-invariant sheaf of abelian groups generated by $Spec(k)_+$ is $\mathbb{Z}$, not $\mathbf{K}_0^{MW}$. To see a analogous presentation of $\mathbf{K}_0^{MW}$ see Theorem 3.46 below. □

Roughly, the idea of the proof is to first use Lemma 3.4 to show that $\phi : (\mathbb{G}_m)^{\wedge n} \to M$ induces on fields $F \in \mathcal{F}_k$ a morphism $K_n^{MW}(F) \to M(F)$ and then to use our work on unramified sheaves in Chap. 2 to observe this induces a morphism of sheaves.

Theorem 3.39. *Let M be a strongly $\mathbb{A}^1$-invariant sheaf, let $n \geq 1$ be an integer, and let $\phi : (\mathbb{G}_m)^{\wedge n} \to M$ be a morphism of pointed sheaves. For any* *field $F \in \mathcal{F}_k$, there is unique morphism*

$$\Phi(F) : K_n^{MW}(F) \to M(F)$$

such that for any $(u_1, \ldots, u_n) \in (F^\times)^n$, $\Phi_n(F)([u_1, \ldots, u_n]) = \phi(u_1, \ldots, u_n)$.

Preliminaries. We will freely use some notions and some elementary results from [59].

Let M be a sheaf of groups on Sm_k. Recall that we denote by M_{-1} the sheaf $M^{(\mathbb{G}_m)}$, and for $n \geq 0$, by M_{-n} the n-th iteration of this construction. To say that M is strongly $\mathbb{A}^1$-invariant is equivalent to the fact that $K(M, 1)$ is $\mathbb{A}^1$-local [59]. Indeed from *loc. cit.*, for any pointed space $\mathcal{X}$, we have $Hom_{\mathcal{H}_\bullet(k)}(\mathcal{X}; K(M, 1)) \cong H^1(\mathcal{X}; M)$ and $Hom_{\mathcal{H}_\bullet(k)}(\Sigma(\mathcal{X}); K(M, 1)) \cong \tilde{M}(X))$. Here we denote for M a strongly $\mathbb{A}^1$-invariant sheaf of abelian groups and $\mathcal{X}$ a pointed space by $\tilde{M}(\mathcal{X})$ the kernel of the evaluation at the base point of $M(\mathcal{X}) \to M(k)$, so that $M(\mathcal{X})$ splits as $M(k) \oplus \tilde{M}(\mathcal{X})$.

We also observe that because M is assumed to be abelian, the map (from "pointed to base point free classes")

$$Hom_{\mathcal{H}_\bullet(k)}(\Sigma(\mathcal{X}); K(M, 1)) \to Hom_{\mathcal{H}(k)}(\Sigma(\mathcal{X}); K(M, 1))$$

is a bijection.

From Lemma 2.32 and its proof we know that in that case, $R\mathbf{Hom}_\bullet(\mathbb{G}_m; K(M, 1))$ is canonically isomorphic to $K(M_{-1}, 1)$ and that M_{-1} is also strongly $\mathbb{A}^1$-invariant. We also know that $R\Omega_s(K(M, 1) \cong M$.

As a consequence, for a strongly $\mathbb{A}^1$-invariant sheaf of abelian groups M, the evaluation map

$$Hom_{\mathcal{H}_\bullet(k)}(\Sigma((\mathbb{G}_m)^{\wedge n}), K(M, 1)) \to M_{-n}(k)$$

is an isomorphism of abelian groups.

Now for $\mathcal{X}$ and $\mathcal{Y}$ pointed spaces, the cofibration sequence $\mathcal{X} \vee \mathcal{Y} \to \mathcal{X} \times \mathcal{Y} \to \mathcal{X} \wedge \mathcal{Y}$ splits after applying the suspension functor Σ. Indeed, as $\Sigma(\mathcal{X} \times \mathcal{Y})$ is a co-group object in $\mathcal{H}_\bullet(k)$ the (ordered) sum of the two morphism $\Sigma(\mathcal{X} \times \mathcal{Y}) \to \Sigma(\mathcal{X}) \vee \Sigma(\mathcal{Y}) = \Sigma(\mathcal{X} \vee \mathcal{Y})$ gives a left inverse to $\Sigma(\mathcal{X}) \vee \Sigma(\mathcal{Y}) \to \Sigma(\mathcal{X} \times \mathcal{Y})$. This left inverse determines an $\mathcal{H}_\bullet(k)$-isomorphism $\Sigma(\mathcal{X}) \vee \Sigma(\mathcal{Y}) \vee \Sigma(\mathcal{X} \wedge \mathcal{Y}) \cong \Sigma(\mathcal{X} \times \mathcal{Y})$.

We thus get canonical isomorphisms:

$$\tilde{M}(\mathcal{X} \times \mathcal{Y}) = \tilde{M}(\mathcal{X}) \oplus \tilde{M}(\mathcal{Y}) \oplus \tilde{M}(\mathcal{X} \wedge \mathcal{Y})$$

and analogously

$$H^1(\mathcal{X} \times \mathcal{Y}; M) = H^1(\mathcal{X}; M) \oplus H^1(\mathcal{Y}; M) \oplus H^1(\mathcal{X} \wedge \mathcal{Y}; M)$$

As a consequence, the product $\mu : \mathbb{G}_m \times \mathbb{G}_m \to \mathbb{G}_m$ on $\mathbb{G}_m$ induces in $\mathcal{H}_\bullet(k)$ a morphism $\Sigma(\mathbb{G}_m \times \mathbb{G}_m) \to \mathbb{Z}(\mathbb{G}_m)$ which using the above splitting decomposes as

$$\Sigma(\mu) = \langle Id_{\Sigma(\mathbb{G}_m)}, d_{\Sigma(\mathbb{G}_m)}, \eta \rangle : \Sigma(\mathbb{G}_m) \vee \Sigma(\mathbb{G}_m) \vee \Sigma((\mathbb{G}_m)^{\wedge 2}) \to \Sigma(\mathbb{G}_m)$$

The morphism $\Sigma((\mathbb{G}_m)^{\wedge 2}) \to \Sigma(\mathbb{G}_m)$ so defined is denoted η. It can be shown to be isomorphic in $\mathcal{H}_\bullet(k)$ to the Hopf map $\mathbb{A}^2 - \{0\} \to \mathbb{P}^1$.

Let M be a strongly $\mathbb{A}^1$-invariant sheaf of abelian groups. We will denote by

$$\eta : M_{-2} \to M_{-1}$$

the morphism of strongly $\mathbb{A}^1$-invariant sheaves of abelian groups induced by η.

In the same way let $\Psi : \Sigma(\mathbb{G}_m \wedge \mathbb{G}_m) \cong \Sigma(\mathbb{G}_m \wedge \mathbb{G}_m)$ be the twist morphism and for M a strongly $\mathbb{A}^1$-invariant sheaf of abelian groups, we still denote by

$$\Psi : M_{-2} \to M_{-2}$$

the morphism of strongly $\mathbb{A}^1$-invariant sheaves of abelian groups induced by Ψ.

Lemma 3.40. *Let M be a strongly $\mathbb{A}^1$-invariant sheaf of abelian groups. Then the morphisms $\eta \circ \Psi$ and η*

$$M_{-2} \to M_{-1}$$

are equal.

Proof. This is a direct consequence of the fact that μ is commutative. □

As a consequence, for any $m \geq 1$, the morphisms of the form

$$M_{-m-1} \to M_{-1}$$

obtained by composing m times morphisms induced by η doesn't depend on the chosen ordering. We thus simply denote by $\eta^m : M_{-m-1} \to M_{-1}$ this canonical morphism.

Proof of Theorem 3.39 By Lemma 3.6 1), the uniqueness is clear. By a base change argument analogous to [52, Corollary 5.2.7], we may reduce to the case $F = k$.

From now on we fix a morphism of pointed sheaves $\phi : (\mathbb{G}_m)^{\wedge n} \to M$, with M a strongly $\mathbb{A}^1$-invariant sheaf of abelian groups. We first observe that ϕ determines and is determined by the $\mathcal{H}_\bullet(k)$-morphism $\phi : \Sigma((\mathbb{G}_m)^{\wedge n}) \to K(M, 1)$, or equivalently by the associated element $\phi \in M_{-n}(k)$.

For any symbol $(u_1, \ldots, u_r) \in (k^\times)^r$, $r \in \mathbb{N}$, we let $S^0 \to (\mathbb{G}_m)^{\wedge r}$ be the (ordered) smash-product of the morphisms $[u_i] : S^0 \to \mathbb{G}_m$ determined by u_i. For any integer $m \geq 0$ such that $r = n + m$, we denote by $[\eta^m, u_1, \ldots, u_r] \in M(k) \cong Hom_{\mathcal{H}_\bullet(k)}(\Sigma(S^0), K(M, 1))$ the composition

$$\eta^m \circ \Sigma([u_1, \ldots, u_n]) : \Sigma(S^0) \to \Sigma((\mathbb{G}_m)^{\wedge r}) \xrightarrow{\eta^m} \Sigma((\mathbb{G}_m)^{\wedge n}) \xrightarrow{\phi} K(M, 1)$$

The theorem now follows from the following:

Lemma 3.41. *The previous assignment* $(m, u_1, \ldots, u_r) \mapsto [\eta^m, u_1, \ldots, u_r] \in M(k)$ *satisfies the relations of Definition 3.3 and as a consequence induce a morphism*

$$\Phi(k) : K_n^{MW}(k) \to M(k)$$

Proof. The proof of the Steinberg relation $\mathbf{1_n}$ will use the following stronger result by P. Hu and I. Kriz:

Lemma 3.42. *(Hu–Kriz [33]) The canonical morphism of pointed sheaves* $(\mathbb{A}^1 - \{0, 1\})_+ \to \mathbb{G}_m \wedge \mathbb{G}_m$, $x \mapsto (x, 1 - x)$ *induces a trivial morphism* $\tilde{\Sigma}(\mathbb{A}^1 - \{0, 1\}) \to \Sigma(\mathbb{G}_m \wedge \mathbb{G}_m)$ *(where* $\tilde{\Sigma}$ *means unreduced suspension[1]) in* $\mathcal{H}_\bullet(k)$.

For any $a \in k^\times - \{1\}$ the suspension of the morphism of the form $[a, 1-a] : S^0 \to (\mathbb{G}_m)^{\wedge 2}$ factors in $\mathcal{H}_\bullet(k)$) through $\tilde{\Sigma}(\mathbb{A}^1 - \{0, 1\}) \to \Sigma(\mathbb{G}_m \wedge \mathbb{G}_m)$ as the morphism $Spec(k) \to \mathbb{G}_m \wedge \mathbb{G}_m$ factors itself through $\mathbb{A}^1 - \{0, 1\}$. This implies the Steinberg relation in our context as the morphism of the form $\Sigma([u_i, 1 - u_i]) : \Sigma(S^0) \to \Sigma((\mathbb{G}_m)^{\wedge 2})$ appears as a factor in the morphism which defines the symbol $[\eta^m, u_1, \ldots, u_r]$, with $u_i + u_{i+1} = 1$, in $M(k)$.

Now, to check the relation $\mathbf{2}_n$, we observe that the pointed morphism $[ab] : S^0 \to \mathbb{G}_m$ factors as $S^0 \xrightarrow{[a][b]} \mathbb{G}_m \times \mathbb{G}_m \xrightarrow{\mu} \mathbb{G}_m$. Taking the suspension and using the above splitting which defines η, yields that

$$\Sigma([ab]) = \Sigma([a]) \vee \Sigma([b]) \vee \eta([a][b]) : \Sigma(S^0) \to \Sigma(\mathbb{G}_m)$$

in the group $Hom_{\mathcal{H}_\bullet(k)}(\Sigma(S^0), \Sigma(\mathbb{G}_m))$ whose law is denoted by $\vee$. This implies relation $\mathbf{2}_n$.

Now we come to check the relation $\mathbf{4}_n$. For any $a \in k^\times$, the morphism $a : \mathbb{G}_m \to \mathbb{G}_m$ given by multiplication by a is not pointed (unless $a = 1$). However the pointed morphism $a_+ : (\mathbb{G}_m)_+ \to \mathbb{G}_m$ induces after suspension $\Sigma(a_+) : S^1 \vee \Sigma(\mathbb{G}_m) \cong \Sigma((\mathbb{G}_m)_+) \to \Sigma(\mathbb{G}_m)$. We denote by $< a >: \Sigma(\mathbb{G}_m) \to \Sigma(\mathbb{G}_m)$ the morphism in $\mathcal{H}_\bullet(k)$ induced on the factor $\Sigma(\mathbb{G}_m)$. We need:

---

[1] Observe that if $k = \mathbb{F}_2$, $\mathbb{A}^1 - \{0, 1\}$ has no rational point.

Lemma 3.43. *1) For any $a \in k^{\times}$, the morphism $M_{-1} \to M_{-1}$ induced by $< a >: \Sigma(\mathbb{G}_m) \to \Sigma(\mathbb{G}_m)$ is equal to $Id + \eta \circ [a]$.*

2) The twist morphism $\Psi \in Hom_{\mathcal{H}_{\bullet}(k)}(\Sigma(\mathbb{G}_m \wedge \mathbb{G}_m), \Sigma(\mathbb{G}_m \wedge \mathbb{G}_m))$ and the inverse, for the group structure, of $Id_{\mathbb{G}_m} \wedge < -1 > \cong < -1 > \wedge Id_{\mathbb{G}_m}$ have the same image in the set $Hom_{\mathcal{H}(k)}(\Sigma(\mathbb{G}_m \wedge \mathbb{G}_m), \Sigma(\mathbb{G}_m \wedge \mathbb{G}_m))$.

Remark 3.44. In fact the map

$$Hom_{\mathcal{H}_{\bullet}(k)}(\Sigma(\mathbb{G}_m \wedge \mathbb{G}_m), \Sigma(\mathbb{G}_m \wedge \mathbb{G}_m)) \to Hom_{\mathcal{H}(k)}(\Sigma(\mathbb{G}_m \wedge \mathbb{G}_m), \Sigma(\mathbb{G}_m \wedge \mathbb{G}_m))$$

is a bijection. Indeed we know that $\Sigma(\mathbb{G}_m \wedge \mathbb{G}_m))$ is $\mathbb{A}^1$-equivalent to $\mathbb{A}^2 - \{0\}$ and also to SL_2 because the morphism $SL_2 \to \mathbb{A}^2 - \{0\}$ (forgetting the second column) is an $\mathbb{A}^1$-weak equivalence. As SL_2 is a group scheme, the classical argument shows that this space is $\mathbb{A}^1$-simple. Thus for any pointed space $\mathcal{X}$, the action of $\pi_1^{\mathbb{A}^1}(SL_2)(k)$ on $Hom_{\mathcal{H}_{\bullet}(k)}(\mathcal{X}, SL_2)$ is trivial. We conclude because as usual, for any pointed spaces $\mathcal{X}$ and $\mathcal{Y}$, with $\mathcal{Y}$ $\mathbb{A}^1$-connected, the map $Hom_{\mathcal{H}_{\bullet}(k)}(\mathcal{X}, \mathcal{Y}) \to Hom_{\mathcal{H}(k)}(\mathcal{X}, \mathcal{Y})$ is the quotient by the action of the group $\pi_1^{\mathbb{A}^1}(\mathcal{Y})(k)$.

Proof. 1) The morphism $a : \mathbb{G}_m \to \mathbb{G}_m$ is equal to the composition $\mathbb{G}_m \xrightarrow{[a] \times Id} \mathbb{G}_m \times \mathbb{G}_m \xrightarrow{\mu} \mathbb{G}_m$. Taking the suspension, the previous splittings give easily the result.

2) Through the $\mathcal{H}_{\bullet}(k)$-isomorphism $\Sigma(\mathbb{G}_m \wedge \mathbb{G}_m) \cong \mathbb{A}^2 - \{0\}$, the twist morphism becomes the opposite of the permutation isomorphism $(x, y) \mapsto (y, x)$. This follows easily from the definition of this isomorphism using the Mayer–Vietoris square

$$\begin{array}{ccc}
\mathbb{G}_m \times \mathbb{G}_m & \subset & \mathbb{A}^1 \times \mathbb{G}_m \\
\cap & & \cap \\
\mathbb{G}_m \times \mathbb{A}^1 & \subset & \mathbb{A}^2 - \{0\}
\end{array}$$

and the fact that our automorphism on $\mathbb{A}^2 - \{0\}$ permutes the top right and bottom left corner.

Consider the action of $GL_2(k)$ on $\mathbb{A}^2 - \{0\}$. As any matrix in $SL_2(k)$ is a product of elementary matrices, the associated automorphism $\mathbb{A}^2 - \{0\} \cong \mathbb{A}^2 - \{0\}$ is the identity in $\mathcal{H}(k)$. As the permutation matrix $\begin{pmatrix} 0 & 1 \\ 1 & 0 \end{pmatrix}$ is congruent to $\begin{pmatrix} -1 & 0 \\ 0 & 1 \end{pmatrix}$ or $\begin{pmatrix} 1 & 0 \\ 0 & -1 \end{pmatrix}$ modulo $SL_2(k)$, we get the result. □

Proof of Theorem 3.37 By Lemma 3.45 below, we know that for any smooth irreducible X with function field F, the restriction map $M(X) \subset M(F)$ is injective.

As $\mathbf{K}_n^{MW}$ is unramified, the Remark 2.15 of Sect. 1.1 shows that to produce a morphism of sheaves $\Phi : \mathbf{K}_n^{MW} \to M$ it is sufficient to prove that for any discrete valuation v on $F \in \mathcal{F}_k$ the morphism $\Phi(F) : K_n^{MW}(F) \to M(F)$ maps $\mathbf{\underline{K}}_n^{MW}(\mathcal{O}_v)$ into $M(\mathcal{O}_v)$ and in case the residue field $\kappa(v)$ is separable, that some square is commutative (see Remark 2.15).

But by Theorem 3.22, we know that the subgroup $\mathbf{\underline{K}}_n^{MW}(\mathcal{O}_v)$ of $K_n^{MW}(F)$ is the one generated by symbols of the form $[u_1, \ldots, u_n]$, with the $u_i \in \mathcal{O}_v^\times$. The claim is now trivial: for any such symbol there is a smooth model X of $\mathcal{O}_v$ and a morphism $X \to (\mathbb{G}_m)^{\wedge n}$ which induces $[u_1, \ldots, u_n]$ when composed with $(\mathbb{G}_m)^{\wedge n} \to \mathbf{\underline{K}}_n^{MW}$. But now composition with $\phi : (\mathbb{G}_m)^{\wedge n} \to M$ gives an element of $M(X)$ which lies in $M(\mathcal{O}_v) \subset M(F)$ which is by definition the image of $[u_1, \ldots, u_n]$ through $\Phi(F)$. A similar argument applies to check the commutativity of the square of the Remark 2.15: one may choose X so that there is a closed irreducible $Y \subset X$ of codimension 1, with $\mathcal{O}_{X, \eta_Y} = \mathcal{O}_v \subset F$. Then the restriction of $\Phi([u_1, \ldots, u_n]) \subset M(\mathcal{O}_v)$ is just induced by the composition $Y \to X \to (\mathbb{G}_m)^{\wedge n} \to M$, and this is also compatible with the s_v in Milnor–Witt K-theory.   □

Lemma 3.45. *Let M be an $\mathbb{A}^1$-invariant sheaf of pointed sets on Sm_k. Then for any smooth irreducible X with function field F, the kernel of the restriction map $M(X) \subset M(F)$ is trivial.*

In case M is a sheaf of groups, we see that the restriction map $M(X) \to M(F)$ is injective.

Proof. This follows from [52, Lemma 6.1.4] which states that $L_{\mathbb{A}^1}(X/U)$ is always 0-connected for U non-empty dense in X. Now the kernel of $M(X) \to M(U)$ is covered by $Hom_{\mathcal{H}_\bullet(k)}(X/U, M)$, which is trivial as M is his own π_0 and $L_{\mathbb{A}^1}(X/U)$ is 0-connected.   □

We know deal with $\mathbf{\underline{K}}_0^{MW}$. We observe that there is a canonical morphism of sheaves of sets $\mathbb{G}_m/2 \to \mathbf{\underline{K}}_0^{MW}$, $U \mapsto\, <U>$, where $\mathbb{G}_m/2$ means the cokernel in the category of sheaves of abelian groups of $\mathbb{G}_m \xrightarrow{2} \mathbb{G}_m$.

Theorem 3.46. *The canonical morphism of sheaves $\mathbb{G}_m/2 \to \mathbf{\underline{K}}_0^{MW}$ is the universal morphism of sheaves of sets to a strongly $\mathbb{A}^1$-invariant sheaf of abelian groups. In other words $\mathbf{\underline{K}}_0^{MW}$ is the free strongly $\mathbb{A}^1$-invariant sheaf on the space $\mathbb{G}_m/2$.*

Proof. Let M be a strongly $\mathbb{A}^1$-invariant sheaf of abelian groups. Denote by $\mathbb{Z}[\mathcal{S}]$ the free sheaf of abelian groups on a sheaf of sets $\mathcal{S}$. When $\mathcal{S}$ is pointed, then the latter sheaf splits canonically as $\mathbb{Z}[\mathcal{S}] = \mathbb{Z} \oplus \mathbb{Z}(\mathcal{S})$ where $\mathbb{Z}(\mathcal{S})$ is the free sheaf of abelian groups on the pointed sheaf of sets $\mathcal{S}$, meaning the quotient $\mathbb{Z}[\mathcal{S}]/\mathbb{Z}[*]$ (where $* \to \mathcal{S}$ is the base point). Now a morphism of sheaves of sets $\mathbb{G}_m/2 \to M$ is the same as a morphism of sheaves of abelian groups $\mathbb{Z}[\mathbb{G}_m] = \mathbb{Z} \oplus \mathbb{Z}(\mathbb{G}_m) \to M$. By the Theorem 3.37 a morphism $\mathbb{Z}(\mathbb{G}_m) \to M$ is the same as a morphism $\mathbf{\underline{K}}_1^{MW} \to M$.

Thus to give a morphism of sheaves of sets $\mathbb{G}_m/2 \to M$ is the same as to give a morphism of sheaves of abelian groups $\mathbb{Z} \oplus \mathbf{\underline{K}}_1^{MW} \to M$ together with extra conditions. One of this conditions is that the composition $\mathbb{Z} \oplus \underline{\mathbf{K}}_1^{MW} \overset{[2]}{\to}$ $\mathbb{Z} \oplus \underline{\mathbf{K}}_1^{MW} \to M$ is equal to $\mathbb{Z} \oplus \underline{\mathbf{K}}_1^{MW} \overset{[*]}{\to} \mathbb{Z} \oplus \underline{\mathbf{K}}_1^{MW} \to M$. Here $[*]$ is represented by the matrix $\begin{pmatrix} Id_{\mathbb{Z}} & 0 \\ 0 & 0 \end{pmatrix}$ and $[2]$ by the matrix $\begin{pmatrix} Id_{\mathbb{Z}} & 0 \\ 0 & [2]_1 \end{pmatrix}$. The morphism $[2]_1 : \underline{\mathbf{K}}_1^{MW} \to \underline{\mathbf{K}}_1^{MW}$ is the one induced by the square map on $\mathbb{G}_m$. From Lemma 3.14, we know that this map is the multiplication by $2_\epsilon = h$. recall that we set $\underline{\mathbf{K}}_1^{W} := \underline{\mathbf{K}}_1^{MW}/h$. Thus any morphism of sheaves of sets $\mathbb{G}_m/2 \to M$ determines a canonical morphism $\mathbb{Z} \oplus \underline{\mathbf{K}}_1^{W} \to M$. Moreover the morphism $\mathbb{Z}[\mathbb{G}_m] \to \mathbb{Z} \oplus \underline{\mathbf{K}}_1^{W}$ factors through $\mathbb{Z}[\mathbb{G}_m] \to \mathbb{Z}[\mathbb{G}_m/2]$; this morphism is induced by the map $U \mapsto (1, <U>)$. □

We have thus proven that given any morphism $\phi : \mathbb{Z}[\mathbb{G}_m/2] \to M$, there exists a unique morphism $\mathbb{Z} \oplus \underline{\mathbf{K}}_1^{W} \to M$ such that the composition $\mathbb{Z}[\mathbb{G}_m/2] \to \mathbb{Z} \oplus \underline{\mathbf{K}}_1^{W} \to M$ is ϕ. As $\mathbb{Z} \oplus \underline{\mathbf{K}}_1^{W}$ is a strongly $\mathbb{A}^1$-invariant sheaf of abelian groups, it is the free one on $\mathbb{G}_m/2$.

Our claim is now that the canonical morphism $i : \mathbb{Z} \oplus K_1^{W} \to \underline{\mathbf{K}}_0^{MW}$ is an isomorphism.

We know proceed closely to proof of Theorem 3.37. We first observe that for any $F \in \mathcal{F}_k$, the canonical map $\mathbb{Z}[F^\times/2] \to \mathbb{Z} \oplus K_1^{W}(F)$ factors through $\mathbb{Z}[F^\times/2] \twoheadrightarrow K_0^{MW}(F)$. This is indeed very simple to check using the presentation of $K_0^{MW}(F)$ given in Lemma 3.9. We denote by $j(F) : K_0^{MW}(F) \to \mathbb{Z} \oplus K_1^{W}(F)$ the morphism so obtained.

Using Theorem 3.22 and the same argument as in the end of the proof of Theorem 3.37 we see that the $j(F)$'s actually come from a morphism of sheaves $j : \underline{\mathbf{K}}_0^{MW} \to \mathbb{Z} \oplus \underline{\mathbf{K}}_1^{W}$. It is easy to check on $F \in \mathcal{F}_k$ that i and j are inverse morphisms to each other. □

The following corollary is immediate from the Theorem and its proof:

Corollary 3.47. *The canonical morphism*

$$K_1^{W}(F) \to I(F)$$

is an isomorphism.

We now give some applications concerning abelian sheaves of the form M_{-1}, see Sect. 2.2. From Lemma 2.32 if M is strongly $\mathbb{A}^1$-invariant, so is M_{-1}. Now we observe that there is a canonical pairing:

$$\mathbb{G}_m \times M_{-1} \to M$$

In case M is a sheaf of abelian groups, as opposed to simply a sheaf of groups, we may view $M_{-1}(X)$ for $X \in Sm_k$ as fitting in a short exact sequence:

$$0 \to M(X) \to M(\mathbb{G}_m \times X) \to M_{-1}(X) \to 0 \tag{3.8}$$

Given $\alpha \in \mathcal{O}(X)^\times$ that we view as a morphism $X \to \mathbb{G}_m$, we may consider the evaluation at $\alpha\ ev_\alpha : M(\mathbb{G}_m \times X) \to M(X)$, that is to say the restriction map through $(\alpha, Id_X) \circ \Delta_X : X \to \mathbb{G}_m \times X$. Now $ev_\alpha - ev_1 : M(\mathbb{G}_m \times X) \to M(X)$ factor through $M_{-1}(X)$ and induces a morphism $\alpha\cup : M_{-1}(X) \to M(X)$. This construction define a morphism of sheaves of sets $\mathbb{G}_m \times M_{-1} \to M$ which is our pairing.

Iterating this process gives a pairing

$$(\mathbb{G}_m)^{\wedge n} \times M_{-n} \to M$$

for any $n \geq 1$.

Lemma 3.48. *For any $n \geq 1$ and any strongly $\mathbb{A}^1$-invariant sheaf, the above pairing induces a bilinear pairing*

$$\underline{K}_n^{MW} \times M_{-n} \to M \quad , \quad (\alpha, m) \mapsto \alpha.m$$

Proof. Let's us prove first that for each field $F \in \mathcal{F}_k$, the pairing $(F^\times)^{\wedge n} \times M_{-n}(F) \to M(F)$ factors through $\mathbb{Z}(F^\times) \times M_{-1}(F) \to K_n^{MW}(F) \times M_{-n}(F)$. Fix $F_0 \in \mathcal{F}_k$ and consider an element $u \in M_{-n}(F_0)$. We consider the natural morphism of sheaves of abelian groups on Sm_{F_0}, $\mathbb{Z}((\mathbb{G}_m)^{\wedge n}) \to M|_{F_0}$ induced by the cup product with u, where $M|_{F_0}$ is the "restriction" of M to Sm_{F_0}. It is clearly a strongly $\mathbb{A}^1$-invariant sheaf of groups (use an argument of passage to the colimit in the H^1) and by Theorem 3.37, this morphism $\mathbb{Z}((\mathbb{G}_m)^{\wedge n}) \to M|_{F_0}$ induces a unique morphism $\underline{K}_n^{MW} \to M|_{F_0}$. Now the evaluation of this morphism on F_0 itself is a homomorphism $K_n^{MW}(F_0) \to M(F_0)$ and it is induced by the product by u. This proves that the pairing $(F^\times)^{\wedge n} \times M_{-n}(F) \to M(F)$ factors through $\mathbb{Z}(F^\times) \times M_{-1}(F) \to K_n^{MW}(F) \times M_{-n}(F)$. Now to check that this comes from a morphisms of sheaves

$$\underline{K}_n^{MW} \times M_{-n} \to M$$

is checked using the techniques from Sect. 2.1. The details are left to the reader.   □

Now let us observe that the sheaves of the form M_{-1} are endowed with a canonical action of $\mathbb{G}_m$. We start with the short exact sequence (3.8):

$$0 \to M(X) \to M(\mathbb{G}_m \times X) \to M_{-1}(X) \to 0$$

We let $\mathcal{O}(X)^\times$ act on the middle term by translations, through $(u, m) \mapsto U^*(m)$ where $U : \mathbb{G}_m \times X \cong \mathbb{G}_m \times X$ is the automorphism multiplication by the unit $u \in \mathcal{O}(X)^\times$. The left inclusion is equivariant if we let $\mathcal{O}(X)^\times$ act trivially on $M(X)$. Thus M_{-1} gets in this way a canonical and functorial structure of $\mathbb{G}_m$-module.

Lemma 3.49. *If M is strongly $\mathbb{A}^1$-invariant, the canonical structure of $\mathbb{G}_m$-modules on M_{-1} is induced from a $\mathbf{K}_0^{MW}$-module structure on M_{-1} through the morphism of sheaves (of sets) $\mathbb{G}_m \to \mathbf{K}_0^{MW}$ which maps a unit u to its symbol $<u>= \eta[u] + 1$. Moreover the pairing of Lemma 3.48, for $n \geq 2$*

$$\mathbf{K}_n^{MW} \times M_{-n+1} \to M_{-1}$$

is $\mathbf{K}_0^{MW}$-bilinear: for units u, v and an element $m \in M_{-2}(F)$ one has:

$$<u> ([v].m) = (<u> [v]).m = [v].(<u>.m)$$

Proof. The sheaf $X \mapsto M(\mathbb{G}_m \times X)$ is the internal function object $M^{\mathbb{Z}(\mathbb{G}_m)}$ in the following sense: it has the property that for any sheaf of abelian groups N one has a natural isomorphism of the form

$$Hom_{\mathcal{A}b_k}(N \otimes \mathbb{Z}(\mathbb{G}_m), M) \cong Hom_{\mathcal{A}b_k}(N, M^{\mathbb{Z}(\mathbb{G}_m)})$$

where $\mathcal{A}b_k$ is the abelian category of sheaves of abelian groups on Sm_k and $\otimes$ is the tensor product of sheaves of abelian groups. The above exact sequence corresponds to the adjoint of the split short exact sequence

$$0 \to \tilde{\mathbb{Z}}(\mathbb{G}_m) \to \mathbb{Z}(\mathbb{G}_m) \to \mathbb{Z} \to 0$$

This short exact sequence is an exact sequence of $\mathbb{Z}(\mathbb{G}_m)$-modules (but non split as such !) and this structure induces exactly the structure of $\mathbb{Z}(\mathbb{G}_m)$-module on $M^{\mathbb{G}_m}$ and M_{-1} that we used above.

In other words, the functional object $M^{\tilde{\mathbb{Z}}(\mathbb{G}_m)}$ is isomorphic to M_{-1} as a $\mathbb{Z}(\mathbb{G}_m)$-module, where the structure of $\mathbb{Z}(\mathbb{G}_m)$-module on the sheaf $\tilde{Z}(\mathbb{G}_m)$ is induced by the tautological one on $\mathbb{Z}(\mathbb{G}_m)$.

Now as M is strongly $\mathbb{A}^1$-invariant the canonical morphism

$$M^{\mathbf{K}_1^{MW}} \to M_{-1} = M^{\tilde{Z}(\mathbb{G}_m)}$$

induced by $\tilde{Z}(\mathbb{G}_m) \to \mathbf{K}_1^{MW}$, is an isomorphism. Indeed given any N a morphism $N \otimes \tilde{Z}(\mathbb{G}_m) \to M$ factorizes uniquely through $N \otimes \mathbb{Z}(\mathbb{G}_m) \to N \otimes \mathbf{K}_1^{MW}$ as the morphism $\tilde{Z}(\mathbb{G}_m) \to \mathbf{K}_1^{MW}$ is the universal one to a strongly $\mathbb{A}^1$-invariant sheaf by Theorem 3.37.

Now the morphism $\tilde{Z}(\mathbb{G}_m) \to \mathbf{K}_1^{MW}$ is $\mathbb{G}_m$-equivariant where $\mathbb{G}_m$ acts on $\mathbf{K}_1^{MW}$ through the formula on symbols $(u, [x]) \mapsto [ux] - [u]$. Now this action factors through the canonical action of $\mathbf{K}_0^{MW}$ by the results of Sect. 3.1 as in $\mathbf{K}_1^{MW}$ one has $[ux] - [u] =<u> [x]$.

The last statement is straightforward to check. □

For $n \geq 2$ we thus get also on M_{-n} a structure of $\mathbf{K}_0^{MW}$-module by expressing M_{-n} as $(M_{-n+1})_{-1}$. However there are several ways to express it this way, one for each index in $i \in \{1, \ldots, n\}$, by expressing $M_{-n}(X)$ as a quotient of $M((\mathbb{G}_m)^n \times X)$ and letting $\mathbb{G}_m$ acts on the given i-th factor. One shows using the results from Sect. 3.1 that this action doesn't depend on the factor one chooses. Indeed given $F_0 \in \mathcal{F}_k$ and $u \in M_{-n}(F_0)$, we may see u as a morphism of pointed sheaves (over F_0) $u : (\mathbb{G}_m)^{\wedge n} \to M|_{F_0}$ and Theorem 3.37 tells us that u induces a unique $u' : \underline{\mathbf{K}}_n^{MW} \to M|_{F_0}$. Now the action of a unit $\alpha \in (F_0)^{\times}$ on u through the i-th factor of $M((\mathbb{G}_m)^n \times X)$ corresponds to letting α acts through the i-th factor $\mathbb{Z}(\mathbb{G}_m)$ of $(\mathbb{Z}(\mathbb{G}_m))^{\otimes n}$ and compose with $(\mathbb{Z}(\mathbb{G}_m))^{\otimes n} \to \underline{\mathbf{K}}_n^{MW} \to M$. A moment of reflexion shows that this action of α on a symbol $[a_1, \ldots, a_n] \in K_n^{MW}(F_0)$ is explicitly given by $[a_1, \ldots, \alpha.a_i, \ldots, a_n] - [a_1, \ldots, \alpha, \ldots, a_n] \in K_n^{MW}(F_0)$. Now the formulas in Milnor–Witt K-theory from Sect. 3.1 show that this is equal to

$$[a_1] \ldots (< \alpha > .[a_i]) \ldots [a_n] = < \alpha > [a_1, \ldots, a_n]$$

which doesn't depend on i.

This structure of $\underline{\mathbf{K}}_0^{MW} = \underline{\mathbf{GW}}$-module on sheaves of the form M_{-1} will play an important role in the next sections. We may emphasize it with the following observation. Let F be in $\mathcal{F}_k$ and let v be a discrete valuation on F, with valuation ring $\mathcal{O}_v \subset F$. For any strongly $\mathbb{A}^1$-invariant sheaf of abelian groups M, each non-zero element μ in $\mathcal{M}_v/(\mathcal{M}_v)^2$ determines by Corollary 2.35 a canonical isomorphism of abelian groups

$$\theta_\mu : M_{-1}(\kappa(v)) \cong H_v^1(\mathcal{O}_v; M)$$

Lemma 3.50. *We keep the previous notations. Let $\mu' = u.\mu$ be another non zero element of $\mathcal{M}_y/(\mathcal{M}_y^2)$ and thus $u \in \kappa(y)^{\times}$. Then the following diagram is commutative:*

$$
\begin{array}{ccc}
M_{-1}(\kappa(v)) & \overset{<u>}{\cong} & M_{-1}(\kappa(v)) \\
\theta_\mu \downarrow & & \theta_{\mu'} \downarrow \\
H_v^1(\mathcal{O}_v; M) & = & H_v^1(\mathcal{O}_v; M)
\end{array}
$$

The proof is straightforward and we leave the details to the reader.

Chapter 4
Geometric Versus Canonical Transfers

In this section M denotes a strongly $\mathbb{A}^1$-invariant sheaf of abelian groups in the Nisnevich topology on Sm_k and unless otherwise stated, the cohomology groups are always computed in the Nisnevich topology.

4.1 The Gersten Complex in Codimension 2

Let X be a smooth k-scheme. The coniveau spectral sequence, see [12], for X with coefficients in M is a cohomological spectral sequence of the form

$$E_1^{p,q} = \oplus_{x \in X^{(p)}} H_x^{p+q}(X; M) \Rightarrow H_{Nis}^{p+q}(X; M)$$

where for $x \in X$ a point the group $H_x^n(X; M)$ is the colimit $colim_\Omega H_{Nis}^n$ $(\Omega/\Omega - \overline{x} \cap \Omega; M)$ over the ordered set of open neighborhood Ω of x. For instance if $x \in X$ is a closed point $H_x^n(X; M) = H^n(X/X - x; M)$. If $U \subset X$ is an open subset, the coniveau spectral sequence for X maps to the one of U by functoriality. Moreover the induced morphism on the E_1-term is easy to analyse: it maps $H_x^{p+q}(X; M)$ to 0 if $x \notin U$ and if $x \in U$, then it maps $H_x^{p+q}(X; M)$ isomorphically onto $H_x^{p+q}(U; M)$. In the sequel we will often use this fact, and will identify both $H_x^{p+q}(X; M)$ and $H_x^{p+q}(U; M)$.

Remark 4.1. It is convenient to extend all the definitions to essentially smooth k-schemes in the obvious way, by taking the corresponding filtering colimits. If X is essentially smooth and X_α is a projective system in Sm_k representing X, and $x \in X$ a point, we mean by $H_x^n(X; M)$ the filtering colimit of the $H_{x_\alpha}^n(X_\alpha; M)$, where x_α is the image of x in X_α. In the sequel we will freely use this extension of notions and notations to essentially smooth k-schemes. For instance if $x \in X$ is a closed point in an essentially smooth k-scheme we also "have" $H_x^n(X; M) = H^n(X/X - x; M)$. If X is local and $x \in X$ is the closed point, and if $y \in X$ is the generic point

F. Morel, $\mathbb{A}^1$-*Algebraic Topology over a Field*, Lecture Notes in Mathematics 2052, DOI 10.1007/978-3-642-29514-0_4, © Springer-Verlag Berlin Heidelberg 2012

of a 1-dimensional integral closed subscheme containing x then we have
$H_y^n(X; M) = H^n(X - x/X - \overline{y}; M)$. etc... $\qquad\qquad\qquad\qquad\qquad\qquad$ □

Now back to the coniveau spectral sequence observe that the E_1 term
vanishes for $q > 0$ for cohomological dimension reasons.

The *Gersten complex* $C^*(X; M)$ of X with coefficients in M is the
horizontal line $q = 0$ of the E^1-term; it is also called the Cousin complex
in [19, Sect. 1]. This complex extends to the right the complex $C^*(X; \mathcal{G})$ that
we previously used for $\mathcal{G}$ a sheaf of groups in Sect. 2.2; its term $C^n(X; M)$ in
degree n is thus isomorphic to

$$C^n(X; M) \cong \oplus_{x \in X^{(n)}} H_x^n(X; M)$$

Observe that given an open subset $U \to X$, we get by functoriality of the
coniveau spectral sequence above a morphism of chain complexes

$$C^*(X; M) \to C^*(U; M)$$

Form what we said above, it is an epimorphism with kernel in degree n the
direct sum of the $H_x^n(X; M)$ over the point x of codimension n in X not in U.
The kernel will be denoted by $C_Z^*(X; M)$ where Z is the closed complement of
U. For instance if $Z = \{z\}$ is the complement of a closed point of codimension
d, $C_z^*(X; M)$ is the group $H_z^d(X; M)$ viewed as a complex concentrated in
degree d.

In degree ≤ 2 the complex $C^*(X; M)$ coincides with the complex $C^*(X; \mathcal{G})$
that we previously used for $\mathcal{G}$ a sheaf of groups in Sect. 2.2. Because of the
abelian group structure, the weak-product Π' becomes a usual direct sum.
Our notations are then compatible.

For X a smooth k-scheme, $z \in X^{(d+1)}$ and $y \in X^{(d)}$ we denote by

$$\partial_z^y : H_y^d(X; M) \to H_z^{d+1}(X; M)$$

the component corresponding to the pair (y, z) in the differential of the chain
complex $C^*(X; M)$. It is easy to check that if $z \notin \overline{y}$ then $\partial_z^y = 0$.

If $z \in \overline{y}$ then ∂_z^y can be described as follows. First we may replace X by
its localization at z, and z is now a closed point. Now let $Y \subset X$ be the
closure of y in X. As we observed above, one has canonical isomorphisms
$H_z^i(X; M) = H^i(X/(X - z); M)$ and $H_y^i(X; M) = H^i(X - z/(X - Y); M)$.
Then ∂_z^y is the connecting homomorphism in the long exact sequence

$$\cdots \to H^d(X/X - Y) \to H^d(X - z/(X - Y); M) \to H^{d+1}(X/X - z; M) \to \cdots$$

Lemma 4.2. *Let X be the localization of a smooth k-scheme at a point z of
codimension 2 and let y be a point of codimension 1 in X. Then the following
sequence is exact:*

$$0 \to H^1(X/X - \overline{y}; M) \to H^1_y(X; M) \xrightarrow{\partial^y_z} H^2_z(X; M)$$

Proof. From the description of the differential given above we see that ∂^y_z is the connecting homomorphism ∂ in the cohomology long exact sequence

$$\cdots \to H^1(X/X - \overline{y}; M) \to H^1(X - z/X - \overline{y}; M) \xrightarrow{\partial} H^i(X/X - x; M) \to \cdots$$

of the triple $(X - z/X - \overline{y}) \subset (X/X - \overline{y}) \to (X/X - z)$. The kernel of $H^1(X/X - \overline{y}; M) \to H^1_y(X; M)$ is thus the image of $H^1(X/X - z; M) \to H^1(X/X - \overline{w})$; but $H^1(X/X - z; M) = H^1_z(X; M)$ vanishes by (4.1) below. $\square$

Let $i : Y \subset X$ be a closed immersion between smooth k-schemes. The quotient $X/(X - Y)$ is called the Thom space of the closed immersion i and by the $\mathbb{A}^1$-homotopy purity Theorem [59] there exists a canonical $\mathbb{A}^1$-weak equivalence

$$X/(X - Y) \cong Th(\nu_i)$$

where $Th(\nu_i)$ denotes the Thom space $E(\nu_i)/(E(\nu_i)^\times)$ of the normal bundle ν_i of i.

We assume now that the normal bundle of i is of rank n and is trivialized, that is to say that we assume that we choose an isomorphism $\nu_i \cong \theta^n_Z$ between the normal bundle of i and the trivial bundle of rank n on Z. In that case, following [59], we introduce $T = \mathbb{A}^1/\mathbb{G}_m$ and we have a canonical isomorphism $T^{\wedge n} \wedge (Y_+) \cong Th(\theta^n_Y)$.

If $Z \subset Y$ is an irreducible closed subscheme, we may consider the following cofibration sequence of pointed spaces, that is to say an exact sequence of pointed spaces

$$(X - Z)/(X - Y) \to X/(X - Y) \to X/(X - Z)$$

We thus get a cofibration sequence in the pointed $\mathbb{A}^1$-homotopy category of spaces

$$Th(\nu_i|_{Y-Z}) \to Th(\nu_i) \to X/(X - Z)$$

And using the given trivialization the above cofibration sequence takes the form

$$T^{\wedge n} \wedge (Y - Z)_+ \to T^{\wedge n} \wedge Y_+ \to X/(X - Z)$$

and produces an $\mathbb{A}^1$-equivalence of the form

$$T^{\wedge n} \wedge (Y/Y - Z)_+ \cong X/(X - Z)$$

Now we apply this to the situation where $Z = \overline{z}$ is irreducible with generic point z of codimension 2 in X, and where $Y = \overline{y}$ is irreducible with generic point of codimension 1. The long exact sequence in cohomology takes the form

$$\ldots \to H^1(X/X - Z; M) \to H^1(X/X - Y; M) \to H^1(X - Z/(X - Y); M)$$
$$\to H^2(X/X - Z; M) \to \ldots$$

As the Eilenberg–MacLane space BM is $\mathbb{A}^1$-local, using the identifications above induced by the trivialization gives

$$H^1(X/X - Z; M) \cong H^1(T \wedge (Y/(Y - Z)); M) = H^0(Y/(Y - Z); M_{-1}) = 0$$
$$(4.1)$$

and identifications $H^1(X/X - Y; M) \cong H^1(T \wedge (Y_+); M) \cong H^0(Y; M_{-1}) = M_{-1}(Y)$ and similarly $H^1(X - Z/(X - Y); M) \cong M_{-1}(Y - Z)$. Thus we get an exact sequence of the form:

$$0 \to M_{-1}(Y) \to M_{-1}(Y - Z) \to H^2(X/X - Z; M) \qquad (4.2)$$

By an easy argument of passing to filtering colimits, the above exact sequence extends to the situation where $i : Y \subset X$ is a closed immersion between essentially smooth k-schemes and a ν_i is trivialized.

We will need more generally the notion of *orientation* of a vector bundle:

Definition 4.3. Let X be an essentially smooth k-scheme and ξ an algebraic vector bundle over X, of rank $r \geq 0$. An *orientation* ω of ξ is a pair consisting of a line bundle Λ over X and an isomorphism $\Lambda^{\otimes 2} \cong \Lambda^r(\xi)$. Two orientations ω and ω' are said to be equivalent if there is an isomorphism $\Lambda' \cong \Lambda$ (with obvious notations) which takes ω to ω'. We let $Q(\xi)$ be the set of equivalence classes of orientations of ξ.

For instance a trivialization of ξ, that is to say an isomorphism of line bundles over Y: $\theta_Y^r \cong \xi$, with θ_Y^r the trivial rank r vector bundle, defines an orientation.

Let X be an essentially smooth k-scheme of dimension 2 and let $i : Y \subset X$ be an irreducible closed scheme of dimension 1 which is essentially k-smooth. Let y be its generic point. Given an orientation ω of ν_i we obtain at the generic point y of Y an isomorphism $\lambda_y^{\otimes 2} \cong (\nu_i)_y = \mathcal{M}_y/(\mathcal{M}_y)^2$ where $\mathcal{M}_y$ is the maximal ideal of the discrete valuation ring $\mathcal{O}_{X,y}$. A choice of a generator of the $\kappa(y)$-vector space (of dimension 1) λ_y determines through the above isomorphism a uniformizing element of $\mathcal{O}_{X,y}$. If we choose an other generator of λ_y we get the previous uniformizing element multiplied by the square of a unit in $\mathcal{O}_{X,y}$. By Lemma 3.50 we thus see that such an orientation defines a canonical isomorphism

$$\theta_\omega : M_{-1}(\kappa(y)) \cong H^1_y(X; M)$$

Lemma 4.4. *Let X be the localization of a point z of codimension 2 in a smooth k-scheme and let $y \in X$ be a point of codimension 1. Assume* *that $Y := \bar{y}$ is essentially smooth. Let i be the normal bundle of the closed immersion $Y \subset X$. Then given an orientation $\omega : \lambda^{\otimes 2} \cong \nu_i$ of ν_i there exists a unique commutative diagram of the form*

$$
\begin{array}{ccccccccc}
0 \to & M_{-1}(Y) & \subset M_{-1}(\kappa(y)) & \to H^1_z(Y; M_{-1}) & \to 0 \\
& \downarrow \wr & \downarrow \wr & \cap \\
0 \to & H^1(X/X - Y; M) & \subset H^1_y(X; M) & \to H^2_z(X; M)
\end{array}
$$

in which the top horizontal line is the tautological short exact sequence and the middle vertical isomorphism is the isomorphism θ_ω mentioned above.

Proof. As λ is trivial over the local scheme Y, we may choose a trivialization of λ. This induces a trivialization of ν_i as well. Changing the trivialization of λ changes the trivialization of ν_i by the square of a unit in Y. The claim is a consequence of the exact sequence (4.2), observing that this exact sequence remains identical if we change the trivialization of ν_i by the square of a unit defined on Y and also from the fact that M_{-1} is a strongly $\mathbb{A}^1$-invariant sheaf by Lemma 2.32 which gives the top horizontal exact sequence and Lemma 4.2 which gives the bottom exact sequence. $\square$

Lemma 4.5. *Let X be the localization of a point z of codimension 2 in a smooth k-scheme. Let $X' \to X$ be a Nisnevich neighborhood of z, that is to say $f : X' \to X$ is étale, has only one point lying over z (in particular X' is also the localization of a smooth k-scheme at a point of codimension 2), and this point has the same residue field as z. We denote still by $z \in X'$ this point. Then the morphism*

$$
H^2_z(X; M) \to H^2_z(X'; M)
$$

is an isomorphism.

Let $y \in X$ be a point of codimension 1. Assume that $Y := \bar{y}$ is essentially k-smooth. We denote by y' the only point of codimension 1 in X' lying over y. Then in the commutative diagram

$$
\begin{array}{ccc}
H^1_y(X; M) & \overset{\partial^y_z}{\to} & H^2_z(X; M) \\
\downarrow & & \downarrow \wr \\
H^1_{y'}(X'; M) & \overset{\partial^{y'}_z}{\to} & H^2_z(X'; M)
\end{array}
$$

the two horizontal morphisms have the same image (where we identify $H^2_z(X; M)$ and $H^2_z(X'; M)$).

Proof. Choose a non zero element of $\mathcal{M}_y/(\mathcal{M}^2_y)$. This determines also a non zero element in $\mathcal{M}_{y'}/(\mathcal{M}^2_{y'})$ because $X' \to X$ is étale. Set $Y' := \bar{y'}$. As Y is essentially smooth, so is $Y' \cong Y \times_X X'$ and the morphism $Y' \to Y$ a Nisnevich neighborhood of z. Our claim now follows at once from the previous

Lemma and the fact that under our assumptions the induced morphism

$$H_z^1(Y; M_{-1}) \to H_z^1(Y'; M_{-1})$$

is an isomorphism, which in turns, follows from the fact that M_{-1} is a strongly $\mathbb{A}^1$-invariant sheaf, see Sect. 2.2. $\qquad\qquad\square$

Corollary 4.6. *Let X be the localization of a point z of codimension 2 in a smooth k-scheme, and let y be a point of codimension 1 in X, essentially k-smooth. Then the image of the morphism*

$$H_y^1(X; M) \overset{\partial_z^y}{\to} H_z^2(X; M)$$

doesn't depend on y. We denote it by I_z.

Proof. Let X^h be the henselization of X (at z) and y^h be the unique point of codimension 1 of X^h lying over y. Then by passing to the filtering colimit we see from the previous Lemma that the image of the two horizontal morphisms in the diagram

$$
\begin{array}{ccc}
H_y^1(X; M) & \overset{\partial_z^y}{\to} & H_z^2(X; M) \\
\downarrow & & \downarrow \wr \\
H_{y^h}^1(X^h; M) & \overset{\partial_z^{y^h}}{\to} & H_z^2(X^h; M)
\end{array}
$$

are the same. As $Y^h := \overline{y^h}$ is henselian and in fact is the henselization of $Y := \overline{y}$ at z, we see that it suffices to prove that given Y_0 and Y_1 irreducible closed subschemes of codimension 1 in X, and essentially k-smooth, there exists a k-automorphism of X^h preserving z, which takes Y_0^h to Y_1^h. As k is perfect, there is always a Nisnevich neighborhood $\Omega \to X$ of z which is also a Nisnevich neighborhood $\Omega \to \mathbb{A}^1 \times Y_0$ of z in $\mathbb{A}^1 \times Y_0$. Thus the pair (X^h, z) is isomorphic to the pair $((\mathbb{A}^1 \times Y_0)^h, z)$ which is also isomorphic to the pair $((\mathbb{A}^1 \times (Y_0^h)^h, z)$. As the henselizations Y_0^h and Y_1^h of Y_0 and Y_1 at z are z-preserving isomorphic, the Corollary is proven. $\qquad\square$

Let now S be the localization of a smooth k-scheme at a point s of codimension 1; let κ be the residue field of s. Let $\eta \in (\mathbb{A}_S^1)^{(1)}$ be the generic point of $\mathbb{A}_\kappa^1 \subset \mathbb{A}_S^1$. We will study the morphism

$$\partial^\eta : H_\eta^1(\mathbb{A}_S^1; M) \to \oplus_{z \in (\mathbb{A}_s^1)^{(2)}} H_z^2(\mathbb{A}_S^1; M)$$

sum of each ∂_z^η where z runs over the set of points of codimension 2 in $\mathbb{A}_S^1$, that is to say the set of closed point in $\mathbb{A}_\kappa^1$.

Lemma 4.7. *Let S the localization of a point s of codimension 2 in a smooth k-scheme, whose residue field κ. Let $\eta \in \mathbb{A}_S^1$ be the generic point of the irreducible curve $\mathbb{A}_\kappa^1 \subset \mathbb{A}_S^1$. Then the image of the morphism*

$$\partial^\eta : H^1_\eta(\mathbb{A}^1_S; M) \to \oplus_{z \in (\mathbb{A}^1_\kappa)^{(1)}} H^2_z(\mathbb{A}^1_S; M)$$

is the direct sum of the images of each of the $\partial^\eta_z : H^1_\eta(\mathbb{A}^1_S; M) \to H^2_z$
$(\mathbb{A}^1_S; M)$ *'s.*

Proof. Choose a non zero element of the $\kappa(T)$-vector space $\mathcal{M}_\eta/(\mathcal{M}_\eta)^2$.
It follows from Lemma 4.4 that for each $z \in (\mathbb{A}^1_\kappa)^{(1)}$ the image of ∂^η_z
is isomorphic to $H^1_z(\mathbb{A}^1_\kappa; M_{-1})$ through the epimorphism $M_{-1}(\kappa(T)) \to$
$H^1_z(\mathbb{A}^1_\kappa; M_{-1})$. Now our claim follows from the fact that the morphism

$$H^1_\eta(\mathbb{A}^1_S; M) \cong M_{-1}(\kappa(T)) \to \oplus_{z \in (\mathbb{A}^1_\kappa)^{(1)}} H^1_z(\mathbb{A}^1_\kappa; M_{-1})$$

is onto, which on the other hand follows from the fact that M_{-1} is a
strongly $\mathbb{A}^1$-invariant sheaf by Lemma 2.32: the cokernel of this morphism is
$H^1(\mathbb{A}^1_\kappa; M_{-1})$ and vanishes for each strongly $\mathbb{A}^1$-invariant sheaf, see Sect. 2.2.
$\square$

Lemma 4.8. *Let X be the localization of a point z of codimension 2 in a
smooth k-scheme, and let y be a point of codimension 1 in X, essentially
k-smooth. Then the image of*

$$H^1_y(X; M) \xrightarrow{\partial^y_z} H^2_z(X; M)$$

is contained in I_z.

Proof. Let $y_0, y_1 \in X^{(1)}$ be points of codimension 1 and with y_0 is smooth
at z. To prove that the image of

$$\partial^{y_1}_z : H^1_{y_1}(X; M) \to H^2_z(X; M)$$

is contained in the image I_z of

$$\partial^{y_0}_z : H^1_{y_0}(X; M) \to H^2_z(X; M)$$

we may replace X by its henselization X^h_z at z, as this follows from Lemma
4.5. By Gabber's presentation Lemma 1.15 below one may choose a pro-étale
morphism $X^h_z \to \mathbb{A}^1_S$, such that both compositions $\overline{y_0} \to \mathbb{A}^1_S$ and $\overline{y_1} \to \mathbb{A}^1_S$
are closed immersions, where S is the henselization of a smooth k-scheme at
some point (of codimension 1). Observe that then for any point $x \in \overline{y_0} \cup \overline{y_1}$
of codimension i in X (or $\mathbb{A}^1_S$) the morphism

$$H^i_x(\mathbb{A}^1_S; M) \to H^i_x(X; M)$$

As $\overline{1_0}$ and $\mathbb{A}^1_{\kappa(s)}$ are both essentially k-smooth, we conclude by Corollary 4.6
that the images of

$$\partial_z^{y_0} : H_{y_0}^1(X; M) \cong H_{y_0}^1(\mathbb{A}_S^1; M) \to H_z^2(X; M)$$

and of

$$\partial_z^\eta : H_\eta^1(X; M) \cong H_\eta^1(\mathbb{A}_S^1; M) \to H_z^2(X; M)$$

are equal, that is to say both I_z. But from Lemma 4.7 above, the image of $\partial_z^{y_1} : H_{y_1}^1(X; M) \cong H_{y_1}^1(\mathbb{A}_S^1; M) \to H_z^2(X; M)$ is contained in that of ∂_z^η which is also I_z.   □

Lemma 4.9. *Let X be an essentially smooth k-scheme of dimension ≤ 2. Then the Gersten complex $C^*(X; M)$ computes the Nisnevich cohomology of X with coefficients in M.*

Proof. One uses the coniveau spectral sequence for X with coefficients in M. One observes that the spectral sequence has to collapse from the E_2-term, as M is strongly $\mathbb{A}^1$-invariant, thus unramified. This indeed implies that no differential can starts from $E_2^{0,0} = M(X)$. The result follows as the spectral sequence converges strongly to $H_{Nis}^*(X; M)$.   □

Lemma 4.10. *Let X be the localization of a point z of codimension 2 in a smooth k-scheme, and let y be a point of codimension 1 in X, essentially k-smooth. Then*

$$H_y^1(X; M) \overset{\partial_z^y}{\to} H_z^2(X; M)$$

is surjective. In particular $H_{Nis}^2(X; M) = 0$ and it follows that for any essentially smooth k-scheme X of dimension ≤ 2, the comparison map

$$H_{Zar}^*(X; M) \to H_{Nis}^*(X; M)$$

is an isomorphism.

Proof. We first treat the case of the henselization X^h of X at z. From the previous Lemma it follows that the last differential

$$\oplus_{y \in (X^h)^{(1)}} H_y^1(X^h; M) \overset{\partial_z^y}{\to} H_z^2(X^h; M)$$

is surjective. As the image of this morphism is contained in I_z by Lemma 4.8 (observe that there is always at least a $y \in (X^h)^{(1)}$ which is essentially k-smooth), this shows that $I_z = H_z^2(X^h; M)$. Now by Lemma 4.5, we see that $I_z = H_z^2(X; M)$ as well.

The rest of the statement is straightforward to deduce as any such X contains a smooth such y (observe that the last statement is already know for $* \leq 1$).   □

Corollary 4.11. *Let X be the localization of a point z of codimension 2 in a smooth k-scheme, and let $y \in X$ be a point of codimension 1. Assume that $Y := \overline{y}$ is essentially smooth over k and let $i : Y \subset X$ be the induced closed* *immersion. Then for any orientation ω of ν_i in the commutative diagram of Lemma 4.4*

$$
\begin{array}{ccccc}
0 \to & M_{-1}(Y) & \subset M_{-1}(\kappa(y)) \to & H^1_z(Y; M_{-1}) \to 0 \\
& \theta_\omega \downarrow \wr & \theta_\omega \downarrow \wr & \phi_{\omega,z} \downarrow \wr \\
0 \to & H^1(X/X - Y; M) \subset & H^1_y(X; M) \twoheadrightarrow & H^2_z(X; M) \to 0
\end{array}
$$

the right vertical morphism, which we denote by $\phi_{\omega,z}$, is also an isomorphism.

The statement still holds if we only assume X to be an local essentially smooth k-scheme of dimension 2 with closed point z.

Proof. The first claim follows from Lemma 4.4 and the previous Lemma. To deduce it for a general local essentially smooth k-scheme X of dimension 2 follows from general results on inverse limit of schemes [30], as such an X is the inverse limit of a system of affine étale morphisms between localizations of smooth k-schemes at points of codimension 2. □

Let X be an essentially smooth k-scheme of dimension 2, and let $i : Y \subset X$ be an irreducible closed scheme of dimension 1 which is essentially k-smooth and let y be its generic point.

We saw above that an orientation ω of ν_i induces an isomorphism

$$
\theta_\omega : M_{-1}(\kappa(y)) \cong H^1_y(X; M)
$$

Now for any closed point z in Y by the previous corollary there is an induced isomorphism of the form $\phi_{\omega,z} : H^1_z(Y; M_{-1}) \cong H^2_z(X; M)$. The commutative diagram of the same Corollary implies that these morphisms induce altogether a canonical isomorphism of complexes, only depending on ω:

$$
\phi_\omega : C^*(Y; M_{-1})[1] \cong C^*_Y(X; M) \tag{4.3}
$$

between the complex $C^*(Y; M_{-1})$ shifted by $+1$ (in cohomological degrees) and the complex $C^*_Y(X; M) \subset C^*(X; M)$.

If z is a closed point of X we may apply what precedes to the essentially k-smooth scheme $X - z$ and the closed immersion $Y - z \subset X - z$ we obviously get a compatible isomorphism of complexes still denoted by ϕ_ω:

$$
\phi_\omega : C^*(Y - z; M_{-1})[1] \cong C^*_{Y-z}(X - z; M)
$$

Altogether we get a commutative diagram

$$
\begin{array}{ccccc}
0 \to H^1_z(Y; M_{-1})[1] & \subset C^*(Y; M_{-1})[1] & \twoheadrightarrow C^*(Y - z; M_{-1})[1] \to 0 \\
\downarrow \wr & \downarrow \wr & \downarrow \wr \\
0 \to H^2_z(X; M)[2] & \subset C^*_Y(X; M) & \twoheadrightarrow C^*_{Y-z}(X - z; M) \to 0
\end{array} \tag{4.4}
$$

in which the vertical are the isomorphisms of the form ϕ_ω and the horizontal are the canonical short exact sequences of complexes. This diagram induces a canonical commutative square by evaluation of cohomology on the left square:

$$
\begin{array}{ccc}
H^1_z(Y; M_{-1}) & \to & H^1(Y; M_{-1}) \\
\phi_{\omega, z} \downarrow \wr & & \phi_\omega \downarrow \wr \\
H^2_z(X; M) & \to & H^2_Y(X; M)
\end{array}
\tag{4.5}
$$

Corollary 4.12. *Let X be a local essentially smooth k-scheme of dimension 2 with closed point z, and let $y_0 \in X^{(1)}$ be a point of codimension 1. Assume that $Y_0 := \overline{y} \subset X$ is essentially smooth over k. Then the canonical morphism*

$$\oplus_{y \in X^{(1)} - \{y_0\}} \partial_y : M(X) \to \oplus_{y \in X^{(1)} - \{y_0\}} H^1_y(X; M)$$

is onto.

Proof. The morphism in question is just the differential of the complex $C^*(X - Y_0; M)$. Thus we have to prove $H^1_{Nis}(X - Y_0; M) = 0$. We consider the subcomplex $C^*_{Y_0}(X; M) \subset C^*(X; M)$, which is the kernel of the epimorphism $C^*(X; M) \to C^*(X - Y_0; M)$. We may choose a trivialization of the normal bundle of $Y_0 \subset X$ and thus using the isomorphism of complexes (4.3) we see that $H^2(X/X - Y_0; M) \cong H^1(Y_0; M_{-1}) = 0$ (as Y_0 is also local). The long exact sequence gives then the result. □

The following result is one of the main results of this section:

Theorem 4.13. *Let M be a strongly $\mathbb{A}^1$-invariant sheaf. Then for any field $K \in \mathcal{F}_k$, one has $H^2(\mathbb{A}^2_K; M) = 0$.*

Proof. From Lemma 4.9 we may use $C^*(\mathbb{A}^2_K; M)$ to compute $H^2(\mathbb{A}^2_K; M)$. We then analyze $C^*(\mathbb{A}^2_K; M)$ as follows. Let $\mathbb{A}^1_{K(X)} \to \mathbb{A}^2_K$ the morphism induced by the morphism of k-algebras $K[X, Y] \to K(X)[Y]$. This is the inverse limit of open immersions and thus there is an induced morphism of Gersten complexes

$$C^*(\mathbb{A}^2_K; M) \to C^*(\mathbb{A}^1_{K(X)}; M)$$

This morphism is clearly surjective. Let $\mathcal{K}$ denote its kernel. As $H^2(\mathbb{A}^1_{K(X)}; M) = 0$, to prove that $H^2(\mathbb{A}^2_K; M) = 0$ it suffices to prove that $H^2(\mathcal{K}) = 0$.

The complex $\mathcal{K}$ in dimension 1 is $\oplus_y H^1_y(\mathbb{A}^2_K; M)$ where y runs over the set of points of codimension 1 in $\mathbb{A}^2_K$ which do not dominate $\mathbb{A}^1_K$ through the first projection $pr_1 : \mathbb{A}^2_K \to \mathbb{A}^1_K$ which forgets Y. These points y are exactly in one to one correspondence with the closed points t in $\mathbb{A}^1_K$ by taking their image, as automatically $y = \mathbb{A}^1_{\kappa(t)} \subset \mathbb{A}^2_K$, with $t = pr_1(y)$, for such y's. As any points z of codimension 2 in $\mathbb{A}^2_K$ lie exactly in one and only one of these, indeed in $\mathbb{A}^1_{pr_1(z)}$, we see that $\mathcal{K}$ is the direct sum over the $t \in (\mathbb{A}^1_F)^{(1)}$ of the complexes of the form:

$$0 \to H^1_{\eta_t}(\mathbb{A}^2_K; M) \to \oplus_{z \in (\mathbb{A}^1_{\kappa(t)})^{(1)}} H^2_z(\mathbb{A}^2_K; M)$$

where η_t is the generic point of $\mathbb{A}^1_{\kappa(t)} \subset \mathbb{A}^2_K$ and with $H^1_{\eta_t}(\mathbb{A}^2_K; M)$ placed in degree 1. We have to prove that each of these complexes has trivial H^2 that is to say that the morphism $H^1_{\eta_t}(\mathbb{A}^2_K; M) \to \oplus_{z \in (\mathbb{A}^1_{\kappa(t)})^{(1)}} H^2_z(\mathbb{A}^2_K; M)$ is onto for each t. The minimal polynomial of t defines a non zero element in $\mathcal{M}_t/(\mathcal{M}_t)^2$ and induces a non zero element in $\mathcal{M}_{\eta_t}/(\mathcal{M}_{\eta_t})^2$ as well. By Corollary 4.11 and its consequence the isomorphism (4.3), we see that for fixed t the above morphism may be identified with

$$M_{-1}(\kappa(t)(Y)) \to \oplus_{z \in (\mathbb{A}^1_{\kappa(t)})^{(1)}} H^1_z(\mathbb{A}^1_{\kappa(t)}; M_{-1})$$

which is onto as M_{-1} is strongly $\mathbb{A}^1$-invariant. The Theorem is proven. $\square$

Corollary 4.14. *Let K be in $\mathcal{F}_k$. Then*

$$H^2(\mathbb{P}^1 \times \mathbb{A}^1_K; M) = 0$$

Proof. We use the covering of $\mathbb{P}^1 \times \mathbb{A}^1_K$ by the two open subsets isomorphic to $\mathbb{A}^2_K$ with intersection $\mathbb{G}_m \times \mathbb{A}^1_K$. By the Theorem 4.13 above we have $H^1(\mathbb{A}^2_K; M) = 0$ and the Mayer–Vietoris sequence produces an isomorphism

$$H^1(\mathbb{G}_m \times \mathbb{A}^1; M) \cong H^2(\mathbb{P}^1 \times \mathbb{A}^1_K; M)$$

and we conclude as $H^1(\mathbb{G}_m \times \mathbb{A}^1_K; M) \cong H^1((\mathbb{G}_m)_K; M) = 0$ because M is strongly $\mathbb{A}^1$-invariant. $\square$

Corollary 4.15. *Let K be in $\mathcal{F}_k$. Then for any K-rational point z of $(\mathbb{P}^1)^2_K$ the morphism*

$$H^2_z((\mathbb{P}^1)^2_K; M) \to H^2((\mathbb{P}^1)^2_K; M)$$

is an isomorphism.

Proof. It suffices to treat the case $z = (0,0)$. To do this, we analyze the restriction epimorphism

$$C^*((\mathbb{P}^1)^2_K; M) \twoheadrightarrow C^*(\mathbb{P}^1 \times \mathbb{A}^1_K; M)$$

of Gersten complexes where $\mathbb{P}^1 \times \mathbb{A}^1_K$ is the complement of $\mathbb{P}^1 \times \{\emptyset\} \subset (\mathbb{P}^1)^2_K$. From Corollary 4.11 its kernel $C^*_{\mathbb{P}^1 \times \{\emptyset\}}((\mathbb{P}^1)^2_K; M)$ is canonically isomorphic to the complex $C^*(\mathbb{P}^1_K; M_{-1})$ shifted by one; this provides an isomorphism

$$H^1(\mathbb{P}^1_K; M_{-1}) \cong H^2_{\mathbb{P}^1 \times \{\infty\}}((\mathbb{P}^1)^2_K; M)$$

where $H^*_{\mathbb{P}^1 \times \{\infty\}}((\mathbb{P}^1)^2_K; M)$ denotes the cohomology of $C^*_{\mathbb{P}^1 \times \{\infty\}}((\mathbb{P}^1)^2_K; M)$.

On the other hand, the long exact sequence associated to the short exact sequence of complexes

$$0 \to C^*_{\mathbb{P}^1 \times \{\infty\}}((\mathbb{P}^1)^2_K; M) \subset C^*((\mathbb{P}^1)^2_K; M) \twoheadrightarrow C^*(\mathbb{P}^1 \times \mathbb{A}^1_K; M) \to 0$$

gives that the morphism

$$H^2_{\mathbb{P}^1 \times \{\infty\}}((\mathbb{P}^1)^2_K; M) \to H^2((\mathbb{P}^1)^2_K; M)$$

is an isomorphism, taking into account the vanishing $H^2(\mathbb{P}^1 \times \mathbb{A}^1_K; M) = 0$ of Corollary 4.14 and the fact that $H^1((\mathbb{P}^1)^2_K; M) \to H^1(\mathbb{P}^1 \times \mathbb{A}^1_K; M) \cong H^1(\mathbb{P}^1_K; M)$ is onto as $\mathbb{P}^1 \times \{0\} \to (\mathbb{P}^1)^2_K$ admits a retraction.

Now the morphism $H^2_\infty((\mathbb{P}^1)^2_K; M) \to H^2((\mathbb{P}^1)^2_K; M)$ factorizes as

$$H^2_\infty((\mathbb{P}^1)^2_K; M) \to H^2_{\mathbb{P}^1 \times \{\infty\}}((\mathbb{P}^1)^2_K; M) \cong H^2((\mathbb{P}^1)^2_K; M)$$

where the factorization is induced by the factorization of morphisms of complexes

$$C^*_\infty((\mathbb{P}^1)^2_K; M) \to C^*_{\mathbb{P}^1 \times \{\infty\}}((\mathbb{P}^1)^2_K; M) \to C^*((\mathbb{P}^1)^2_K; M)$$

Now $C^*_\infty((\mathbb{P}^1)^2_K; M) = H^2_\infty((\mathbb{P}^1)^2_K; M)$ and by the diagram (4.4) we see that it suffices to prove that

$$H^1_\infty(\mathbb{P}^1_K; M_{-1}) \to H^1(\mathbb{P}^1_K; M_{-1})$$

is an isomorphism which follows from the next Lemma. $\square$

Lemma 4.16. *For any rational point z in $\mathbb{P}^1_K$ and any strongly $\mathbb{A}^1$-invariant sheaf of abelian groups M the morphism*

$$H^1_z(\mathbb{P}^1_K; M) \to H^1(\mathbb{P}^1_K; M)$$

is an isomorphism.

Proof. This follows from the fact that the epimorphism of Gersten complexes (of length 1 here) $C^*(\mathbb{P}^1_K; M) \twoheadrightarrow C^*(\mathbb{A}^1_K; M)$ has kernel $H^1_z(\mathbb{P}^1_K; M)$ placed in degree 1 and the fact that $H^1(\mathbb{A}^1_K; M) = 0$ and $H^0(\mathbb{P}^1_K; M) = H^0(\mathbb{A}^1_K; M)$. $\square$

Remark 4.17. The analogue statement as in the Theorem with $\mathbb{P}^2_K$ instead of $(\mathbb{P}^1)^2_K$ is wrong in general. One finds for z a rational K-point of $\mathbb{P}^2_K$ an exact sequence of the form:

$$H^1_z(\mathbb{P}^1_K; M_{-1}) \to H^2_z(\mathbb{P}^2_K; M) \twoheadrightarrow H^2(\mathbb{P}^2_K; M) \to 0$$

where the left morphism is some "Hopf map", non trivial in general. The role of the choice of the compactification $(\mathbb{P}^1)^2_K$ of $\mathbb{A}^2_K$ is thus important in the sequel. □

Remark 4.18. Of course most of the previous computations in the 2-dimensional case would be much easier to perform if we already knew that M is strictly $\mathbb{A}^1$-invariant, or even only that $K(M,2)$ is $\mathbb{A}^1$-local. □

Orientations and the Gersten complex in codimension 2

Let X be an essentially smooth k-scheme of dimension 2 and let $i : Y \subset X$ be an irreducible closed scheme which is essentially k-smooth. Let τ be a trivialization of the normal bundle ν_i; we have constructed above an isomorphism

$$\theta_\tau : M_{-1}(\kappa(y)) \cong H^1_y(X; M)$$

which induces for each closed point z in Y an isomorphism

$$\phi_\tau : H^1_z(Y; M_{-1}) \cong H^2_z(X; M)$$

(see the diagram (4.5)). Given a non-zero element $\mu \subset \mathcal{N}_z/(\mathcal{N}_z)^2$, where $\mathcal{N}_z$ is the maximal ideal corresponding to z in the dvr $\mathcal{O}_{Y,z}$, we know from Corollary 2.35 that we get a further isomorphism

$$\theta_\mu : M_{-2}(\kappa(z)) \cong H^1_z(Y; M_{-1})$$

Thus in the above situation, a pair (τ, μ) defines a canonical isomorphism

$$\Phi_{\tau,\mu} := \phi_\tau \circ \theta_\mu : M_{-2}(\kappa(z)) \cong H^2_z(X; M) \tag{4.6}$$

We want to study the dependence of this isomorphism on the pair (τ, μ). Observe that the definition of this isomorphism is local and we may assume $X = Spec(A)$ is a local scheme with closed point z and residue field κ. $\mathcal{M}$ or sometimes $\mathcal{M}_z$ will denote the maximal ideal of A.

Let (π, ρ) be a regular system of parameters in A, that is to say (π, ρ) generates the maximal ideal $\mathcal{M}$ of A or equivalently their classes form a basis of the 2-dimensional κ-vector space $\mathcal{M}/\mathcal{M}^2$. Set $Y_0 := Spec(A/\pi)$ and $Y_1 := Spec(A/\rho)$; these are irreducible regular (thus essentially k-smooth) closed subschemes of X. π defines a trivialization of the normal bundle of $Y_0 \subset X$ and ρ a trivialization of the normal bundle of $Y_1 \subset X$. The class of π in A/ρ is a uniformizing element of the d.v.r. A/ρ and the class of ρ in A/π is also a uniformizing element. We thus get two isomorphisms still denoted by

$$\Phi_{\pi,\rho} : M_{-2}(\kappa) \cong H^2_z(X; M)$$

and

$$\Phi_{\rho,\pi} : M_{-2}(\kappa) \cong H_z^2(X; M)$$

Recall from the end of Sect. 3.3 that there is a canonical $\underline{\mathbf{K}}_0^{MW}$-module structure on the sheaves of the form M_{-n}, for $n \geq 1$ and M a strongly $\mathbb{A}^1$-invariant sheaf of abelian groups. We denote by $< -1 > \in K_0^{MW}(k) = GW(k)$ the class of the form $(A, B) \mapsto -AB$. It thus acts on M_{-2}. In the proof of the Theorem we will also make use of the pairing

$$\underline{\mathbf{K}}_n^{MW} \times M_{-n} \to M$$

considered in Lemma 3.48, for $n \in \{1,2\}$. If $[x_1, \ldots, x_n]$ is a symbol in $\underline{\mathbf{K}}_n^{MW}(F)$ and $m \in M_{-n}(F)$ we let $[x_1, \ldots, x_n] \cup m \in M(F)$ be the product of the above pairing.

Theorem 4.19. *Keeping the previous assumptions and notations, we have the equality*

$$\Phi_{\rho,\pi_1} = \Phi_{\pi,\rho_0} \circ < -1 > : M_{-2}(\kappa) \cong H_z^2(X; M)$$

Proof. If $X' \to X$ is a Nisnevich neighborhood of z, that is to say an étale morphism with exactly one point lying over z with the same residue field, to prove the statement for X and (π, ρ) is equivalent to prove it for X' and (π', ρ') where π' is the image of π and ρ' that of ρ. In this way we may reduce to assume that X is henselian, and thus admits a structure of $\kappa(z)$-smooth scheme, and then by standard techniques to reduce further to the case $X = \mathbb{A}_\kappa^2$, with κ, and $(\pi, \rho) = (X, Y)$, the two coordinate functions on $\mathbb{A}_\kappa^2$.

From the Lemma 4.21 below we have (with the notations of the Lemmas 4.21 and 4.22) for $m \in M_{-2}(\kappa)$

$$\Phi_{X,Y}(m) = \partial_0 \circ \partial_\beta \circ ([X,Y] \cup m)$$

and from the same Lemma applied by permuting X and Y,

$$\Phi_{Y,X}(m) = \partial_0 \circ \partial_\alpha \circ ([Y,X] \cup m)$$

The claim now follows from this and the Lemma 4.22, together with the fact that $[X,Y] = - < -1 > [Y,X] \in K_2^{MW}(\kappa(X,Y))$ by Lemma 3.8. $\quad\square$

Lemma 4.20. *Let K be in $\mathcal{F}_k$, let C be a smooth K-curve with function field $K(C)$, let z be a K-rational point locally defined on C by the function $\pi \in K(C)$. Then the following diagram is commutative:*

$$
\begin{array}{ccc}
M_{-1}(K) & \subset & M_{-1}(K(C)) \\
\theta_\pi \downarrow \wr & & [\pi]\cup \downarrow \\
H_z^1(C; M) & \xleftarrow{\partial_z} & M(K(C))
\end{array}
$$

Proof. This follows from the definition of the isomorphism θ_π (see Corollary 2.35), the definition of M_{-1} and an inspection in the commutative square:

$$
\begin{array}{ccc}
Spec(k(C)) & \subset & Spec(\mathcal{O}_z) \\
\pi \downarrow & & \pi \downarrow \\
(\mathbb{G}_m)_K & \subset & \mathbb{A}^1_K
\end{array}
$$

$\square$

Lemma 4.21. *Let K be in $\mathcal{F}_k$ and let $\mathbb{A}^2_K = Spec(K[X,Y]$, let η be the generic point of $\mathbb{A}^2_K$, let $C_X \subset \mathbb{A}^2_K$ be the closed subscheme defined by $Y = 0$ and α be its generic point, and let $0 \in \mathbb{A}^2_K$ be the closed point defined by $X = Y = 0$. Then the following diagram is commutative:*

$$
\begin{array}{ccc}
M_{-2}(K) & \subset & M_{-2}(K(X,Y)) \\
\Phi_{Y,X} \downarrow \wr & & [Y,X] \cup \downarrow \\
H^2_0(\mathbb{A}^2_K; M) & \xleftarrow{\partial^\alpha_0 \circ \partial^\eta_\alpha} & M(K(X,Y))
\end{array}
$$

Proof. This follows from the previous lemma, applied twice, and the definition of the isomorphism $\Phi_{Y,X} = \Phi_Y \circ \theta_X$. $\square$

Lemma 4.22. *We keep the same notations as above and we further denote by β the generic point of the curve in $\mathbb{A}^2_K$ defined by $X = 0$. Then for any $m \in M_{-2}(K)$ one has*

$$
\partial_0 \circ \partial_\beta \circ ([X,Y] \cup m) + \partial_0 \circ \partial_\alpha \circ ([X,Y] \cup m) = 0
$$

Proof. Indeed in the Gersten complex for $\mathbb{A}^2_K$, one check that the only points $y \in (\mathbb{A}^2_K)^{(1)}$ where $[X,Y] \cup m$ is not defined are α and β, as in any other case, X and Y are in the units of the local ring $\mathcal{O}_{\mathbb{A}^2_K,y}$. Now the formula of the Lemma is just the fact that in a complex $\partial \circ \partial = 0$ applied to $[X,Y] \cup m$. $\square$

We keep the previous notations and assumptions. Let $\lambda \in A$ be a regular function on X. The pair $(\pi, \rho + \lambda\pi)$ of elements of $\mathcal{M}$ is still a regular system of parameters and clearly one has the equality $\Phi_{\pi,\rho+\lambda\pi} = \Phi_{\pi,\rho}$ between isomorphisms $M_{-2}(\kappa) \cong H^2_z(X; M)$ as the reduction of $\rho + \lambda\pi$ and of ρ modulo π are equal. From the Theorem 4.19 we also have the equality

$$
\Phi_{\rho+\lambda\pi,\pi} = \Phi_{\pi,\rho+\lambda\pi} < -1 >= \Phi_{\pi,\rho} < -1 >= \Phi_{\rho,\pi}
$$

We let $\bar{\pi} \wedge \bar{\rho} \in \Lambda_2(\mathcal{M}/\mathcal{M}^2)$ be the exterior product of the reductions of π and ρ respectively in $\mathcal{M}/\mathcal{M}^2$. Observe finally that, almost by construction, if $\lambda \notin \mathcal{M}$ is a unit of A, then

$$
\Phi_{\lambda\pi,\rho} = \Phi_{\pi,\rho} < \lambda >
$$

We let $\Lambda_2(\mathcal{M}/\mathcal{M}^2)^{\times}$ be the set of non-zero elements; observe that $\kappa^{\times}$ acts freely and transitively on this set.

The next Corollary follows from what we have just done and from the classical fact that the matrices of the form $\begin{pmatrix} 1 & \lambda \\ 0 & 1 \end{pmatrix}$, $\begin{pmatrix} 1 & 0 \\ \lambda & 1 \end{pmatrix}$ and $\begin{pmatrix} \lambda & 0 \\ 0 & \mu \end{pmatrix}$ (λ and μ both units) generate $GL_2(\kappa)$:

Corollary 4.23. *For (π, ρ) as above, and for $M = \begin{pmatrix} \lambda_{11} & \lambda_{12} \\ \lambda_{21} & \lambda_{22} \end{pmatrix}$ an invertible matrix in $GL_2(A)$, whose determinant is $d \in A^{\times}$, one has the equality*

$$\Phi(\pi, \rho) = \Phi(\lambda_{11}\pi + \lambda_{21}\rho, \lambda_{12}\pi + \lambda_{22}\rho)\circ <d>$$

It follows that the isomorphism

$$\Phi_{\rho,\pi} : M_{-2}(\kappa) \cong H^2_z(X; M)$$

only depends on the class $\overline{\pi} \wedge \overline{\rho} \in \Lambda_2(\mathcal{M}/\mathcal{M}^2)$ up to multiplication by a square.

Given a vector space V of dimension n over a field κ, an equivalence class of non zero element in $\Lambda_n(V)$ modulo the multiplication by a square of a unit of κ can be checked to be exactly an *orientation* of V in the sense of Definition 4.3 above. In other words $Q(V) = \Lambda^n(V) - \{0\}/(\kappa^{\times})^2$.

In other words, the previous Corollary states that the isomorphism $\Phi_{\pi,\rho}$ only depends on the orientation $\overline{\pi \wedge \rho} \in Q(\mathcal{M}/\mathcal{M}^2)$.

Remark 4.24. Observe that if the structure of **GW**-module on M_{-2} is trivial, the previous result proves that $\Phi(\pi, \rho)$ doesn't depend on (π, ρ), and thus doesn't depend on any choice.

4.2 Geometric Versus Cohomological Transfers on M_{-n}

Geometric transfers on M_{-1}

We now show that M_{-1} inherits canonical geometric transfers morphisms for finite monogenous fields extensions in $\mathcal{F}_k$ and then show that for sheaves of the form M_{-n}, $n \geq 2$, these transfers for monogenous extensions can be conveniently extended to canonical transfer morphisms for any finite extension in $\mathcal{F}_k$. Our construction is inspired by [9], conveniently adapted, and was also the basic inspiration to the axioms of strictly $\mathbb{A}^1$-invariant sheaf with generalized transfers of [56–58].

The construction of transfers for monogenous extensions given in [9] is what we call the "geometric transfers". In general given a finite extension $K \subset L$ with a set of generators, the compositions of these geometric transfers using the successive generators (as in Bass–Tate) will depend on these choices. However, in characteristic $\neq 2$, these transfers can be turned into canonical ones, for sheaves of the form M_{-2}, by twisting the geometric ones conveniently. These are what we call the "cohomological transfers". In the Sect. 5.1 we also introduce a variant, called the "absolute transfers", which exists in any characteristic.

Recall that the construction M_{-1} on M is also a strongly $\mathbb{A}^1$-invariant sheaf of abelian groups by Lemma 2.32, and so is it for all the iterations M_{-n} of that construction on M.

Let K be in $\mathcal{F}_k$. As in the classical case of Milnor K-theory, we start with the short exact sequence, which holds for any strongly $\mathbb{A}^1$-invariant sheaf M:

$$0 \to M(K) \subset M(K(X)) \to \oplus_{z \in (\mathbb{A}^1_K)^{(1)}} H^1_z(\mathbb{A}^1_K; M) \to 0$$

For each such closed point z in $\mathbb{A}^1_K$, the minimal polynomial P_z defines a uniformizing element and thus by Corollary 2.35 a canonical isomorphism

$$\theta_{\overline{P_z}} : M_{-1}(\kappa(z)) \cong H^1_z(\mathbb{A}^1_K; M)$$

depending only on the class of P_z in $\mathcal{M}_z / (\mathcal{M}_z)^2$, where $\mathcal{M}_z$ is the maximal ideal in the local ring (a d.v.r.) of $\mathbb{A}^1_K$ at z. We may thus rewrite the previous exact sequence as

$$0 \to M(K) \subset M(K(X)) \to \oplus_{z \in (\mathbb{A}^1_K)^{(1)}} M_{-1}(\kappa(z)) \to 0$$

using at each z the residue homomorphism $\partial_z^{\overline{P_z}} : M(K(X)) \to M_{-1}(\kappa(z))$ discussed after Corollary 2.35.

This also holds for the valuation at infinity of $K(X)$ and we thus get the morphism

$$\partial_\infty^\eta : M(K(X)) \to M_{-1}(K)$$

by using the uniformizing element $P_\infty = \frac{-1}{X}$ at ∞; see below Lemma 4.25 (and its proof) to explain this choice. As $\partial_\infty^{P_\infty}$ is zero on $M(K) \subset M(K(X))$ we see that the previous exact sequence and the morphism

$$- \partial_\infty^\eta \tag{4.7}$$

induces a morphism

$$\oplus_{z \in (\mathbb{A}^1_K)^{(1)}} M_{-1}(\kappa(z)) \to M_{-1}(K)$$

In other words, given a finite extension $K \subset L$ and a generator $z \in L$ of L over K, we have defined a canonical morphism:

$$\tau_K^L(z) : M_{-1}(L) \to M_{-1}(K)$$

which we call the *geometric transfer*.

This is exactly the approach of Bass and Tate in [9]. The sign $-$ appearing in the morphism (4.7) is there to guaranty the formula

$$\sum_{z \in \mathbb{P}_K^1} \tau_K^{\kappa(z)}(z) \circ \partial_z^{P_z} = 0 : M(K(T)) \to M_{-1}(K) \qquad (4.8)$$

in which we take for $\tau_K^K(\infty)$ the isomorphism $\theta_{P_\infty} : M_{-1}(K) \cong H^1_\infty(\mathbb{P}_K^1; M) \cong H^1(\mathbb{P}_K^1; M)$. We will call this isomorphism $\tau_K^K(\infty) = \theta_{P_\infty}$ the *canonical isomorphism* and will denote it simply by $\gamma : M_{-1}(K) \cong H^1(\mathbb{P}_K^1; M)$. For any rational K-point z of $\mathbb{P}_K^1$ different from ∞, that is to say contained in $\mathbb{A}_K^1$ we have the isomorphism $\theta_{X-z} : M_{-1}(K) \cong H^1(\mathbb{P}_K^1; M)$. The following Lemma explains our choices:

Lemma 4.25. *Let z be a K-rational point of $\mathbb{P}_K^1$ distinct from ∞. Then*

$$\theta_{X-z} = \gamma : M_{-1}(K) \cong H^1(\mathbb{P}_K^1; M)$$

Proof. As the morphism $\mathbb{P}_K^1 \to \mathbb{P}_K^1$, $y \mapsto y - z$ is $\mathbb{A}^1$-homotopic to the identity (take $(T, y) \mapsto y - Tz$) and as M is strongly $\mathbb{A}^1$-invariant, the induced morphism $H^1(\mathbb{P}_K^1; M) \to H^1(\mathbb{P}_K^1; M)$ is the identity. One deduces from that $\theta_z = \theta_0$ for any $z \in A_K^1$. It remains to prove that $\theta_0 = \gamma$. For this we take the morphism $\rho : \mathbb{P}_K^1 \to \mathbb{P}_K^1$, $y \mapsto \frac{-1}{y}$. It takes the point 0 to the point at infinity and the uniformizing element X to $\frac{-1}{X}$; thus the diagram

$$
\begin{array}{ccc}
M_{-1}(K) & \xrightarrow{\gamma} & H^1(\mathbb{P}_K^1; M) \\
\| & & \downarrow \rho \\
M_{-1}(K) & \xrightarrow{\theta_X} & H^1(\mathbb{P}_K^1; M)
\end{array}
$$

Now it thus suffices to prove that ρ induces the identity on $H^1(\mathbb{P}_K^1; M)$. This follows from the fact that $\rho([x, y]) = [-y, x]$ in homogenous coordinates and this automorphism comes from the action the matrix $\begin{pmatrix} 0 & -1 \\ 1 & 0 \end{pmatrix}$ of $SL_2(K)$. As such a matrix is a product of elementary matrices, it is well-known that the morphism ρ is $\mathbb{A}^1$-homotopic to the identity morphism and we conclude by the fact that M is strongly $\mathbb{A}^1$-invariant. $\square$

One may reformulate the construction of the geometric transfers above in the following way. We see a closed point z in $\mathbb{A}_K^1$ as a closed point in $\mathbb{P}_K^1$ and we thus get a canonical morphism of the form

$$M_{-1}(L) \overset{\theta_{P_z}}{\simeq} H^1_z(\mathbb{P}^1_K; M) \to H^1(\mathbb{P}^1_K; M) \overset{\gamma^{-1}}{\simeq} M_{-1}(K)$$

which is seen to be the same as the geometric transfer morphism above. Using the $\mathbb{A}^1$-purity theorem in the same way as in the beginning of the previous section we may even reformulate the construction as follows. We consider the closed embedding $i_z : Spec(L) \subset \mathbb{P}^1_K$ defined by z and we thus get a cofiber sequence:

$$\mathbb{P}^1_K - \{z\} \subset \mathbb{P}^1_K \to Th(\nu_z)$$

where ν_z is the normal bundle of i_z. Of course it is trivial, and in fact the polynomial P_z defines a trivialization of ν_z so that the previous cofibration is equivalent to

$$\mathbb{P}^1_K - \{z\} \subset \mathbb{P}^1_K \to T \wedge (Spec(L)_+)$$

with $T := \mathbb{P}^1/\mathbb{A}^1$. Now using the isomorphisms γ, $H^1(T \wedge (Spec(L)_+); M) \cong H^1_\infty(\mathbb{P}^1_L; M)$ and γ_L we get finally a morphism which is equal to the geometric transfer

$$\tau^L_K(z) : M_{-1}(L) \overset{\gamma_L}{\simeq} H^1(T \wedge (Spcc(L)_+); M) \to H^1(\mathbb{P}^1_K; M) \overset{\gamma}{\simeq} M_{-1}(K)$$

This last description will have the advantage to make the Lemma 4.40 below relatively obvious.

The cohomological transfer. Now let us recall from the end of the Sect. 3.3 that the sheaves of the form M_{-1} (and M_{-n}, $n \geq 1$) are endowed with a canonical action of $\mathbb{G}_m$ which in fact is induced from a $\mathbf{K}^{MW}_n$-module structure on M_{-1}, see Lemma 3.49. We denote the action of a unit u by $< u >$.

Given a field K of characteristic exponent p and a monic irreducible polynomial $P \in K[X]$ we may write canonically P as $P_0(X^{p^i})$ where P_0 is a monic irreducible separable polynomial and $i \in \mathbb{N}$ an integer (which we take equal to 0 in characteristic 0 where $p = 1$). This corresponds to the canonical factorization of $K \subset L = k[X]/P$ as $K \subset L_{sep} \subset L$ where $K \subset L_{sep} = K[x^{p^i}]$ is the separable closure of K in L, and $L_{sep} \subset L$ purely inseparable; i is thus the smallest integer with $x^{p^i} \in L_{sep}$ and $[L : L_{sep}] = p^i$ is the inseparable degree of x over K. P_0 is then the minimal polynomial of x^{p^m} over K. We will denote by P'_0 the derivative of P_0; thus $P'_0 \neq 0$. Observe that $P'_0(x^{p^m}) \in L^\times_{sep} \subset L^\times$ in any characteristic.

Definition 4.26. Given a monogenous extension $K \subset L$ with a generator x we set

$$\omega_0(x) := P'_0(x^{p^m}) \in L^\times$$

and the composed morphism

$$Tr_K^L(x) : M_{-1}(L) \overset{<\omega_0(x)>}{\simeq} M_{-1}(L) \overset{\tau_K^L(x)}{\longrightarrow} M_{-1}(K)$$

is called the *cohomological transfer* for the monogenous extension with generator $(K \subset L, x)$.

The following result justifies the previous definition:

Theorem 4.27. *Let* $K \subset L$ *be a finite extension between finite type extensions of* k. *Assume one of the following assumptions holds:*

a) $char(k) \neq 2$;
b) *the extension* $K \subset L$ *is separable;*
c) *the structure of* $\underline{\mathbf{GW}}$-*modules on* M_{-2} *is trivial.*

Choose an increasing sequence $K = L_0 \subset L_1 \subset L_2 \subset \cdots \subset L_r = L$ *for which any of the intermediate extensions* $L_{i-1} \subset L_i$ *is monogenous and choose for each* i *a generator* x_i *of* L_i *over* L_{i-1}. *Then the composition of each of the cohomological transfers morphisms previously constructed*

$$M_{-2}(L) \overset{Tr_K^L(x_r)}{\longrightarrow} M_{-2}(L_{r-1}) \to \cdots \to M_{-2}(L_1) \overset{Tr_K^{L_1}(x_1)}{\longrightarrow} M_{-2}(K)$$

only depends on the extension $K \subset L$.

We give now the following definition:

Definition 4.28. Let $K \subset L$ be a finite extension between finite type extensions of k. In case one of the assumptions (a), (b) or (c) holds, the morphism

$$M_{-2}(L) \to M_{-2}(K)$$

obtained from the previous Theorem and any choice of increasing sequence as in the statement is called *the cohomological transfer morphism* for the extension $K \subset L$ and is denoted by

$$Tr_K^L : M_{-2}(L) \to M_{-2}(K)$$

Remark 4.29. Observe that for a purely inseparable extension generated by an element $x \in L$, one has $Tr_K^L = \tau_K^L(x)$. In characteristic $p > 2$ the Theorem says that this morphism is independent of the choice of x.

The case of characteristic 2 is different and tricky. The transfers depends in general on the generator and on the dependence on a generator x is parametrised by a certain L-vector space of Kähler differential forms of L over K. See Corollary 5.2 below.   □

Remark 4.30. It follows from Lemma 4.34 below that in characteristic $p > 2$ or in characteristic 2 and L separable over K, that one has the "projection formula" for any unit $u \subset K$ and element $m \in M_{-2}(L)$:

$$Tr_K^L(<u|_L>.m) = <u>.Tr_K^L(m)$$

In characteristic 2 for a monogenous purely inseparable extension $K \subset L$ with generator x it follows from the same Lemma as well as Lemma 4.41 that one has the projection formula for each $\tau_K^L(x)$:

$$\tau_K^L(x)(<u|_L>.m) = <u>.\tau_K^L(x)_K^L(m)$$

Remark 4.31. One can show more generally that the statement of the Theorem holds as well for M_{-1} in place of M_{-2}. This requires for the moment quite a bit more work; we hope to come back on this point elsewhere.   □

Remark 4.32. In the case of the sheaf $\underline{\mathbf{K}}_n^M$ of unramified Milnor K-theory in weight n, which was constructed in the previous Sect. 3.2 as a strongly $\mathbb{A}^1$-invariant sheaf with the property that $(\underline{\mathbf{K}}_{n+1}^M)_{-1} \cong \underline{\mathbf{K}}_n^M$, it is clear that the $\underline{\mathbf{K}}_0^{MW} = \underline{\mathbf{GW}}$-module structure is the trivial one and that the transfers

$$\tau_K^L(x) : K_n^M(L) \to K_n^M(K)$$

obtained by the above method for a finite extension $K \subset L$ with generator x is the same as the one constructed in [9]. The previous Theorem thus also reproves in a new way the Theorem of Kato [37] that these transfers morphisms are independent of any choices of an increasing sequence as in the Theorem. Observe indeed that we didn't use any transfer in Sect. 3.2. Our new proof relies on the use of the contractible chain complex $C^*(\mathbb{A}_F^2; M)$ (see the details below).   □

Remark 4.33. For any surjective finite morphism $Y \to X$ between smooth k-schemes with induced function fields extension $K \subset L$, together with a closed embedding $Y \subset \mathbb{A}_X^1$ such that the composition $Y \to \mathbb{A}_X^1 \to X$ is the given morphism, one may observe (compare with the Lemma 4.40 below) that the *geometric transfer* morphism $\tau_K^L(x) : M_{-2}(L) \to M_{-2}(K)$ on the fields extension level (where $x \in L$ is the generator of L over K corresponding to the function $Y \to \mathbb{A}^1$ of the embedding) induces a canonical morphism $M_{-2}(Y) \to M_{-2}(X)$ such that the obvious diagram

$$M_{-2}(Y) \to M_{-2}(X)$$
$$\cap \qquad\qquad \cap$$
$$\tau_K^L(x) : M_{-2}(L) \to M_{-2}(K)$$

commutes. Observe that the sheaves M_{-2} being unramified, the vertical morphisms are indeed injective. However it is not true that the cohomological transfer morphism induces a morphism $M_{-2}(Y) \to M_{-2}(X)$. Consequently the previous theorem can't be extended for a general finite dominant morphism $Y \to X$ in Sm_k, and there is no way in general to define a transfer $M_{-2}(Y) \to M_{-2}(X)$. This is possible however, if the morphism $Y \to X$ is finite and étale, or if the structure of $\mathbf{GW}$-module on M_{-2} happens to be trivial. This is exactly the case of Voevodsky's homotopy invariant sheaves with transfers, for example with the sheaves $\mathbf{K}_n^M$.

We will also see below in Sect. 5.2 that one may always define a canonical transfer by conveniently *twisting* the transfer $M_{-2}(L) \to M_{-2}(K)$.   □

4.3   Proof of the Main Theorem

In this section we give the proof of Theorem 4.27. We start with the following Lemma in which p denotes the exponential characteristic of k:

Lemma 4.34. *Let $K \subset L$ be an extension and let $x \in L$ be a generator of L over K. For any unit $u \in K^\times$ the following diagram is commutative*

$$
\begin{array}{ccc}
M_{-1}(L) & \xrightarrow{\tau_K^L(x)} & M_{-1}(K) \\
\downarrow \wr <u^n> & & \downarrow \wr <u> \\
M_{-1}(L) & \xrightarrow{\tau_K^L(u.x)} & M_{-1}(K)
\end{array}
$$

For odd exponential characteristic p or for $p = 2$ and $K \subset L$ separable, the following diagram commutes:

$$
\begin{array}{ccc}
M_{-1}(L) & \xrightarrow{Tr_K^L(x)} & M_{-1}(K) \\
\downarrow \wr <u> & & \downarrow \wr <u> \\
M_{-1}(L) & \xrightarrow{Tr_K^L(u.x)} & M_{-1}(K)
\end{array}
$$

and that for $p = 2$ and a purely inseparable extension the following diagram commutes:

$$
\begin{array}{ccc}
M_{-1}(L) & \xrightarrow{Tr_K^L(x)} & M_{-1}(K) \\
\| & & \downarrow \wr <u> \\
M_{-1}(L) & \xrightarrow{Tr_K^L(u.x)} & M_{-1}(K)
\end{array}
$$

Proof. Let us denote by P the minimal polynomial of x over K and by Q that of ux. The K-automorphism $f_u : \mathbb{P}_K^1 \cong \mathbb{P}_K^1$, $[x,y] \mapsto [ux,y]$ induces an automorphism of complexes $<u>: C^*(\mathbb{P}_K^1; M) \cong C^*(\mathbb{P}_K^1; M)$ such that obvious diagram

$$
\begin{array}{ccccc}
M_{-1}(L) & \xrightarrow{\theta_P} & H_x^1(\mathbb{P}_K^1; M) & \subset & C^*(\mathbb{P}_K^1; M) \\
\downarrow \wr <u^n> & & \downarrow \wr & & \downarrow \wr <u> \\
M_{-1}(L) & \xrightarrow{\theta_Q} & H_x^1(\mathbb{P}_K^1; M) & \subset & C^*(\mathbb{P}_K^1; M)
\end{array}
$$

commutes, as clearly $Q(Y) = u^n P(\frac{Y}{u})$. Moreover one has (using our standard conventions) $Q_0(Y) = u^n P_0(\frac{1}{u^{p^i}} Y)$ and thus $Q_0'((ux)^{p^m}) = (u^n \cdot u^{-p^m})$ $P_0'(x^{p^m})$. If p is odd u^{-p^m} is congruent to u mod the square $u^{p^m - 1}$ and if $p = 2$ but $m = 0$ we thus obtain $< \omega_0(ux) > = < u^n > < u > < \omega_0(x) >$. If $p = 2$ and $m > 0$ then $< \omega_0(ux) > = < u^n > < \omega_0(x) >$. These facts imply the last claims. □

We now observe that if $[L : K] = 1$ and $x \in L$ is any element then $Tr_K^L(x) = \tau_K^L(x) : M_{-2}(L) \to M_{-2}(K)$ is nothing but the inverse to the canonical isomorphism $M_{-2}(K) \cong M_{-2}(L)$ by using Lemma 4.25 and in particular doesn't depend on x.

Using this fact and an easy induction on r, we see that to prove Theorem 4.27 it suffices to prove the statement of the theorem in the case $r = 1$ and 2. In fact the case $r = 2$ suffices because this case also implies the case $r = 1$: apply the case $r = 2$ to the extension $K \subset L$ with two generators x and y of L over K, and the previous observation.

We fix now $K \subset L$ a finite extension in $\mathcal{F}_k$ and (x, y) generators of $L|K$. Set $E := K[x] \subset L$ and $F := K[y] \subset L$. Observe that x is a generator of L over F and y of L over E. We have to prove that the following diagram

$$
\begin{array}{ccc}
M_{-2}(L) & \xrightarrow{Tr_F^L(x)} & M_{-2}(F) \\
\downarrow Tr_E^L(y) & & \downarrow Tr_K^F(y) \\
M_{-2}(E) & \xrightarrow{Tr_K^E(x)} & M_{-2}(K)
\end{array}
\qquad (4.9)
$$

is commutative. Our method is to "embed everything" in $(\mathbb{P}^1)_K^2$ and to use the results at the end of the previous Sect. 4.1. More precisely let us denote by $(\infty, \infty) \in (\mathbb{P}^1)_K^2$ the K rational point given by the two points at ∞. For this point the pair of functions $(\frac{-1}{X}, \frac{-1}{Y})$ is a system of local parameters and we get the canonical isomorphism

$$
\gamma_{X,Y} : M_{-2}(K) \overset{\Phi_{\frac{-1}{X}, \frac{-1}{Y}}}{\underset{\simeq}{\longrightarrow}} H_{\infty,\infty}^2((\mathbb{P}^1)_K^2; M) \cong H^2((\mathbb{P}^1)_K^2; M)
$$

where the right isomorphism is given by Corollary 4.15. Observe that if we interchange the order of X and Y (or rather $\frac{-1}{X}$ and $\frac{-1}{Y}$) then $\gamma_{Y,X} = \gamma_{X,Y} \circ < -1 >$ by Theorem 4.19. By abuse we will also sometimes denote by $\gamma_{X,Y}$ the isomorphism $M_{-2}(K) \overset{\Phi_{\frac{-1}{X}, \frac{-1}{Y}}}{\underset{\simeq}{\longrightarrow}} H_{\infty,\infty}^2((\mathbb{P}^1)_K^2; M)$.

We will denote by p is the exponential characteristic of k, by P the minimal polynomial of x, by Q that of y, both over K. We write $P(X) = P_0(X^{p^i})$ where $P_0 \in K[X]$ is irreducible and separable, and analogously $Q(Y) = Q_0(Y^{p^j})$ for y. We let z denote the closed point of $\mathbb{A}_K^2$ (or $(\mathbb{P}^1)_K^2$) corresponding to L with the generators (x, y) in this ordering; $\mathcal{M}_z$ will denote the maximal ideal at z (in either $\mathbb{A}_K^2$ or $(\mathbb{P}^1)_K^2$). Each of our arguments below will amount to use and analyze the morphism

$$H_z^2((\mathbb{P}^1)_K^2; M) \to H^2((\mathbb{P}^1)_K^2; M) \overset{\gamma_{X,Y}^{-1}}{\simeq} M_{-2}(K)$$

Lemma 4.35. *Assume that* $(P, Q) \in \mathcal{M}_z$ *is a regular system of parameters.*

1) *We set* $\omega(P) = \omega_0(x)P(X) = P_0'(X^{p^i})P(X)$ *and* $\omega(Q) = \omega_0(y)Q(Y) = Q_0'(Y^{p^j})Q(Y)$. *Then* $(\omega(P), \omega(Q))$ *is also a regular system of parameters in* $\mathcal{M}_z$. *We let* P_F *(resp.* Q_E*) be the minimal polynomial of* x *(resp.* y*) over* F *(resp.* E*). Then one has:*

$$\Phi_{\omega(P), \omega(Q_E)} = \Phi_{\omega(P), \omega(Q)} = \Phi_{\omega(P_F), \omega(Q)}$$

2) *The diagram*

$$
\begin{array}{ccc}
M_{-2}(L) & \overset{Tr_F^L(-x)}{\longrightarrow} & M_{-2}(F) \\
\downarrow Tr_E^L(y) & & \downarrow Tr_K^F(y) \\
M_{-2}(E) & \overset{Tr_K^E(-x)}{\longrightarrow} & M_{-2}(K)
\end{array}
$$

commutes.

Remark 4.36. For instance $(P, Q) \in \mathcal{M}_z$ is a regular system of parameters if one of x or y is separable over K. Indeed, the quotient ring $K[X, Y]/(P, Q)$ is always isomorphic to $E \otimes_K F$. If one of the extension is separable, then $E \otimes_K F$ is a product of fields, one of these being L itself, this implies the claim.   □

Proof. 1) It suffices to prove the first equality. In $E[Y]$ one has a decomposition $Q(Y) = Q_E(Y)R(Y)$. The assumption that (P, Q) is a regular system of generators implies (it is in fact equivalent) that $R(Y)$ is prime to $Q(Y)$, that is to say that Q_E has multiplicity 1 in Q. Recall that $Q(Y) = Q_0(Y^{p^j})$ in $K[Y]$ with Q_0 irreducible and separable, that is to say the minimal polynomial of y^{p^j} over K. Write $Q_{E,0}(Y)$ for the minimal polynomial of y^{p^j} over E. We get in $E[Y]$ the decomposition $Q_0(Y) = Q_{E,0}(Y)R_0(Y)$ with $R_0(Y)$ prime to $Q_{E,0}$. Now we get the equality in $E[Y]$:

$$Q(Y) = Q_0(Y^{p^j}) = Q_{E,0}(Y^{p^j})R_0(Y^{p^j})$$

As $Q(y) = 0$ and $R_{E,0}(y^{p^j}) \neq 0$, the minimal polynomial Q_E of y over E divides $Q_{E,0}(Y^{p^j})$. Now $Q_{E,0}$ being irreducible, $Q_{E,0}(Y^{p^j})$ is of the form $Q_1(Y^{p^{j'}})^{p^{j''}}$ with $Q_1(Y^{p^{j'}})$ irreducible. Thus as $Q_E(Y)$ divides $Q(Y)$ in $E[Y]$ with multiplicity 1, we see that $Q_E(Y) = Q_1(Y^{p^j})$, $j'' =$

0 and $R(Y) = R_0(Y^{p^j})$. It follows that in $E[Y]$: $Q(Y) = Q_0(Y^{p^j}) = Q_{E,0}(Y^{p^j} R_0(Y^{p^j})) = Q_E(Y)R_0(Y^{p^j})$ and $Q_0'(Y^{p^j}) = Q_{E,0}'(Y^{p^j})R_0(Y^{p^j}) + Q_E(Y)R_0'(Y^{p^j})$. Now (see the definition of the isomorphism Φ (4.6)) this implies that

$$\Phi_{\omega(P),\omega(Q)} = \Phi_{\omega(P),u^2\,\omega(Q_E)} = \Phi_{\omega(P),\omega(Q_E)}$$

with $u = \overline{R_0(Y^{p^j})} \in E^\times$, the last equality coming from Corollary 4.23; the point (1) is established.

2) We introduce the composed morphism

$$Tr_K^L(x,y) : M_{-2}(L) \overset{\Phi_{\omega(P),\omega(Q)}}{\cong} H_z^2((\mathbb{P}^1)_K^2; M)$$

$$\to H^2((\mathbb{P}^1)_K^2; M) \overset{\gamma_{X,Y}^{-1}}{\simeq} M_{-2}(K)$$

We claim now and prove below that the morphism $Tr_K^L(x,y)$ is equal to

$$M_{-2}(L) \overset{Tr_E^L(y)}{\longrightarrow} M_{-2}(E) \overset{Tr_K^E(-x)}{\longrightarrow} M_{-2}(K)$$

This fact implies the Lemma because by interchanging X and Y we see that on the other hand that

$$Tr_K^L(y,x) : M_{-2}(L) \overset{\Phi_{\omega(Q),\omega(P)}}{\cong} H_z^2((\mathbb{P}^1)_K^2; M)$$

$$\to H^2((\mathbb{P}^1)_K^2; M) \overset{\gamma_{Y,X}^{-1}}{\simeq} M_{-2}(K)$$

is equal to

$$M_{-2}(L) \overset{Tr_F^L(x)}{\longrightarrow} M_{-2}(F) \overset{Tr_K^F(-y)}{\longrightarrow} M_{-2}(K)$$

By Theorem 4.19

$$\gamma_{X,Y} = \gamma_{Y,X} \circ <-1>$$

and $\Phi_{\omega(Q),\omega(P)} = \Phi_{\omega(P),\omega(Q)} \circ <-1>$. Thus we get the equation between morphisms $M_{-2}(L) \to M_{-2}(K)$:

$$Tr_K^E(-x) \circ Tr_E^L(y) =<-1> Tr_K^F(-y) \circ Tr_F^L(x) \circ <-1>$$

The Lemma 4.34 for $u = -1$ (observe that for $p = 2$ this Lemma for $u = -1 = 1$ is trivial) implies that $<-1> Tr_K^F(-y) = Tr_K^F(y) \circ <-1>$ and also $<-1> Tr_F^L(x) = Tr_F^L(-x) \circ <-1>$. As $<-1> \circ <-1>=<1>$ is the identity the Lemma is proven, modulo the claim. $\square$

To prove this claim, that $Tr_K^L(x,y)$ is the composition:

$$M_{-2}(L) \xrightarrow{Tr_E^L(y)} M_{-2}(E) \xrightarrow{Tr_K^E(-x)} M_{-2}(K)$$

we proceed as follows. Write $\mathbb{P}_E^1 = \{x\} \times \mathbb{P}^1 \subset (\mathbb{P}^1)_K^2$ the closed subscheme defined by the product immersion of $x : Spec(E) \subset \mathbb{P}_K^1$ corresponding to x and $Id_{\mathbb{P}_K^1}$ over K and we write $P_\infty^1 = \mathbb{P}^1 \times \{\infty\} \subset (\mathbb{P}^1)_K^2$ the closed subscheme defined by the product immersion of $Id_{\mathbb{P}_K^1}$ and the point at infinity $\infty : Spec(K) \to \mathbb{P}_K^1$ over K. We denote in the same way $(x,\infty) \in \mathbb{P}_E^1$ the E-rational point defined by the product of $x : Spec(E) \to \mathbb{P}_K^1$ and the point at infinity $Spec(K) \to \mathbb{P}_K^1$. The main observation is that $H_z^2((\mathbb{P}^1)_K^2; M) \to H^2((\mathbb{P}^1)_K^2; M)$ factorizes canonically as

$$H_z^2((\mathbb{P}^1)_K^2; M) \to H_{\mathbb{P}_E^1}^2((\mathbb{P}^1)_K^2; M) \to H^2((\mathbb{P}^1)_K^2; M)$$

We analyze this decomposition through the various identifications we have in hands, to prove our claim. First we have the following diagram where the morphisms without name are the "obvious one":

$$
\begin{array}{ccccc}
H_z^2((\mathbb{P}^1)_K^2; M) & \to & H_{\mathbb{P}_E^1}^2((\mathbb{P}^1)_K^2; M) & & \\
\phi_{\omega_P} \uparrow \wr & & \phi_{\omega_P} \uparrow \wr & \nwarrow & \\
H_z^1(\mathbb{P}_E^1; M_{-1}) & \to & H^1(\mathbb{P}_E^1; M_{-1}) & & H_{(x,\infty)}^2((\mathbb{P}^1)_K^2; M) \\
\theta_{\omega_{Q_E}} \uparrow \wr & & \gamma \uparrow \wr & \nwarrow & \phi_{\omega_P} \uparrow \wr \\
M_{-2}(L) & \xrightarrow{Tr_E^L(y)} & M_{-2}(E) & \xrightarrow{\simeq}{} & H_{(x,\infty)}^1(\mathbb{P}_E^1; M_{-1})
\end{array}
$$

It is commutative by all what we have seen so far: the point (1) of the Lemma that established ($\Phi_{\omega(P),\omega(Q_E)} = \Phi_{\omega(P),\omega(Q)}$) show that the left vertical composition is indeed $\Phi_{\omega(P),\omega(Q)}$. The left bottom square commutes by the very definition of $Tr_E^L(y)$. The top left square and the right square involving ω_P are commutative diagrams by (4.5) for ω_P, that we use for two points z and (x,∞). The remaining triangle involving γ is commutative by Lemma 4.25.

Now we concatenate on the left side of the previous diagram the following one:

$$
\begin{array}{ccccc}
H_{\mathbb{P}_E^1}^2((\mathbb{P}^1)_K^2; M) & \longrightarrow & H^2((\mathbb{P}^1)_K^2; M) & & \\
& \nwarrow & & \nwarrow & \\
& & H_{(x,\infty)}^2((\mathbb{P}^1)_K^2; M) & \to & H_{P_\infty^1}^2((\mathbb{P}^1)_K^2; M) \\
& \phi_{\omega_P} \nearrow & \uparrow \wr \phi_{\frac{-1}{\gamma}} & & \uparrow \wr \phi_{\frac{-1}{\gamma}} \\
H_{(x,\infty)}^1(\mathbb{P}_E^1; M_{-1}) & & H_{(x,\infty)}^1(\mathbb{P}_\infty^1; M) & \to & H^1(\mathbb{P}_\infty^1; M) \\
\uparrow \wr \gamma & & \uparrow \wr \theta_{\omega_P} & & \uparrow \wr \gamma \\
M_{-2}(E) & \xrightarrow[\simeq]{<\frac{-1}{}>} & M_{-2}(E) & \xrightarrow{Tr_K^E(x)} & M_{-2}(K)
\end{array}
$$

The latter commutes for the following reasons: the top square commutes for trivial reasons. The left bottom diagram commutes by Theorem 4.19. The middle right square commutes by (4.5) for $\frac{-1}{Y}$ and the bottom left square commutes by definition.

Altogether the commutativity of the diagram obtained from the union of the previous two ones gives, after a moment of reflexion, the fact that $Tr_K^L(x,y)$ is the composition

$$M_{-2}(L) \xrightarrow{Tr_E^L(y)} M_{-2}(E) \xrightarrow{<\frac{-1}{\simeq}>} M_{-2}(E) \xrightarrow{Tr_K^E(x)} M_{-2}(K) \xrightarrow{<\frac{-1}{\simeq}>} M_{-2}(K)$$

The last right multiplication by $< -1 >$ comes from the fact that the vertical right morphism of the previous diagram is $\Phi_{\frac{-1}{Y},\frac{-1}{X}}$ which is equal to $\Phi_{\frac{-1}{X},\frac{-1}{Y}} \circ < -1 >$ and we conclude using Lemma 4.34 and $u = -1$. The Lemma is established. $\square$

Corollary 4.37. *Assume that $K \subset L$ is separable. Then the diagram (4.9) is commutative and in particular the Theorem 4.27 holds in characteristic 0 or in case (b).*

Proof. Indeed apply the previous Lemma and conclude by the induction on r that we mentioned above. $\square$

Corollary 4.38. *Assume that E and F are linearly disjoint, that is to say $E \otimes_K F \to L$ is an isomorphism. Then the diagram (4.9) commutes. In particular, this holds if x is separable over K and y is purely inseparable.*

Proof. Indeed this conditions implies that (P,Q) is a regular system of parameters at z, and we just have to apply the previous Lemma. $\square$

It remains to establish the Theorem in finite characteristic p.

Lemma 4.39. *Assume that $y = x^{p^m}$ for some integer m. Then the following diagram commutes:*

$$
\begin{array}{ccc}
M_{-2}(L) & = & M_{-2}(L) \\
\downarrow \tau_F^L(-x) & & \downarrow \tau_K^L(-x) \\
M_{-2}(F) & \xrightarrow{\tau_K^F(y)} & M_{-2}(K)
\end{array}
$$

In particular, for p odd, the diagram

$$
\begin{array}{ccc}
M_{-2}(L) & = & M_{-2}(L) \\
Tr_F^L(-\omega_0(x).x) \downarrow & & \downarrow Tr_K^L(-x) \\
M_{-2}(F) & \xrightarrow{Tr_K^F(y)} & M_{-2}(K)
\end{array}
$$

is commutative.

For $p = 2$ the following diagram

$$
\begin{array}{ccc}
M_{-2}(L) & = & M_{-2}(L) \\
\downarrow \tau_F^L(\omega_0(x).x) & & \downarrow \tau_K^L(x) \\
M_{-2}(F) & \overset{Tr_K^F(y)}{\longrightarrow} & M_{-2}(K)
\end{array}
$$

Proof. Observe that in that case $E = L$ and that $F = K[y] = K[x^{p^m}]$. P and Q denoting (as usual) the minimal polynomial of x and y over K respectively, it is clear that $P(X) = Q(X^{p^m})$ and also that $\omega_0(x) = \omega_0(y)$. We consider the morphism

$$
H_z^2((\mathbb{P}^1)_K^2; M) \to H^2((\mathbb{P}^1)_K^2; M) \overset{\gamma_{X,Y}^{-1}}{\simeq} M_{-2}(K)
$$

corresponding to the closed point z. We claim that on one hand the composition

$$
M_{-2}(L) \overset{\Phi_{P(X), Y - X^{p^m}}}{\simeq} H_z^2((\mathbb{P}^1)_K^2; M) \to H^2((\mathbb{P}^1)_K^2; M) \overset{\gamma_{X,Y}^{-1}}{\simeq} M_{-2}(K)
$$

is $Tr_K^L(-x)$ and the other hand that the composition obtained by interchanging the role of X and Y

$$
M_{-2}(L) \overset{\Phi_{Q(Y), X^{p^m} - Y}}{\simeq} H_z^2((\mathbb{P}^1)_K^2; M) \to H^2((\mathbb{P}^1)_K^2; M) \overset{\gamma_{Y,X}^{-1}}{\simeq} M_{-2}(K)
$$

is the composition $Tr_K^{K[x^{p^m}]}(x^{p^m}) \circ Tr_{K[x^{p^m}]}^L(-x)$.

The first claim follows by using the same techniques as in the proof of Lemma 4.35. Observe that $(P(X), Y - X^{p^m})$ is a system of regular parameters at z and we may factorize the above morphism as

$$
\begin{array}{ccc}
H_z^2((\mathbb{P}^1)_K^2; M) & \to H_{\mathbb{P}_L^1}^2((\mathbb{P}^1)_K^2; M) \to H^2((\mathbb{P}^1)_K^2; M) \\
\wr \uparrow \; \Phi_{P(X), Y - X^{p^m}} & \downarrow \wr \gamma_{X,Y}^{-1} \\
M_{-2}(L) & M_{-2}(K)
\end{array}
$$

where $\mathbb{P}_L^1 \subset (\mathbb{P}^1)_K^2$ is the closed immersion corresponding to $P(X) = 0$. As $E = L$, $Y - X^{p^m}$ is equal in the maximal ideal of z in $\mathbb{P}_L^1$ to polynomial $Y - y$, so that the isomorphism $\theta_{Y - X^{p^m}} : M_{-2}(L) \cong H_z^1(\mathbb{P}_L^1; M_{-1})$ is the canonical one (corresponding to the rational point y in L). The rest of the claim is straightforward, copying the reasoning used in the proof of Lemma 4.35.

For the second claim we observe that $(Q(Y), X^{p^m} - Y)$ is also a regular system of parameters at z and that the curve defined by the first equation $Q(Y) = 0$ is $\mathbb{P}_F^1 = \mathbb{P}^1 \times Spec(F) \subset (\mathbb{P}^1)_K^2$. The second parameter $X^{p^m} - Y$ is a local parameter for the closed point $x := Spec(L) \subset \mathbb{P}_F^1$, as $X^{p^m} - y$ is thc minimal polynomial of x over F. The rest of the claim follows from an analogous reasoning as in the proof of Lemma 4.35.

Once this is done we claim that

$$\Phi_{P(X),Y-X^{p^m}} = \Phi_{Q(Y),X^{p^m}-Y} \circ < -1 >$$

which suffices to imply the Lemma as $\gamma_{X,Y} = \gamma_{Y,X} \circ < -1 >$.

To prove the above claim, by Corollary 4.23, it suffices to prove in the set $\omega(\mathcal{M}_z/(\mathcal{M}_z))$ of orientations that one has the equality

$$\overline{P(X)} \wedge \overline{(Y - X^{p^m})} = \overline{Q(Y)} \wedge \overline{-(X^{p^m} - Y)}$$

But $P(X) = Q(X^{p^m})$ and $Q(X^{p^m}) - Q(Y)$ is divisible by $Y - X^{p^m}$ Thus

$$\overline{P(X)} \wedge \overline{(Y - X^{p^m})} = \overline{Q(Y)} \wedge \overline{(Y - X^{p^m})}$$

which establishes the claim. The rest of the Lemma is straightforward.   □

Lemma 4.40. *If $K \subset L = E = F$ is purely inseparable, and if $\mathrm{char}(k) \neq 2$ or if the $\mathbf{GW}$-module structure on M_{-2} is trivial, one has the equality*

$$\tau_K^L(x) = \tau_K^L(y) : M_{-2}(L) \to M_{-2}(K)$$

Proof. Let $[L : K] = p^i > 0$ be the degree (the case $K = L$ is trivial). We consider again the morphism

$$H_z^2((\mathbb{P}^1)_K^2; M) \to H^2((\mathbb{P}^1)_K^2; M)$$

corresponding to the closed point $z = (x, y)$.

There is a unique polynomial $R(X) \in K[X]$ with degree $< p^i$ such that $R(x) = y$ and there is a unique polynomial $S(Y) \in K[Y]$ with degree $< p^i$ such that $S(y) = x$.

We claim that on one hand the composition

$$M_{-2}(L) \overset{\Phi_{P(X),Y-R(X)}}{\cong} H_z^2((\mathbb{P}^1)_K^2; M) \to H^2((\mathbb{P}^1)_K^2; M) \overset{\gamma_{X,Y}^{-1}}{\cong} M_{-2}(K)$$

is $\tau_K^L(-x)$ and on the other hand that

$$M_{-2}(L) \overset{\Phi_{Q(Y),X-S(Y)}}{\cong} H_z^2((\mathbb{P}^1)_K^2; M) \to H^2((\mathbb{P}^1)_K^2; M) \overset{\gamma_{Y,X}}{\cong} M_{-2}(K)$$

is $\tau_K^L(-y)$. Of course the second claim follows from the first one by interchanging X and Y.

Now the first claim follows by using the same techniques as in the proof of Lemma 4.35. Observe that $(P(X), Y - R(X))$ is a system of regular parameters at z and the first factorization is $z \in \mathbb{P}^1_L \subset (\mathbb{P}^1)^2_K$ with $\mathbb{P}^1_L = Spec(L) \times \mathbb{P}^1_K$ corresponding to $P(X) = 0$. As $E = F = L$, y is a rational point of $\mathbb{P}^1_L$ and the isomorphism $\theta_{Y-R(X)} = \theta_{Y-y} : M_{-2}(L) \cong H^1_z(\mathbb{P}^1_L; M_{-1})$ is the canonical one corresponding to any rational point. The rest is straightforward.

Now we prove that

$$\Phi_{P(X),Y-R(X)} = \Phi_{Q(Y),X-S(Y)} \circ < -1 > \qquad (4.10)$$

at least in odd characteristic. Taking into account that $\gamma_{X,Y} = \gamma_{Y,X} \circ < -1 >$ this implies the Lemma as we get (using Lemma 4.34) for any generators x, y the equality $\tau^L_K(-x) = \tau^L_K(-y)$. Observe that if the **GW**-structure is trivial, by Remark 4.24 (or Corollary 4.23) the isomorphisms $\Phi_{\pi,\rho}$ do not depend on any choice of the regular system of parameters. The Lemma in that case is thus proven.

It remains to compare $\Phi_{P(X),Y-R(X)}$ and $\Phi_{Q(Y),X-S(Y)}$. By Corollary 4.23 and Remark 4.24 it suffices to compare in $\omega(\mathcal{M}_z/(\mathcal{M}_z))$ the induced orientations $\overline{P(X)} \wedge \overline{(Y - R(X))}$ and $\overline{Q(Y)} \wedge \overline{-(X - S(Y))}$.

Set $a := x^{p^i} \in K$ and $b := y^{p^i} \in K$. Observe that $P(X) = X^{p^i} - a$ and $Q(Y) = Y^{p^i} - b$. We have $R(S(Y)) \equiv Y[Y^{p^i} - b]$ and that $S(R(X)) \equiv X$ $[X^{p^i} - a]$ as $S(y) = x$ and $R(x) = y$. Now $X^{p^i} - a = X^{p^i} - S(Y)^{p^i} + S(Y)^{p^i} - a$. The polynomial

$$X^{p^i} - S(Y)^{p^i} = (X - S(Y))^{p^i}$$

lies in $(\mathcal{M}_z)^2$ (because $p^i \geq 2$) so that $\overline{P(X)} = \overline{S(Y)^{p^i} - a}$ in the L-vector space $\mathcal{M}_z/(\mathcal{M}_z)^2$. Now the Taylor development in $L[Y]$:

$$S(Y) - S(y) = (Y - y)S'(y) + (Y - y)^2 T(Y)$$

implies by raising to the p^i-th power the equality in $K[Y]$ (observe that $a = x^{p^i} = S(y)^{p^i}$):

$$S(Y)^{p^i} - a = (Y^{p^i} - b).(S'(y))^{p^i} + (Y^{p^i} - b)^2 (T(Y))^2$$

This implies that $\overline{S(Y)^{p^i} - a} = (S'(y))^{p^i}.\overline{Q(Y)}$. So we have established

$$\overline{P(X)} \wedge \overline{(Y - R(X))} = (S'(y))^{p^i}.\overline{Q(Y)} \wedge \overline{(Y - R(X))}$$

Now $Y - R(X) = Y - R(S(Y)) + R(S(Y)) - R(X)$. As the minimal polynomial $Q(Y)$ of y divides $Y - R(S(Y))$, we see that $\overline{Q(Y)} \wedge \overline{(Y - R(X))} = \overline{Q(Y)} \wedge \overline{(R(S(Y)) - R(X))}$. Now the Taylor formula tells us that $R(S(Y)) - R(X) = R'(X).(S(Y) - X) + (S(Y) - X)^2.U(X,Y)$ in $K[X,Y]$. Thus again $\overline{(Y - R(X))} = R'(x).\overline{S(Y) - X}$ and we get that

$$\overline{P(X)} \wedge \overline{(Y - R(X))} = (S'(y))^{p^i}.R'(x).\overline{Q(Y)} \wedge \overline{(S(Y) - X)}$$

As $S'(y)).R'(x) = 1$ because $R(S(Y)) \equiv Y[Y^{p^i} - b]$, we obtain at the end:

$$\overline{P(X)} \wedge \overline{(Y - R(X))} = (S'(y))^{p^m - 1}.\overline{Q(Y)} \wedge \overline{(S(Y) - X)}$$

In characteristic $p > 2$, $S'(y)^{p^m - 1}$ is a square so the claim (4.10) follows. □
If $p = 2$, in the previous proof we actually got:

$$\overline{P(X)} \wedge \overline{(Y - R(X))} = < S'(y) > .\overline{Q(Y)} \wedge \overline{(S(Y) - X)} \qquad (4.11)$$

Observe that $< S'(y) > = < R'(x) >$. It follows that in that case we proved:

Lemma 4.41. *Assume $char(k) = 2$ and let $K \subset L$ be purely inseparable and monogenous generated by x. Let $R \in K[X]$ be such that $R(x)$ is a generator of $L|K$ (that is to say $R' \neq 0 \in K[X]$). Then one has the equality*

$$\tau_K^L(y) = \tau_K^L(x) \circ < R'(x) > : M_{-2}(L) \to M_{-2}(K)$$

Remark 4.42. We will use below the previous Lemma to define the Rost–Schmid complex in characteristic 2. In fact using the notations which will be introduced later in characteristic 2, we will see in Corollary 5.2 that there is a canonical "Transfer morphism" for any finite dimensional purely inseparable extension $K \subset L$ of the form

$$Tr_K^L : M_{-2}(L) \times_{L^\times} \Omega^{max}(L|K)^\times \to M_{-2}(K)$$

where $\Omega^{max}(L|K)$ is the maximal external power of the finite dimensional L-vector space $\Omega(L|K)$ of Kähler differential forms of L over K. This morphism has the property that given any sequence of generators $(x_1, \ldots, x_r)$ of L over K, where $r = dim_L \Omega(L|K)$, one has for $m \in M_2(L)$ and $L_i := K[x_1, \ldots, x_i] \subset L$:

$$Tr_K^L(m; dx_1 \wedge dx_2 \wedge \cdots \wedge dx_r) = \tau_K^{L_1}(x_1) \circ \cdots \circ \tau_{L_{r-1}}^L(x_r)(m)$$

Corollary 4.43. *Assume that L is purely inseparable over K (and so are E and F). Then the diagram (4.9) commutes if characteristic $\neq 2$ or if the* **GW***-structure is trivial.*

Proof. Let $K' \subset L$ be $E \cap F$. One may check that if $p^{i'} := [L : F]$ is the degree of x over F and $p^{j'} := [L : E]$ that of y over E then $\alpha = x^{p^{i'}} \in K'$ and $\beta = y^{p^{j'}} \in K'$ are both generators of K' and that the extensions $K' \subset E$ and $K' \subset F$ are linearly disjoint. By Lemma 4.40 we know that $\tau_K^{K'}(\alpha) = \tau_K^{K'}(\beta)$. Using the commutativity given by Corollary 4.38 and repeated use of Lemma 4.39 we obtain the result. □

Now we may finish the proof of Theorem 4.27 as follows. We introduce the following commutative diagram of intermediate fields, in which E_{sep} is the separable closure of K in E and F_{sep} the separable closure of K in F:

$$
\begin{array}{ccccc}
E & \subset & E.F_{sep} & \subset & L \\
\cup & & \cup & & \cup \\
E_{sep} & \subset & E_{sep}.F_{sep} & \subset & E_{sep}.F \\
\cup & & \cup & & \cup \\
K & \subset & F_{sep} & \subset & F
\end{array}
$$

The case (b) of the Theorem was done in Corollary 4.37. We treat both remaining cases (a) and (c) at the same time. We observe that each of the extension appearing in the diagram is monogenous and either separable or purely inseparable. In each of the case we know that the cohomological transfer doesn't depend on any choice by Corollary 4.37 and Lemma 4.40.

We start with $Tr_K^E(x) \circ Tr_E^L(y)$. We write $p^{j'} = [L : E.F_{sep}]$ and $p^{i'} = [L : E_{sep}.F]$. From Lemma 4.39 (and Lemma 4.40 in case of characteristic 2 and trivial **GW**-structure) we see that $Tr_E^L(y) = Tr_E^{E.F_{sep}}(y^{p^{j'}}) \circ Tr_{E.F_{sep}}^L(y)$ which we simply write (taking into account what we just said) $Tr_E^L(y) = Tr_E^{E.F_{sep}} \circ Tr_{E.F_{sep}}^L$. In the same way $Tr_K^E(x) = Tr_K^{E_{sep}} \circ Tr_{E_{sep}}^E$.

Now the left top square satisfies the assumptions of Corollary 4.38 ($E_{sep} \subset E_{sep}.F_{sep}$ is separable and $E_{sep} \subset E$ is inseparable). Thus in the composition

$$
Tr_K^{E_{sep}} \circ Tr_{E_{sep}}^E \circ Tr_E^{E.F_{sep}} \circ Tr_{E.F_{sep}}^L
$$

we may permute the two central terms to get

$$
Tr_K^{E_{sep}} \circ Tr_{E_{sep}}^{E_{sep}.F_{sep}} \circ Tr_{E_{sep}.F_{sep}}^{E.F_{sep}} \circ Tr_{E.F_{sep}}^L
$$

By Lemma 4.35 for the bottom left square and Corollary 4.43 for the right top square, we get further

$$
Tr_K^{F_{sep}} \circ Tr_{F_{sep}}^{E_{sep}.F_{sep}} \circ Tr_{E_{sep}.F_{sep}}^{E_{sep}.F} \circ Tr_{E_{sep}.F}^L
$$

and from Corollary 4.38 again for the bottom right square we finally get

$$
Tr_K^{F_{sep}} \circ Tr_{F_{sep}}^F \circ Tr_F^{E_{sep}.F} \circ Tr_{E_{sep}.F}^L
$$

which is checked by Lemma 4.39 (and Lemma 4.40 in case of characteristic 2 and trivial **GW**-structure) to be equal to $Tr_L^F(y) \circ Tr_F^L(x)$.

The Theorem is proven.   □

Chapter 5
The Rost–Schmid Complex
of a Strongly $\mathbb{A}^1$-Invariant Sheaf

5.1 Absolute Transfers and the Rost–Schmid Complex

We still fix a strongly $\mathbb{A}^1$-invariant sheaf of abelian groups M on Sm_k.

The absolute transfers. We aim now at unifying, at all the characteristics, including the characteristic 2, the transfers defined previously. This is inspired by [70], and will be needed to define the Rost–Schmid complex below.

Let A be a k-algebra. Recall that one may construct the A-module of Kähler differential forms $\Omega(A|k)$ over k, see [42] for instance. If A is a smooth finite k-algebra, this is a locally free A-module of finite rank. More generally if A is essentially k-smooth, $\Omega(A|k)$ is also a locally free A-module of finite rank (more precisely of rank d if $Spec(A)$ admits a pro-étale morphism to a smooth k-variety of dimension d). It will be convenient in the sequel to use the notation:

$$\omega(A|k) := \Omega^{max}(A|k) = \Lambda_A^{max}(\Omega(A|k))$$

for the maximal exterior power of $\Omega(A|k)$. Sometimes we will even simply write $\omega(A)$ if there no possible confusion about the base field k. Observe that $\omega(A|k)$ is an invertible $\mathbb{A}$-module (locally free module of rank 1). This definition applies for instance to any $F \in \mathcal{F}_k$.

We start with the case of **characteristic $\neq 2$**.

Lemma 5.1. *[70] Assume given a purely inseparable finite extension $i : F \subset E$ in odd characteristic p. Then there is a canonical $Q(E)$-equivariant bijection*

$$Q(i) : Q(\omega(F|k) \otimes_F E) \to Q(\omega(E|k))$$

induced by the (iterations of the) Frobenius homomorphism.

Proof. In *loc. cit.* page 16 this bijection is constructed in case $E^p \subset F$. For a general finite purely inseparable extension $F \subset E$, one may factorize it as $K \subset \ldots E^{p^2} \subset E^p \subset E$, where each intermediary extension satisfies the above property.   □

For a separable finite extension $F \subset E$, the map $\omega(F|k) \otimes_F E \to \omega^{max}(E|k)$ is an isomorphism. Altogether we see that for any finite extension $i : K \subset L$ there is a canonical $Q(L)$-equivariant bijection :

$$Q(i) : Q(\omega(K|k) \otimes_K L) \cong Q(\omega(L|k))$$

With this bijection, we may define for $n \geq 2$, the *twist* (that is to say the product over $Q(K)$) of the transfer morphism of Definition 4.28 to get:

$$Tr_K^L(\omega) : M_{-n}(E; \omega(E|k)) \to M_{-n}(F; \omega(F|k)) \qquad (5.1)$$

In **characteristic** 2 the situation is quite different. We improve our computations related to the transfers in the following form:

Corollary 5.2. *Assume $char(k) = 2$ and let $K \subset L$ be a purely inseparable finite extension. Then there is a unique bilinear map*

$$T_K^L : M_{-2}(L) \times \omega(L|K)^\times \to M_{-2}(K)$$

such that the following property hold: given a factorization $K = L_0 \subset L_1 \subset \cdots \subset L_r = L$ such that, for each $i \in \{1, \ldots, r\}$, $L_{i-1} \subset L_i$ is monogenous and $[L_i : L_{i-1}] > 1$, and given for each $i \in \{1, \ldots, r\}$ a generator $x_i \in L_i$ of L_i over L_{i-1}, one has for $m \in M_{-2}(L)$:

$$Tr_K^L(m, dx_1 \wedge \cdots \wedge dx_r) = \tau_K^{L_1}(x_1) \circ \cdots \circ \tau_{L_{r-1}}^{L_r}(x_r)(m)$$

This map induces a map (actually a group homomorphism):

$$Tr_K^L : M_{-2}(L) \times_{L^\times} \omega(L|K)^\times \to M_{-2}(K)$$

Proof. We prove first the case $r = 1$. It suffices to prove that for a generator $x \in L$, the morphism $\tau_K^L(x) : M_{-2}(L) \to M_{-2}(K)$ only depends on $dx \in \Omega(L|K)$ (observe that in that case $\Omega(L|K)$ is of dimension 1 over L). If $y \in L$ is an other element, there exists a unique polynomial $R(X)$ $inK[X]$ of degree $< [L : K]$ with $y = R(x)$. Observe that y is a generator of L over K if and only if $R'(X) \neq 0$. Thus $R'(x) \neq 0$ and by Lemma 4.41 one has the equality

$$\tau_K^L(y) = \tau_K^L(x) \circ < R'(x) >: M_{-2}(L) \to M_{-2}(K)$$

Of course in $\Omega(L|K)$ we have also $dy = R'(x)dx$, which establishes the case $r = 1$.

For $r > 1$ we proceed by induction on r. Observe that with our assumptions $\Omega(L|K)$ has dimension r. The given factorization produces an isomorphism (with obvious notations)

$$\Omega(L_r|L_{r-1}) \otimes_{L_{r-1}} \Omega(L_{r-1}|L_{r-2}) \otimes_{L_{r-2}} \cdots \otimes_{L_1} \Omega(L_1|K) \cong \Omega^{max}(\Omega(L|K))$$

In this way we reduce to proving the case $r = 2$. The statement means the following: given (x, y) and (x', y') pairs of elements in L which both generates L, that is to say that (dx, dy) and (dx', dy') are both basis of the two-dimensional L-vector space $\Omega(L|K)$ over K then letting $\delta \in L^\times$ be the determinant of (dx', dy') in the base (dx, dy) we have the equality (setting $E = K[x]$ and $E' = K[x']$)

$$\tau_K^{E'}(x') \circ \tau_{E'}^{L}(y') = \tau_K^{E}(x) \circ \tau_E^{L}(y) \circ < \delta >$$

Now clearly one of the pair (dx, dy') or (dx, dx') is also a basis of $\Omega(L|K)$ over K. So to prove the formula we may reduce to the case that both pairs have a common element. For instance (x, y) and (x, y') (the other case can be treated in the same way). Now there is a unique polynomial $R(X, Y) \in K[X, Y]$ such that $y' = R(x, y)$ if we impose moreover that it has only monomials of the form $a_{i,j} X^i Y^j$ with $i < [E : K]$ and $j < [L : E]$. With our conventions we see that the polynomial $R(x, Y) \subset E[Y]$ satisfies the assumption of Lemma 4.41 so that in the group of morphisms $M_{-2}(L) \to M_{-2}(E)$ one has the equality

$$\tau_E^{L}(y') = \tau_E^{L}(y) \circ < \frac{\partial R(x, Y)}{\partial Y}(y) >$$

Now we conclude by observing that $\frac{\partial R(x, Y)}{\partial Y}(y)$ is exactly the determinant δ in this case. □

We now make the following observation: given a purely inseparable extension $F \subset E$ with $E^2 \subset F$, the exact sequence (*) of [70, page 16]

$$0 \to \Omega(E^2/F^2) \otimes_{E^2} E \to \Omega(F|F^2) \otimes_F E \to \Omega(E|E^2) \to \Omega(E|F) \to 0$$

also exists in characteristic 2, using exactly the same arguments. We also have $\Omega(F|F^2) = \Omega(F|k)$ and $\Omega(E|E^2) = \Omega(E|k)$, as in *loc. cit.*. Now in characteristic 2, and this is the main difference, there is no bijection as in the above Lemma 5.1. But we have:

Lemma 5.3. *If $char(k) = 2$, for each purely inseparable extension $F \subset E$ with $E^2 \subset F$, then $\omega(E^2/F^2) \otimes_{E^2} E$ admits a* canonical *orientation.*

Proof. Indeed, if we choose two non-zero elements ω_1 and ω_2 of $\omega(E^2/F^2)$, each defines a non zero element $\omega_i \otimes 1$ in $\omega(E^2/F^2) \otimes_{E^2} E$. But there is a unit u in E^2 such that $\omega_2 = u\omega_1$, and $u \in E^2$ becomes a square v^2 in E by construction. Then

$$\omega_2 \otimes 1 = (u\omega_1) \otimes 1 = \omega_1 \otimes u = (\omega \otimes 1)v^2$$

so that the two elements $\omega_i \otimes 1$ are equivalent in the set $Q(\Omega(E^2/F^2) \otimes_{E^2} E)$ of orientations of $\Omega(E^2/F^2) \otimes_{E^2} E$.   □

The previous lemma and the above exact sequence gives a canonical $Q(F)$-equivariant bijection

$$Q(\omega(E|k)) \cong Q(\omega(F|k)) \otimes_{Q(F)} Q(\omega(E|F))$$

This bijection is checked to extend by composition to a canonical bijection of the same form for any finite purely inseparable extension. Taking into account that the canonical Transfer morphism for purely inseparable extensions $F \subset E$ of Corollary 5.2 (in characteristic 2) has the form:

$$Tr_F^E : M_{-2}(E) \times_{E^\times} \omega(E|F)^\times \to M_{-2}(F)$$

we see that, in characteristic 2 as well, we get a canonical morphism :

$$Tr_F^E(\omega) : M_{-n}(E; \omega(E|k)) \to M_{-n}(F; \omega(F|k)) \qquad (5.2)$$

for any purely inseparable extension $F \subset F$. Now given any finite extension $K \subset L$ we may factorize it canonically as $K \subset L_{sep} \subset L$. We may the also define a morphism

$$M_{-n}(L; \omega(L|k)) \to M_{-n}(K; \omega(K|k))$$

as $M_{-n}(L; \omega(L|k)) \to M_{-n}(L_{sep}; \omega(L_{sep}|k))$ followed by $M_{-n}(L_{sep}; \omega(L_{sep}|k)) \to M_{-n}(K; \omega(K|k))$, where the last one is just the twist of

$$Tr_K^{L_{sep}} : M_{-n}(L_{sep}; \omega(L_{sep}|k)) \to M_{-n}(K; \omega(K|k))$$

(from Definition 4.28) by $\omega(L_{sep}|k)^\times = \omega(K|k) \otimes_K L_{sep}^\times$.

Definition 5.4. Given any finite extension $K \subset L$, for $n \geq 2$, the canonical morphism constructed above (in any characteristic) of the form:

$$Tr_K^L(\omega) : M_{-n}(L; \omega(L|k)) \to M_{-n}(K; \omega(K|k))$$

is called the *absolute transfer morphism*.

The following is now easy to establish, we leave the details to the reader.

Lemma 5.5. *Given any finite extension $K \subset L$, for $n \geq 2$, the absolute transfer morphism is $K^\times$-equivariant for the obvious action.*

The absolute transfer morphism is functorial in the sense that for any composable finite extensions $K \subset E \subset L$ the composition

$$M_{-n}(L; \omega(L|k)) \xrightarrow{Tr_E^L(\omega)} M_{-n}(E; \omega(E|k)) \xrightarrow{Tr_K^E(\omega)} M_{-n}(K; \omega(K|k))$$

equals $Tr_K^L(\omega)$.

Remark 5.6. 1) In odd characteristic this transfer can be canonically "untwisted" using Lemma 5.1 and corresponds to the cohomological transfer $Tr_K^L : M_{-n}(L) \to M_{-n}(K)$ from Definition 4.28. However in characteristic 2, there is no way in general to untwist the canonical transfer in a canonical way.

2) Given a monogenous extension $K \subset L$ with $x \in L$ a given generator, the normal bundle $Spec(L) \subset \mathbb{A}_K^1$ is trivialized by the minimal polynomial of x. This defines an isomorphism $\omega(L|k) \cong \omega(K[X]|k) \cong \omega(K|k)$, thus one gets a trivialization of $\omega(L) \otimes \omega(K)^{-1}$. The twist of the absolute Transfer by $\omega(K)^{-1}$ composed with the previous trivialization gives a morphism $M_{-n}(L) \to M_{-n}(K)$ which is seen to be the geometric transfer $\tau_K^L(x)$.

3) When $\mathcal{L}$ is a one dimensional K-vector space, we will use the notation $Tr(\omega \otimes \mathcal{L})$ for the absolute transfer twisted by $\mathcal{L}$ over $K^\times$, that is to say the obvious morphism

$$M_{-n}(L; \omega(L|k) \otimes_K \mathcal{L}) \to M_{-n}(L; \omega(K|k) \otimes_K \mathcal{L}) \qquad \qquad \square$$

The Rost–Schmid differential. Now we define for any (essentially) smooth k-scheme X a diagram

$$C_{RS}^0(X; M) \to C_{RS}^1(X; M) \to \cdots \to C_{RS}^n(X; M) \to \ldots$$

which we will prove in the next section to be a complex and in the section after that this complex is canonically isomorphic to the Gersten complex $C^*(X; M)$ mentioned in Sect. 4.1.

This complex is constructed along the line of the complex of X with coefficients in a Rost cycle module [68] and its extension to Witt groups [70]. For this reason we will call it the Rost–Schmid complex. The main property of this complex, once constructed, is the *homotopy invariance property*, see Theorem 5.38 below, which comes from an explicit homotopy as in [68] and [70].

Now as in the preceding section we have defined absolute transfers on sheaves of the form M_{-2}, our main observation is that to define the Rost–Schmid complex one actually only needs transfers morphisms on the groups of the form $M_{-n}(F)$ for $n \geq 2$. Thus it must be *possible* to define this complex for any M, always assumed to be strongly $\mathbb{A}^1$-invariant. This is what we are going to do.

Given a field $F \in \mathcal{F}_k$ and an F-vector space Λ of dimension 1 and an integer $n \geq 1$, we set

$$M_{-n}(F; \Lambda) := M_{-n}(F) \otimes_{[F^\times]} \mathbb{Z}[\lambda^\times]$$

where $\Lambda^\times$ means the set of non-zero elements of Λ endowed with its free and transitive action of $F^\times$ through scalar multiplication. Like in [70] we may also observe that as a set $M_{-n}(F; \Lambda)$ is also $M_{-n}(F) \times_{F^\times} \Lambda^\times$; one may explicit the group structure, see *loc. cit.*. Moreover, as the action of $F^\times$ on M_{-n} factors through an action of $Q(F) := F^\times/(F^2)$, we may also see $M_{-n}(F; \Lambda)$, as a set, as defined by the formula

$$M_{-n}(F; Q(\Lambda)) := M_{-n}(F) \times_{Q(F)} Q(\Lambda)$$

with $Q(\Lambda) := \Lambda^\times/(F^\times)^2$ (this is compatible with our previous definitions and with the notations of [70]).

If we let Λ^{-1} denote the dual of Λ as an F-vector space, observe that the map $\Lambda^\times \ni x \mapsto x^* \in \Lambda^{-1}$ ($\{x^*\}$ being the dual basis of $\{x\}$) induces a canonical bijection

$$Q(\Lambda) \cong Q(\Lambda^{-1})$$

Thus in fact we have a canonical isomorphism

$$M_{-n}(F; \Lambda) \cong M_{-n}(F; \Lambda^{-1}) \tag{5.3}$$

In the sequel, we will freely use the identifications discuss above. We don't need to write the exponent -1 appearing for inverse (or dual) line bundles, but we will sometime do it, to keep track of the geometric situation.

Let κ be a field and V be a κ-vector space of finite dimension. We will set $\Lambda_\kappa^{max}(V)$ for its maximal exterior power. This is a one dimensional κ-vector space.

Observe that almost tautologically $Q(V) = Q(\Lambda^{max}(V))$.

Let X be an essentially smooth k-scheme and $z \in X$ be a point of codimension n. The $\kappa(z)$-vector space $\mathcal{M}_z/(\mathcal{M}_z)^2$ has dimension n and its dual $T_z X$ is called the *tangent space* of X at the point z. We set

$$\Lambda_z^X := \Lambda_{\kappa(z)}^{max}(T_z X)$$

We may then define the Rost–Schmid complex as a graded abelian group:

Definition 5.7. Let X be an essentially smooth k-scheme and n an integer. Let $z \in X^{(n)}$ be a point of codimension n. We set:

$$C_{RS}^n(X; M)_z := M_{-n}(\kappa(z); \Lambda_z^X)$$

We set

$$C^n_{RS}(X; M) := \oplus_{z \in X^{(n)}} M_{-n}(\kappa(z); \Lambda^X_z)$$

Remark 5.8. We might as well have defined λ^X_z as

$$\Lambda^X_z := \Lambda^n_{\kappa(z)}(\mathcal{M}_z/(\mathcal{M}_z)^2)$$

and proceed in the same way. We would get the same graded abelian group because of the canonical isomorphism (5.3). The reason for which we made the previous (not so important) choice will appear later. $\square$

The second fundamental exact sequence for the morphisms $k \to A \to \kappa(z)$ (see [42]) yields a short exact sequence of $\kappa(z)$-vector spaces

$$0 \to \mathcal{M}_z/(\mathcal{M}_z)^2 \to \Omega(A|k) \otimes_A \kappa(z) \to \Omega(\kappa(z)|k) \to 0$$

because, k being perfect, $\kappa(z)$ is separable over k. Taking the maximal exterior powers of the dual exact sequence gives a canonical isomorphism of one-dimensional $\kappa(z)$-vector spaces

$$\Lambda^X_z \cong \omega(\kappa(z)|k) \otimes_A \omega(A|k)^{-1} \tag{5.4}$$

Remark 5.9. In the sequel we will often not write over which rings we take the tensor product, as long as it may not lead to a confusion. So with our previous conventions, when no confusion may arise from the base field, etc..., we may simply write the previous group isomorphism as

$$\Lambda^X_z \cong \omega(\kappa(z)) \otimes \omega(A)^{-1} \quad \square$$

We now want to define a boundary morphism for the Rost–Schmid complex

$$\partial_{RS} : C^{n-1}_{RS}(X; M) \to C^n_{RS}(X; M)$$

To do this it suffices to define for $y \in X^{(n-1)}$ and $z \in X^{(n)}$ a morphism

$$\partial^y_z : M_{-n+1}(\kappa(y); \lambda_y) \to M_{-n}(\kappa(z); \lambda_z)$$

We take $\partial^y_z := 0$ if $z \notin \overline{y}$. Now we assume that $z \in \overline{y} = Y$.

We will need the following lemma, which defines the morphism ∂^y_z when z is a point of codimension 1:

Lemma 5.10. *Let $z \in Y$ be a point of codimension 1 in an essentially smooth k-scheme Y, irreducible with generic point y. For any strongly $\mathbb{A}^1$-invariant sheaf N, there exists a canonical isomorphism of the form*

$$\theta_z : N_{-1}(\kappa(z); \lambda^Y_z) \cong H^1_z(Y; N)$$

such that for any choice of a non-zero element $\mu \in \mathcal{O}_{Y,z}$ the morphism

$$\theta(-;\mu) : N_{-1}(\kappa(z)) \cong H_z^1(Y;N)$$

induced by μ is the isomorphism θ_μ defined in Corollary 2.35.

In particular the residue morphism $\partial_z^y : N(\kappa(y)) \to H_z^1(Y;N)$ (see at the beginning of Sect. 4.1) becomes through this isomorphism a canonical morphism of the form

$$\partial_z^y : N(\kappa(y)) \to N_{-1}(\kappa(z); \lambda_z^Y)$$

Given a unit $u \in \mathcal{O}_{Y,z}^\times$ and an element $\nu \in N_{-1}(\kappa(y))$, one has the following formula (where the cup product with $[u]$ comes from the pairing considered in Lemma 3.48)

$$\partial_z^y([u].\nu) = \epsilon[\overline{u}].\partial_z^y(\nu)$$

where $\overline{u} \in \kappa(z)$ is the reduction of u and $\epsilon = - < -1 > \in \mathbf{K}_0^{MW}(\kappa(z))$ is the element introduced in Sect. 3.1. Note that to state the formula we used the fact that $[\overline{u}]. : N_{-2}(\kappa(z)) \to N_{-1}(\kappa(z))$ is a morphism of $\mathbf{K}_0^{MW}(\kappa(z))$-modules by Lemma 3.49 and thus induces a morphism

$$[u]. : N_{-2}(\kappa(z); \lambda_z^Y) \to N_{-1}(\kappa(z); \lambda_z^Y)$$

If N is of the form M_{-m} for some $m > 0$, the residue morphism is moreover equivariant for the obvious action of the group $\mathcal{O}_{Y,z}^\times \subset \kappa(y)^\times$ on both sides.

Proof. First we may shrink Y around z as we wish. We set $Z := \overline{z}$ and we may thus assume that Z is smooth over k. We use the definition of the isomorphisms θ_μ given in the discussion preceding Lemma 2.34. Let μ and μ' be non-zero elements of $\lambda_z^Y = \mathcal{M}_z/(\mathcal{M}_z)^2$. Both define a canonical isomorphism of pointed sheaves $Th(\nu_i) \cong T \wedge (Z_+)$ (at least up to shrinking further Y) where $i : Z \subset Y$ is the canonical closed embedding. Let $u \in \mathcal{O}_{Y,z}^\times$ be a unit such that $\mu' = \overline{u}.\mu$ in $\lambda_z^Y = \mathcal{M}_z/(\mathcal{M}_z)^2$. By the very construction, the isomorphism $\theta'_\mu : N_{-1}(\kappa(z)) \cong H_z^1(Y;N)$ induced by μ' is thus equal to the composition of $\theta_\mu : N_{-1}(\kappa(z)) \cong H_z^1(Y;N)$ with the automorphism of $N_{-1}(\kappa(z))$ induced precisely by the multiplication by u: $T \wedge (Z_+) \cong T \wedge (Z_+)$ and the isomorphism of Lemma 2.34. But this is exactly the multiplication by $\overline{u} \in \kappa(z)^\times$ for the $\kappa(z)^\times$-structure on $N_{-1}(\kappa(z))$ we used to define λ_z^Y as a tensor product. The first claim of the lemma follows from this. The second claim is proven as follows. Given $u : Y \to \mathbb{G}_m$ one gets an induced morphism of the cofibration sequence $(Y - z)_+ \subset Y_+ \to Y/(Y - z)$ to the cofibration sequence $\mathbb{G}_m \wedge (Y - z)_+ \subset \mathbb{G}_m \wedge (Y_+) \to \mathbb{G}_m \wedge (Y/(Y - z))$. We may suppose that Y is henselian local with closed point z, and thus we may assume that there is an étale morphism $f : Y \to \mathbb{A}^1_{\kappa(z)}$ with $f^{-1}(0) = z$; this defines an identification $Y/(Y - z) \cong T \wedge (Spec(\kappa(z)_+)) \cong \Sigma \mathbb{G}_m \wedge (Spec(\kappa(z)_+))$.

Isolating the part related to the connecting homomorphisms (dashed arrows below) the morphism of cofibration sequences above gives us a commutative diagram of the form:

$$
\begin{array}{ccccc}
\mathbb{G}_m \wedge (Spec(\kappa(z)_+)) & \cong & \Sigma^{-1}(Y/(Y-z)) & \overset{\delta}{\dashrightarrow} & (Y-z)_+ \\
\overline{u} \wedge Id \downarrow & & & & u \downarrow \\
\mathbb{G}_m \wedge (\mathbb{G}_m \wedge (Spec(\kappa(z)_+))) & \cong \Sigma^{-1}\mathbb{G}_m \wedge (Y/(Y-z)) & \overset{\delta}{\dashrightarrow} & \mathbb{G}_m \wedge (Y-z)_+
\end{array}
$$

The second claim follows from this and from the geometric interpretation of the residue morphism given in Sect. 4.1. Observe that ϵ appears as one has to permute the two $\mathbb{G}_m$'s in bottom left corner to get the formula in the correct way. By Lemma 3.7, this is the same ϵ and this is indeed $- < -1 >$ as claimed.

The equivariant statement follows just from the naturality of the action of $\mathbb{G}_m$ on sheaves of the form M_{-m} and a straightforward checking.   □

We may now define the morphism $\partial^y_{RS,z}$ in the general case, following Rost [68] and Schmid [70]. We may replace X by its localization at z and thus assume that $X = Spec(A)$ is local with closed point z and residue field κ. We have to define a morphism

$$
\partial^y_z : M_{-n+1}(\kappa(y); \Lambda^X_y) \to M_{-n}(\kappa(z); \Lambda^X_z)
$$

Let $A \twoheadrightarrow B$ be the epimorphism of rings representing the closed immersion $Y = Spec(B) \subset Spec(A) = X$. Let $Spec(\tilde{B}) = \tilde{Y} \to Y$ be its normalization in $\kappa(y)$; $\tilde{B}$ is a principal ring with finitely many maximal ideals. We let $\tilde{z}_i$ be the finitely many closed points in $\tilde{Y}$, each lying over z. We write $\tilde{\kappa}_i$ for the residue field of $\tilde{z}_i$.

By (5.4) above we have that $\Lambda^X_y \cong \omega(\kappa(y)) \otimes \omega(A)^{-1}$. Now as $Spec(\tilde{B})$ is essentially smooth over k, the invertible $\tilde{B}$-module $\omega(\tilde{B})$ is equal to $\omega(\kappa(y))$ at the generic point, by which we mean $\omega(\tilde{B}) \otimes_{\tilde{B}} \kappa(y) = \omega(\kappa(y))$. One gets a canonical map:

$$
M_{-n+1}(\kappa(y)) \times_{\tilde{B}^\times} (\omega(\tilde{B}) \otimes \omega(A)^{-1})^\times \to M_{-n+1}(\kappa(y); \Lambda^X_y) \tag{5.5}
$$

which is automatically an isomorphism.

For each point $\tilde{z}_i$ of $Spec(\tilde{B})$ we use the previous Lemma 5.10 and get a canonical $\tilde{B}^\times$-equivariant morphism

$$
\partial^y_{\tilde{z}_i} : M_{-n+1}(\kappa(y)) \to M_{-n}(\tilde{\kappa}_i); \Lambda^{\tilde{Y}}_{\tilde{z}_i})
$$

One then take the product of this morphism over $\tilde{B}^\times$ with $\omega(\tilde{B}) \otimes \omega(A)^{-1}$ to get, as consequence of the above isomorphism (5.5), a canonical morphism:

$$\tilde{\partial}_i : M_{-n+1}(\kappa(y); \Lambda_y^X) \to M_{-n}(\tilde{\kappa}_i; \Lambda_i) \tag{5.6}$$

where Λ_i is the following one-dimensional $\tilde{\kappa}_i$-vector space:

$$\Lambda_i = \Lambda_{\tilde{z}_i}^{\tilde{Y}} \otimes \omega(\tilde{B}) \otimes \omega(A|k)^{-1})$$

Now we remember that there is a canonical isomorphism $\Lambda_{\tilde{z}_i}^{\tilde{Y}} \cong \omega(\tilde{\kappa}_i) \otimes \omega(\tilde{B})^{-1}$ so that we get at the end a canonical isomorphism of the form

$$\Lambda_i \cong \omega(\tilde{\kappa}_i) \otimes \omega(A)^{-1}$$

To define the morphism ∂_z^y it remains now to define a morphism

$$\tilde{\tau}_i : M_{-n}(\tilde{\kappa}_i; \Lambda_i) \to M_{-n}(\kappa; \Lambda_z^X)$$

As $\Lambda_z^X \cong \omega(\kappa) \otimes \omega(A)^{-1}$ we see that we may take the absolute transfer $Tr_\kappa^{\tilde{\kappa}_i}(\omega) : M_{-n}(\tilde{\kappa}_i; \omega(\tilde{\kappa}_i)) \to M_{-n}(\kappa; \omega(\kappa))$ (defined in Definition 5.4) and then we twist it over $\kappa^\times$ by $\omega(A)^{-1}$ to get $\tilde{\tau}_i$.

For each i the composition of $\tilde{\partial}_i : M_{-n+1}(\kappa(y); \Lambda_y^X) \to M_{-n}(\tilde{\kappa}_i; \Lambda_i)$ and of $\tilde{\tau}_i$ is a morphism

$$M_{-n+1}(\kappa(y); \Lambda_y^X) \to M_{-n}(\kappa; \Lambda_z^X)$$

and we set now:

$$\partial_{RS,z}^y := \sum_i \tilde{\tau}_i \circ \tilde{\partial}_i : M_{-n+1}(\kappa(y); \Lambda_y^X) \to M_{-n}(\kappa; \Lambda_z^X)$$

We may then define the Rost–Schmid complex as a graded abelian group with a morphism ∂_{RS} of degree 1:

Definition 5.11. Let X be a smooth k-scheme and n an integer. The Rost–Schmid complex of X with coefficients in M is the graded abelian group $\{C_{RS}^n(X; M)\}_{n \in \mathbb{N}}$ endowed with the morphism $\partial_{RS} : C_{RS}^*(X; M) \to C_{RS}^{*+1}(X; M)$ defined from the degree $n-1$ as the obvious sum

$$\sum_{y \in X^{(n-1)}, z \in X^{(n)}} \partial_{RS,z}^y$$

Beware that we didn't yet prove that this is a complex, and we consider for the moment $C_{RS}^*(X; M)$ as a diagram, or a quasi-complex:

Definition 5.12. A pair (C^*, ∂) consisting of a graded abelian groups C^* and a morphism $\partial : C^* \to C^{*+1}$ of graded abelian groups of degree $+1$ is called a *quasi-complex*.

Remark 5.13. We will need in the next section, for any line bundle $\mathcal{L}$ over X, the *Rost–Schmid complex twisted by* $\mathcal{L}$: $C_{RS}^*(X; \mathcal{L}; M_{-1})$ (see [70] in the case of the Witt groups). Its definition is as follows. As a graded abelian group we just take:

$$C_{RS}^n(X; \mathcal{L}; M_{-1}) := \bigoplus_{z \in X^{(n)}} M_{-n-1}(\kappa(z); \Lambda_z^X \otimes \mathcal{L})$$

The differential is the sum of the twisted version of the ∂_z^y that we defined above. To do this, we observe that the absolute transfers and residue morphisms that we used are equivariant with respect to the action of the units $\mathcal{O}(X)^\times$ (see Lemmas 5.10, and 5.5). Observe also that we used the action of the units on M_{-1} so that one can't define the twisted Rost–Schmid complex $C_{RS}^*(X; \mathcal{L}; M)$ in general, for M not of the form M_1.

Observe that any orientation of $\mathcal{L}$ in the sense of definition 4.3, that is to say an isomorphism $\omega : \mathcal{L} \cong \mu^{\otimes 2}$ between $\mathcal{L}$ and the square of a line bundle, defines a canonical isomorphism of complexes $C_{RS}^*(X; \mathcal{L}; M) \cong C_{RS}^*(X; M)$. Indeed for any one-dimensional κ-vector spaces λ and μ one has an obvious canonical bijection $\omega(\lambda) \cong \omega(\lambda \otimes (\mu)^{\otimes 2})$ and one checks that this bijection is compatible with the differentials in the Rost–Schmid complex. $\quad\square$

Remark 5.14. Let X be an essentially k-smooth scheme of dimension ≤ 2.

1) Lemma 5.10 defines a canonical isomorphism θ between the Gersten complex and the Rost–Schmid complex in dimension ≤ 1. From Corollary 4.23 we see moreover that the isomorphisms of the form $\Phi_{\pi,\rho}$ induce a canonical isomorphism for any closed point z in X:

$$\theta_z : M_{-2}(\kappa(z); \Lambda_z^X) \cong H_z^2(X; M)$$

Altogether these isomorphisms induces a canonical isomorphism of graded abelian groups denoted by

$$\Theta : C_{RS}^*(X; M) \cong C^*(X; M)$$

whose component corresponding to some point z of X is θ_z. We have seen that Θ commutes to the differential $C^0 \to C^1$. From Corollary 4.11 we see that moreover, for any $y \in X^{(1)}$ such that $Y := \overline{y} \subset X$ is essentially k-smooth, then the diagram

$$\begin{array}{ccc} M_{-1}(\kappa(y); \Lambda_y^X) & \xrightarrow{\partial_{RS}^y} & \bigoplus_{z \in X^{(2)}} M_{-2}(\kappa(z); \Lambda_z^X) \\ \Theta \downarrow \wr & & \Theta \downarrow \wr \\ H_y^1(X; M) & \xrightarrow{\partial^y} & \bigoplus_{z \in X^{(2)}} H_z^2(X; M) \end{array}$$

Thus Θ commutes with the differential in degree 0 and to the differentials in degree 1 starting from the summands corresponding to any essentially k-smooth point of codimension 1.

2) Given a line bundle $\mathcal{L}$ over X, we may also define the twisted Gersten complex $C^*(X; \mathcal{L}; M_{-1})$ analogously as follows. We define on the small Nisnevich site of X a twisted sheaf $M_{-1}(\mathcal{L})$ as follows: it is the associated sheaf to the presheaf which sends an étale morphism $U \to X$ to the abelian group $M_{-1}(U)_{\mathbb{Z}[\mathcal{O}(U)] \times \mathbb{Z}[\mathcal{L}^{\times}(U)]}$ where $\mathcal{L}^{\times}$ is the complement of the zero section. Then the same procedure as in [12] produces a coniveau spectral sequence whose $E_1^{q,*}$ we call the twisted Gersten complex $C^*(X; \mathcal{L}; M_{-1})$. We claim that the isomorphism of graded abelian groups Θ can be extended to define canonically an isomorphism of graded abelian groups

$$\Theta(\mathcal{L}) : C^*_{RS}(X; \mathcal{L}; M_{-1}) \cong C^*(X; \mathcal{L}; M_{-1})$$

If $\mathcal{L}$ is trivialized, this is clear. One has only to check that considering an open covering of X which trivializes $\mathcal{L}$, the untwisted isomorphisms Θ given on the intersections of this open covering may be glued together, to define $\Theta(\mathcal{L})$. We let the details to the reader.

3) In this context, given an essentially k-smooth closed subscheme $Y \subset X$, we may extend the isomorphism (4.5) more generally to a twisted version (which doesn't require any choice) to get a commutative diagram

$$
\begin{array}{ccc}
H_z^1(Y; M_{-1}) & \to & H^1(Y; M_{-1}) \\
\phi_z \downarrow \wr & & \phi \downarrow \wr \\
H_z^2(X; \mathcal{L}; M) & \to & H_Y^2(X; \mathcal{L}; M)
\end{array}
$$

where $\mathcal{L}$ is the line bundle on X corresponding to Y. □

5.2 The Rost–Schmid Complex Is a Complex

To state our next lemmas, we will need the following:

Definition 5.15. Given two quasi-complexes (C^*, ∂) and (D^*, ∂), a morphism $f : (C^*, \partial) \to (D^*, \partial)$ of quasi-complexes is a graded morphism $f^* : C^* \to D^*$ of graded abelian groups of degree 0 which commutes with the differentials in the obvious sense.

Let $f : X' \to X$ be a smooth morphism between smooth k-schemes. Let $z \in X^{(n)}$ be a point of codimension n in X; denote by $Z = \bar{z} \subset X$ it closure and $Z' \subset X'$ it inverse image. The morphism $Z' \to Z$ being smooth, Z' is a finite disjoint sum of integral closed subschemes of codimension n in X'. We denote by z_i' its finitely many irreducible components. Clearly $\overline{z_i'}$ is a smooth $\kappa(z)$-scheme and by definition $\mathcal{M}_z/(\mathcal{M}_z)2 \otimes_{\kappa(z)} \kappa(z_i') \to \mathcal{M}_{z_i'}/(\mathcal{M}_{z_i'})2$ is an isomorphism. In this way we get a canonical morphism for each such z:

$$f_z^* : M_{-n}(\kappa(z); \Lambda_z^X) \to \oplus_i M_{-n}(\kappa(z_i'); \Lambda_{z_i'}^{X'})$$

altogether, by summing up the previous morphisms, we get a morphism of graded abelian groups of degree 0:

$$f^* : C^*_{RS}(X; M) \to C^*_{RS}(X'; M)$$

The following property is straightforward and its proof is left to the reader.

Lemma 5.16. *For any smooth morphism $f : X' \to X$ as above, the morphism $f^* : C^*_{RS}(X; M) \to C^*_{RS}(Y; M)$ of graded abelian groups previously defined is a morphism of quasi-complexes.*

Now, following Rost [68], we want to define a push-forward morphism for (some type of) proper morphisms.

Let $f : X' \to X$ be a morphism between (essentially) smooth k-schemes. The *normal line bundle $\nu(f)$* of f is the line bundle over X' defined as

$$\nu(f) := \omega(X') \otimes f^*(\omega(X))^{-1}$$

Roughly speaking it is the maximal exterior power of the virtual normal bundle $f^*(\Omega(X|k)^\vee) - \Omega(X'|k)^\vee$ of f.

Let $z' \in X'^{(n)}$ and let $z = f(z') \in X$ be its image. If $\kappa(z) \subset \kappa(z')$ is a finite fields extension we define a morphism

$$f_* : M_{-n}(\kappa(z'); \Lambda^{X'}_{z'} \otimes \nu(f)) \to M_{-n}(\kappa(z); \Lambda^X_z)$$

as follows. We consider the sequence of isomorphisms

$$M_{-n}(\kappa(z'); \Lambda^{X'}_{z'} \otimes \nu(f)) \cong M_{-n}(\kappa(z'); \omega(\kappa(z')) \otimes \omega(X')^{-1} \otimes \nu(f))$$

$$\cong M_{-n}(\kappa(z'); \omega(\kappa(z')) \otimes f^*(\omega(X))^{-1})$$

and we compose this isomorphism with the absolute transfer of Definition 5.4 twisted over $\kappa(z)^\times$ with $\omega(X)^{-1}$. We get in this way a morphism of graded abelian groups of degree $d = d(f) = dim(X') - dim(X)$:

$$f^\nu_* : C^*_{RS}(X'; \nu(f); M) \to C^{*-d}_{RS}(X; M_{-d})$$

called the *push-forward morphism*.

Observe that we don't claim that this is a morphism of quasi-complexes. Following the point 2) of Remark 5.14, we may also define, if X' and X have dimension ≤ 2, a morphism of graded abelian groups involving the Gersten complexes

$$f^\nu_* : C^*(X'; \nu(f); M) \to C^{*+d}(X; M_{-d})$$

by using the canonical isomorphism $\Theta(\omega(f)) : C^*_{RS}(X'; \nu(f); M) \cong C^*(X'; \nu(f); M)$ and $\Theta : C^*_{RS}(X; M) \cong C^*(X; M)$ (of graded abelian groups) and the previous formulas. We don't know either that it commutes with the (Gersten) differential.

Remark 5.17. Let ω be an orientation of the normal bundle of f that is to say an orientation of the line bundle $\nu(f)$ over X'; recall that in the sense of definition 4.3 this means an isomorphism of line bundles $\omega :$ $\Lambda^{\otimes 2} \cong \nu(f)$, where Λ is a line bundle over X'. From Remark 5.13 the complex $C_{RS}^*(X'; \nu(f); M)$ is canonically isomorphic to $C_{RS}^*(X'; M)$. The corresponding morphism $C_{RS}^*(X'; M) \to C_{RS}^{*+d}(X'; M_d)$ is denoted by f_*^ω.

We will meet two important cases of proper morphism with an orientable normal bundle: the case of the projection $\mathbb{P}^1 \times X \to X$ and of a finite morphism between local schemes.

(1) If $f : X' \to X$ is a finite morphism with both X' and X local schemes, we get an orientation $\omega(\theta, \theta')$ of the normal bundle of f by choosing a trivialization θ of $\omega(X)$ and θ' of $\omega(X')$.
(2) If f is the projection morphism $\mathbb{P}_X^1 \to X$, we get a canonical orientation $\omega(\mathbb{P}_X^1 | X)$ of the normal bundle of f by contemplating the sequence of canonical isomorphisms

$$\nu(f) \cong \Omega(\mathbb{P}_X^1 | X) \cong \mathcal{O}(-2) \cong (\mathcal{O}(-1))^{\otimes 2} \qquad \square$$

We start with a simple case:

Lemma 5.18. *For any $K \in \mathcal{F}_k$, the projection morphism $f : P_K^1 \to Spec(K)$ together with its canonical orientation induces a morphism of quasi-complexes (indeed complexes):*

$$f_*^{\omega(\mathbb{P}_K^1 | K)} : C_{RS}^*(\mathbb{P}_K^1; M) \to C_{RS}^{*-1}(Spec(K); M_{-1})$$

Proof. This lemma means exactly that the following diagram:

$$M(K(T); \omega(\mathbb{P}_K^1) \otimes \omega(K)^{-1}) \to \oplus_{z \in (\mathbb{P}_K^1)^{(1)}} M_{-1}(\kappa(z); \omega(\kappa(z)) \otimes \omega(K)^{-1})$$
$$\downarrow \qquad\qquad\qquad\qquad\qquad\qquad \downarrow$$
$$0 \qquad\qquad \to \qquad\qquad M_{-1}(K)$$

$$(5.7)$$

in which the right vertical is the sum of the twist of the absolute transfers, commutes. This amounts to proving that the composition of the two morphisms is 0. Now from what we have said above, $\omega(\mathbb{P}_K^1) = \omega(K) \otimes \mathcal{O}(-2)$ and thus $M(K(T); \omega(\mathbb{P}_K^1) \otimes \omega(K)^{-1})$ is canonically isomorphic to $M(K(T))$. In the same way as at the beginning of Sect. 4.2, each $\omega(\kappa(z)) \otimes \omega(K)^{-1}$ is canonically trivialized (by the minimal polynomial for $z \neq \infty$ and by $\frac{-1}{X}$ for $z = \infty$). At the end we have to prove that the following diagram:

$$M(K(T)) \to \oplus_{z \in (\mathbb{P}^1_K)^{(1)}} M_{-1}(\kappa(z))$$
$$\downarrow \qquad\qquad\qquad \downarrow$$
$$0 \quad \to \qquad M_{-1}(K)$$

commutes, which is nothing but the formula (4.8): $\sum_{z \in \mathbb{P}^1_K} \tau_K^{\kappa(z)} \circ \partial_z^{P_z} = 0$:
$M(K(T)) \to M_{-1}(K)$.   □

Now we come to a much more subtle case:

Theorem 5.19. *Let $f : Spec(B) \to Spec(A)$ be a finite morphism, with A and B being d.v.r. essentially smooth over k. If B is monogenous over A, then $f_*^\nu : C_{RS}^*(Spec(B); \nu(f); M_{-1}) \to C_{RS}^*(Spec(A); M_{-1})$ is a morphism of quasi-complexes.*

Remark 5.20. In other words, letting K be the field of fractions of A and L be the field of fractions of B, the following diagram commutes:

$$M_{-1}(L; \omega(B) \otimes \omega(A)^{-1}) \xrightarrow{\partial_z} M_{-2}(\lambda; \omega(\lambda) \otimes \omega(A)^{-1})$$
$$\downarrow \qquad\qquad\qquad\qquad \downarrow \qquad\qquad\qquad\qquad (5.8)$$
$$M_{-1}(K) \qquad \xrightarrow{\partial_s} M_{-2}(\kappa; \omega(\kappa) \otimes \omega(A)^{-1})$$

where the vertical morphisms are both the absolute Transfer twisted by $\omega(A)^{-1}$, and $s \in Spec(A)$ and $z \in Spec(B)$ are the respective closed points.

Note that we will prove below that the statement of the theorem still holds in general, without the assumption that B is monogenous over A, but for sheaves of the form M_{-n}, $n \geq 2$.   □

Remark 5.21. Given a finite morphism $Spec(B) \to Spec(A)$ as in the theorem, we may "untwist" the diagram above to get a commutative diagram of the form

$$M_{-1}(L; \omega(B)) \xrightarrow{\partial_z} M_{-2}(\lambda; \omega(\lambda))$$
$$\downarrow \qquad\qquad\qquad \downarrow$$
$$M_{-1}(K; \omega(A)) \xrightarrow{\partial_s} M_{-2}(\kappa; \omega(\kappa))$$

The induced morphism on the kernel of the horizontal morphisms (the residues):

$$Tr_A^B(\omega) : M_{-1}(B; \omega(B)) \to M_{-1}(A; \omega(A))$$

is called the *absolute* transfer morphism. The previous diagram is commutative in fact without the assumption that B is monogenous, but with sheaves of the form M_{-n}, $n \geq 2$, so that the previous transfer morphism exists in general in this situation.   □

Remark 5.22. Observe that for an essentially smooth k-scheme X of dimension ≤ 1, $\Theta : C_{RS}^*(X; M) \cong C^*(X; M)$ is an isomorphism of complexes! Thus the two previous results are also true for the corresponding Gersten complexes.   □

Proof of the Theorem 5.19. In geometric language x defines a closed immersion $i : Spec(B) \subset \mathbb{P}^1_S$, with $S = Spec(A)$ which doesn't intersect the section at infinity $\infty : S \subset \mathbb{P}^1_S$. Let $P \in A[X]$ denote the minimal polynomial of x over K; then $A[X]/P \cong B$.

We want to prove that the diagram (5.8) is commutative. First we observe that $\omega(B) \otimes \omega(A)^{-1}$ is trivialized using P. Indeed P trivializes the normal bundle of the regular immersion $Spec(B) \subset \mathbb{A}^1_S$, giving an isomorphism $\omega(B) \cong \omega(A[X]) \cong \omega(A)$ (the latter identification comes from the fact that $\Omega(A[X]|A) = A[X].dX$). Through this identification the left vertical morphism from the diagram 5.8 is nothing but the geometric transfer $\tau_K^L(x)$: $M_{-1}(L) \to M_{-1}(K)$.

Let $Q \in \kappa[X]$ denote the minimal polynomial of the image $\overline{x}$ of x in λ. Let e be the ramification index of $A \subset B$. We have thus $\overline{P} = Q^e$ in $\kappa[X]$ (use [71, Lemme 4 p. 29]). We finally choose a uniformizing element π of A. If $e = 1$ we set $\tilde{Q} = P$ and $S(X) = 0$. If $e > 1$ let us choose a lifting $\tilde{Q} \in A[X]$ of Q, of the same degree as Q and monic. Then $P(X) \equiv \tilde{Q}^e[\pi]$ and we may thus write uniquely

$$P(X) = \tilde{Q}^e + \pi S(X) \tag{5.9}$$

for $S(X) \in A[X]$ of degree $\leq [L : K] - 1$ (If $e = 1$ we take $S(X) = 0$). In case $e = 1$ of course π is a uniformizing element of B as well. If $e > 1$ we claim that $t = \tilde{Q}(x) \in B$ is a uniformizing element for B. Indeed the local ring B/π has length e, and its maximal ideal is generated by the image of any uniformizing element of B. But from the expression above, $B/\pi \cong A[X]/(P,\pi)$ is isomorphic to $A[X]/(\tilde{Q}^e, \pi) \cong \kappa[X]/Q^e$. The image of $\tilde{Q}(x)$ in this ring is the class of Q in $\kappa[X]/Q^e$ which generates the maximal ideal. Thus $\tilde{Q}(x)$ is congruent mod π to a uniformizing element of B, and as $e > 1$, the valuation of π in B is > 1 and thus $\tilde{Q}(x)$ itself is a uniformizing element of B. If $e = 1$, let us also write $t := \pi$.

With these choices, we may trivialize $\omega(\lambda) \otimes \omega(A)^{-1}$ as follows: this is the normal line bundle of the morphism $Spec(\lambda) \subset \mathbb{A}^1_S \to Spec(A)$, where $Spec(\lambda) \subset \mathbb{A}^1_S$ corresponds to the quotient map $g : A[X] \twoheadrightarrow \lambda$, $X \mapsto \overline{x}$. The normal bundle of $\mathbb{A}^1_S \to Spec(A)$ is trivial, and the normal bundle of the closed immersion $Spec(\lambda) \subset \mathbb{A}^1_S$ is trivialized by the regular system of parameters $(P, t) \in \mathcal{M}$, where $\mathcal{M} = Ker(g)$ is the maximal ideal defining λ. The diagram (5.8) is now completely "untwisted" and has clearly the form

$$
\begin{array}{ccc}
M_{-1}(L) & \xrightarrow{\partial_z^t} & M_{-2}(\lambda) \\
\tau_K^L(x) \downarrow & & \tau \downarrow \\
M_{-1}(K) & \xrightarrow{\partial_s^\pi} & M_{-2}(\kappa)
\end{array}
\tag{5.10}
$$

τ being the absolute transfers detwisted using the trivialization (P, t) above.

We claim now that $\tau = \tau_\kappa^\lambda(\overline{x}) \circ < -1 >$ if $e = 1$ and that $\tau = \tau_\kappa^\lambda(\overline{x}) \circ < S(\overline{x}) >$ else. Observe that if $e > 1$, $S(x)$ is a unit in B by definition of e as $t = S(x).\pi$.

The trivialization of $\omega(\lambda) \otimes \omega(A)^{-1}$ corresponding to (P, t) only depends on the class $(P \wedge t) \in \Lambda^2_\lambda(\mathcal{M}/(\mathcal{M})^2)$. The above claim for $e > 1$ follows at once from that and from the fact that in $\Lambda^2_\lambda(\mathcal{M}/(\mathcal{M})^2)$ one has for $e > 1$:

$$(P \wedge t) = S(\overline{x}).(\pi \wedge \tilde{Q})$$

Indeed $t \equiv \tilde{Q}[P]$ proves that $(P \wedge t) = (P \wedge \tilde{Q})$ and (5.9) proves the fact that $(P \wedge \tilde{Q}) = (\pi S(x) \wedge \tilde{Q})$. Now in $\mathcal{M}/(\mathcal{M})^2$ one has $S(x) = S(\overline{x})$ and thus one gets the above equation.

In case $e = 1$ one has $P = \tilde{Q}, t = \pi$ which gives trivially $(P \wedge t) = -(\pi \wedge \tilde{Q})$. Moreover in that case one checks analogously that $\tau = \tau^\lambda_\kappa(\overline{x}) \circ < -1 >$.

We introduce the following notation that we will use at the very end of the proof: $\epsilon := S(\overline{x}) \in \lambda^\times$ if $e > 1$ and $\epsilon := -1$ if $e = 1$. We thus have in each case the formulas

$$(P \wedge t) = \epsilon(\pi \wedge Q) \qquad and \qquad \tau = \tau^\lambda_\kappa(\overline{x}) \circ \epsilon \qquad (5.11)$$

Now that we have identified τ in the above diagram, it remains to prove that the diagram is commutative. We will use the geometric language introduced in Sect. 4.1 (and the homotopy purity theorem of [59]). One could also write the argument below using entirely the chain complexes with support of the form $C^*_Y(\mathbb{P}^1_S; M)$ where Y is a closed subscheme of $\mathbb{P}^1_S$ as in Sect. 4.3. These two reasonings are equivalent.

Consider the following diagram of cofibration sequences of the form

$$
\begin{array}{ccccc}
\mathbb{P}^1_S - Spec(B) & \subset & \mathbb{P}^1_S & \to & T \wedge Spec(B)_+ \\
\cup & & \cup & & \uparrow \\
\mathbb{P}^1_K - Spec(L) & \subset & \mathbb{P}^1_K & \to & T \wedge Spec(L)_+
\end{array}
$$

where the Thom space of $Spec(L) \subset \mathbb{P}^1_K$ and of $Spec(B) \subset \mathbb{P}^1_S$ are identified with $T \wedge Spec(L)_+$ and $T \wedge Spec(B)_+$ respectively using the trivialization P of the normal bundle ν_i. We see that $\mathbb{P}^1_K \subset \mathbb{P}^1_S$ is the open complement to the closed immersion $\mathbb{P}^1_\kappa \subset \mathbb{P}^1_S$ and in the same way that $\mathbb{P}^1_K - Spec(L) \subset \mathbb{P}^1_S - Spec(B)$ is the open complement to the closed immersion $\mathbb{P}^1_\kappa - Spec(\lambda) \subset \mathbb{P}^1_S - Spec(B)$.

We use the chosen uniformizing element π of A to trivialize the normal bundle of $\mathbb{P}^1_\kappa \subset \mathbb{P}^1_S$. We may then fill this diagram up as follows:

$$
\begin{array}{ccccc}
T \wedge (\mathbb{P}^1_\kappa - Spec(\lambda))_+ & \to & T \wedge (\mathbb{P}^1_\kappa)_+ & \to & T \wedge Th(j) \\
\uparrow & & \uparrow & & \uparrow \\
\mathbb{P}^1_S - Spec(B) & \subset & \mathbb{P}^1_S & \to & T \wedge Spec(B)_+ \\
\cup & & \cup & & \uparrow \\
\mathbb{P}^1_K - Spec(L) & \subset & \mathbb{P}^1_K & \to & T \wedge Spec(L)_+
\end{array}
\qquad (5.12)
$$

where $j : Spec(\lambda) \subset Spec(B)$ is the tautological closed immersion.

The bottom right horizontal morphism is the one which induces the geometric transfer (see Sect. 4.2). It follows then from the commutativity of the above diagram, using the functoriality of the long exact sequences, that evaluating on $H^2(-; M)$ the two right vertical cofibrations gives us the commutativity of the following square:

$$
\begin{array}{ccc}
M_{-1}(L) & \xrightarrow{\partial_z} & H^1_z(Spec(B); M_{-1}) \\
\tau^L_K(x) \downarrow & & \tau' \downarrow \\
M_{-1}(K) & \xrightarrow{\partial^\pi_s} & M_{-2}(\kappa)
\end{array}
\tag{5.13}
$$

where $\tau' : H^1_z(Spec(B); M_{-1}) \to M_{-2}(\kappa)$ is induced precisely by evaluating the morphism $\mathbb{P}^1_\kappa \to Th(j)$ on $H^1(-; M_{-1})$. Now the diagram (5.12) gives us at the top, a cofibration sequence of the form:

$$
\mathbb{P}^1_\kappa - Spec(\lambda) \to \mathbb{P}^1_\kappa \to Th(j)
$$

or in other words an equivalence $Th(i) \cong Th(j)$ where $i : Spec(\lambda) \subset \mathbb{P}^1_\kappa$ is the closed immersion defined by $\overline{x} \in \lambda$. Thus τ' is the composition

$$
H^1_z(Spec(B); M_{-1}) \overset{\Psi}{\cong} H^1_z(\mathbb{P}^1_\kappa; M_{-1}) \overset{(\theta_Q)^{-1}}{\cong} M_{-2}(\lambda) \xrightarrow{\tau^\lambda_\kappa(\overline{x})} M_{-2}(\kappa)
$$

where the isomorphism Ψ is induced by the equivalence $Th(i) \cong Th(j)$. It follows that we may fill up the diagram (5.13) as follows, so that it remains a commutative diagram of the form:

$$
\begin{array}{ccccc}
M_{-1}(L) & \xrightarrow{\partial_z} & H^1_z(Spec(B); M_{-1}) & \overset{(\theta_t)^{-1}}{\cong} & M_{-2}(\lambda) \\
\tau^L_K(x) \downarrow & & \tau' \downarrow & & \tau'' \downarrow \\
M_{-1}(K) & \xrightarrow{\partial^\pi_s} & M_{-2}(\kappa) & = & M_{-2}(\kappa)
\end{array}
$$

where τ'' is the composition

$$
M_{-2}(\lambda) \overset{\theta_t}{\cong} H^1_z(Spec(B); M_{-1}) \overset{\Psi}{\cong} H^1_z(\mathbb{P}^1_\kappa; M_{-1}) \overset{(\theta_Q)^{-1}}{\cong} M_{-2}(\lambda) \xrightarrow{\tau^\lambda_\kappa(\overline{x})} M_{-2}(\kappa)
$$

As the top horizontal composition is by definition ∂^t_z, to conclude that the diagram (5.10) commutes, and to finish the proof of the theorem, it suffices to prove that we have $\tau'' = \tau$.

A moment of reflection, observing the diagram (5.12), shows that the following diagram of isomorphisms commutes:

$$
\begin{array}{ccccc}
M_{-2}(\lambda) & \overset{\theta_t}{\cong} & H^1_z(Spec(B); M_{-1}) & \overset{\Phi_P}{\cong} & H^2_z(\mathbb{P}^1_S; M) \\
& & \Psi \downarrow \wr & & \| \\
M_{-2}(\lambda) & \overset{\theta_Q}{\cong} & H^1_z(\mathbb{P}^1_\kappa; M_{-1}) & \overset{\Phi_\pi}{\cong} & H^2_z(\mathbb{P}^1_S; M)
\end{array}
$$

Using the notations introduced at the end of Sect. 4.1, we see that the top vertical composition $\Phi_P \circ \theta_t$ is $\Phi_{P,t}$ and the bottom vertical composition is $\Phi_{\pi,Q}$. Applying Corollary 4.23 and the previous formula $(P \wedge t) = \epsilon(\pi \wedge Q)$, we see that the previous diagram can be filled up in a commutative diagram as follows

$$
\begin{array}{ccccc}
M_{-2}(\lambda) & \overset{\theta_t}{\simeq} & H_z^1(Spec(B); M_{-1}) & \overset{\Phi_P}{\simeq} & H_z^2(\mathbb{P}_S^1; M) \\
\epsilon \downarrow \wr & & \Psi \downarrow \wr & & \| \\
M_{-2}(\lambda) & \overset{\theta_Q}{\simeq} & H_z^1(\mathbb{P}_\kappa^1; M_{-1}) & \overset{\Phi_\pi}{\simeq} & H_z^2(\mathbb{P}_S^1; M)
\end{array}
$$

and conclude that τ'' is the composition

$$
M_{-2}(\lambda) \overset{\epsilon}{\simeq} M_{-2}(\lambda) \overset{\tau_\kappa^\lambda(\overline{x})}{\longrightarrow} M_{-2}(\kappa)
$$

which implies the claim by the formulas (5.11). The theorem is proven. □

Lemma 5.23. *Let $S = Spec(A)$ be the spectrum of a essentially k-smooth discrete valuation ring. Then the projection morphism $f : \mathbb{P}_S^1 \to S$ together with its canonical orientation induces a morphism of Gersten complexes:*

$$
f_*^{\omega(\mathbb{P}_K^1 | K)} : C^*(\mathbb{P}_S^1; M) \to C^{*-1}(S; M_{-1})
$$

Proof. The claim for the differential $C^0 \to C^1$ follows directly from Lemma 5.18 applied to K being the field of fractions of A. Now an irreducible closed scheme of codimension 1 in $\mathbb{P}_S^1$ is either finite and surjective over S or is equal to $\mathbb{P}_\kappa^1 \subset \mathbb{P}_S^1$, where $Spec(\kappa) \subset S$ is the closed point of S. To check the claim for the differential starting from $M_{-1}(\kappa(T); \omega(\kappa(T)) \otimes \omega(\mathbb{P}_S^1)^{-1}) \subset C^1(\mathbb{P}_S^1; M)$ is again a clear consequence of the Lemma 5.18 applied to the residue field κ of S.

Now it remains to check the case of an irreducible closed scheme $Spec(B) = Y \subset \mathbb{P}_S^1$ which is finite over S. We let $z : Spec(\lambda) \subset Y$ be the closed point of Y and L be the field of fractions of B. We construct the analogous diagram as (5.12) except that as $Spec(B)$ is not assumed to be essentially k-smooth. We thus write $Th(i)$ for the cone of the morphism $\mathbb{P}_S^1 - Spec(B) \subset \mathbb{P}_S^1$ instead of $T \wedge Spec(B)_+$; $Th(i)$ is the Thom space of the closed immersion $i : Spec(B) \subset \mathbb{P}_S^1$. We write $Th(i_K)$ for the Thom space of $i_K : Spec(L) \subset \mathbb{P}_K^1$, and $Th(i_\kappa)$ for the Thom space of $i_\kappa : Spec(\lambda) \subset \mathbb{P}_\kappa^1$. We get in this way a commutative diagram:

$$
\begin{array}{ccccc}
(\mathbb{P}_S^1 - Spec(B))/(\mathbb{P}_K^1 - Spec(L)) & \to & \mathbb{P}_S^1/\mathbb{P}_K^1 & \to & \mathbb{P}_S^1/(\mathbb{P}_S^1 - \{z\}) \\
\uparrow & & \uparrow & & \uparrow \\
\mathbb{P}_S^1 - Spec(B) & \subset & \mathbb{P}_S^1 & \to & Th(i) \\
\cup & & \cup & & \uparrow \\
\mathbb{P}_K^1 - Spec(L) & \subset & \mathbb{P}_K^1 & \to & Th(i_K)
\end{array}
\qquad (5.14)
$$

Using the functoriality of the long exact sequences in cohomology of the two right vertical cofibration sequences we get the following commutative square:

$$\begin{array}{ccccc}
\Phi_y : M_{-1}(L; \omega(L) \otimes \omega(\mathbb{P}^1_S)^{-1}) & \cong & H^1_y(\mathbb{P}^1_S; M) & \xrightarrow{\partial_z} & H^2_z(\mathbb{P}^1_S; M) \\
& & \tau^L_K \downarrow & & \tau^\lambda_\kappa \downarrow \\
\gamma : M_{-1}(K) & & \cong H^1(\mathbb{P}^1_K; M) & \xrightarrow{\partial_s} & H^2_{\mathbb{P}^1_\kappa}(\mathbb{P}^1_S; M)
\end{array} \qquad (5.15)$$

where s is the closed point of S. Now we claim that through the canonical isomorphisms written in the diagram τ^L_K becomes exactly the convenient twist of the absolute transfer (use the bottom horizontal cofibration sequence of (eq:diagimp1) and that τ^λ_κ becomes through the further isomorphisms Φ_z^{-1} $H^2_z(\mathbb{P}^1_S; M) \cong M_{-2}(\lambda; \omega(\lambda) \otimes \omega(\mathbb{P}^1_S)^{-1})$ and $\phi_\pi^{-1} : H^2_{\mathbb{P}^1_\kappa}(\mathbb{P}^1_S; M) \cong H^1(\mathbb{P}^1_\kappa; M_{-1})$ the convenient twist of the absolute transfer (use a uniformizing element π of A to identify the top horizontal cofibration sequence of (5.14) into $T \wedge (\mathbb{P}^1_\kappa - Spec(\lambda))_+ \to T \wedge (\mathbb{P}^1_\kappa)_+ \to T \wedge Th(i_\kappa))$. $\square$

The following lemma, or rather its proof, contains the geometric explanation for the fact that the Rost–Schmid differential is (at the end) the differential in the Gersten complex:

Lemma 5.24. *Let X be an essentially smooth k-scheme of dimension 2 and let $z \in X$ be a closed point. Let $f : \tilde{X} \to X$ be the blow-up of X at z. Then f induces a morphism of Gersten complexes:*

$$f_* : C^*(\tilde{X}; \nu(f); M) \to C^*(X; M)$$

Proof. We note that $\nu(f)$ is always canonically isomorphic to the line bundle $\mathcal{O}(1)$ corresponding to the divisor $\mathbb{P}^1_\kappa \subset \tilde{X}$. We will freely use this in the proof below. We may assume that X is local with closed point z. Let K be the function field of X and of $\tilde{X}$, κ be the residue field of z. We let $\eta \in X_z^{(1)}$ be the generic point of the exceptional curve $\mathbb{P}^1_\kappa \subset X_z$. We observe that $f_* : C^*(X_z; \nu(f); M) \to C^{*-1}(X; M)$ is an isomorphism in degree 0 and moreover that $\tilde{X}^{(1)} = X^{(1)} \amalg \{\eta\}$. As for $y \in X^{(1)}$ the corresponding point in X_z has the same residue field we conclude that the statement holds for the differential $C^0 \to C^1$. The morphism f_* is zero by definition on the summand $M_{-1}(\kappa(T); \omega(\kappa(T)) \otimes \omega(X)^{-1}) = M_{-1}(\kappa(T); \omega(\kappa(T)))$ (observe that the restriction of $\omega(X)$ to $\mathbb{P}^1_\kappa$ is trivial) of $C^1(\tilde{X}; \nu(f); M)$ corresponding to η. Using the Lemma 5.18, one checks the statement for the differential $C^1 \to C^2$ starting from this summand (we let the reader check that the twists fit well!). Thus it remains to check the commutativity for the differentials starting from a point $\tilde{y}_1 \in \tilde{X}^{(1)} - \{\eta\}$.

Let $\tilde{y}_1$ be such a point, $\tilde{Y}_1 \subset \tilde{X}$ be the corresponding integral closed subscheme of $\tilde{X}$, y_1 the point $f(\tilde{y}_1)$ and $Y_1 = \overline{y_1} \subset X$ the corresponding integral closed subscheme of X; y_1 and $\tilde{y}_1$ have the same residue field κ_1 ($\tilde{y}_1$ is the proper transform of Y_1) and the morphism

$$f_* : M_{-1}(\kappa(\tilde{y}_1); \omega(\kappa(\tilde{y}_1)) \otimes \omega(\tilde{X})^{-1} \otimes \nu(f)) \cong M_{-1}(\kappa(y_1); \omega(\kappa(y_1)) \otimes \omega(X)^{-1})$$

is an isomorphism as well. This corresponds to the fact that $\nu(f)$ is trivialized outside $\mathbb{P}_k^1$ by the previous remark. Let $m \in M_{-1}(\kappa_1; \omega(\kappa_1) \otimes \omega(X)^{-1})$ be an arbitrary element.

By construction $\tilde{Y}_1 \subset \tilde{X}$ is the proper transform of Y_1. We let $\{z_1, \ldots, z_r\}$ be the closed points of $\tilde{Y}_1$, κ_i their residue fields. The z_i's are precisely the closed points appearing in the intersection of $\tilde{Y}$ and the exceptional curve $\mathbb{P}_\kappa^1$.

Let $Y_0 \subset X$ be an essentially k-smooth closed subscheme of codimension 1. Its proper transform $\tilde{Y}_0 \subset \tilde{X}$ is isomorphic to Y_0 itself through the projection $\tilde{Y}_0 \to Y_0$. We will write $z_0 \in \tilde{X}$ for the closed point of $\tilde{Y}_0$, considered in $\tilde{X}$. We write $\tilde{y}_0$ for the generic point of $\tilde{Y}_0$. We may choose Y_0 so that $z_0 \notin \{z_1, \ldots, z_r\}$. (At least if the residue field κ is infinite. In case k is finite, one may pull everything back over $k(T)$, and it suffices to prove the result there.)

By Corollary 4.12 applied to the smooth subscheme $Y_0 \subset X$ the canonical morphism

$$\sum_{y \in X^{(1)} - \{y_0\}} \partial_y : M(K) \to \oplus_{y \in X^{(1)} - \{y_0\}} H_y^1(X; M)$$

is onto. Thus we may find $\alpha \in M(K)$ such that $\partial_y(\alpha) = 0 \in H_y^1(X; M)$ for each $y \in X^{(1)} - \{y_0, y_1\}$ and such that $\partial_{y_1}(\alpha) = m$. We set $m_0 = \partial_{y_0}(\alpha) \in H_{y_0}^1(X; M)$. As the Gersten complex for X is a complex we have

$$\partial_z^y(m) + \partial_z^{y_0}(m_0) \in H_z^2(X; M)$$

Now we see $\alpha \in M(K)$ as defined over the function field of $\tilde{X}$. In that case, for any $y \in \tilde{X}^{(1)}$ different from $\tilde{y}_0$, $\tilde{y}_1$ and η we have $\partial_y(\alpha) = 0$. Write $\mu = \partial_\eta(\alpha) \in H_\eta^1(\tilde{X}; M)$. The closed points of $\tilde{X}$ are exactly the closed points of $\mathbb{P}_\kappa^1$. We apply now the fact that the Gersten complex for $\tilde{X}$ is a complex. The formula $\partial \circ \partial(\alpha) = 0$ gives on the summand corresponding to z_0 the equality in $M_{-2}(\kappa; \omega(\kappa) \otimes \omega(X)^{-1}) \cong H_z^2(X; M)$:

$$\partial_{z_0}^\eta(\mu) + \partial_{z_0}^{y_0}(m_0) = 0 \in H_z^2(X; M)$$

Note that $\partial_{z_0}^{\tilde{y}_0} = \partial_z^{y_0}$ as Y_0 is essentially k-smooth. From what we previously saw, we have $\partial_{z_0}^\eta(\mu) = \partial_z^{y_1}(m)$. By evaluating $\partial \circ \partial = 0$ at each z_i gives

$$\partial_{z_i}^\eta(\mu) + \partial_{z_i}^{\tilde{y}_1}(m) = 0 \in M_{-2}(\kappa_i; \omega(\kappa_i) \otimes \omega(X)^{-1})$$

By the reciprocity formula (Lemma 5.18) for $\mathbb{P}_\kappa^1$ and μ, conveniently twisted by ω^{-1}, we get in $M_{-2}(\kappa; \omega(\kappa) \otimes \omega(X)^{-1}) \cong H_z^2(X; M)$:

$$\partial_{z_0}^\eta(\mu) + \sum_{i=1}^r Tr_\kappa^{\kappa_i}(\omega \otimes \omega(X)^{-1})(\partial_{z_i}^\eta(\mu)) = 0$$

(Note that we used the twisted absolute Transfers defined in Remark 5.6 point 3).) Collecting all our previous equalities we get finally:

$$\partial_z^{y_1}(m) + \sum_{i=1}^r Tr_\kappa^{\kappa_i}(\omega \otimes \omega(X)^{-1})(-\partial_{z_i}^{\tilde{y}_1}(m)) = 0$$

which proves exactly our claim. The lemma is proven.   □

The next result follows by induction:

Corollary 5.25. *Let X be an essentially smooth k-scheme of dimension 2. Let $f : \tilde{X} \to X$ be a composition of finitely many blow-ups at closed points. Then*

$$f_* : C^*(\tilde{X}; \nu(f); M) \to C^*(X; M)$$

is a morphism of Gersten complex.

Now we may prove:

Theorem 5.26. *Let $f : Spec(B) \to Spec(A)$ be a finite morphism, with A and B being essentially k-smooth of dimension 1. Then*

$$f_* : C_{RS}^*(Spec(B); \nu(f); M_{-1}) \to C_{RS}^*(Spec(A); M_{-1})$$

is a morphism of quasi-complexes.

Proof. This results thus generalizes the Theorem 5.19. We make some general comments first. We let $K \subset L$ be the finite fields extension induced by $A \subset B$ of the fraction fields level. Given an intermediary extension $K \subset E \subset L$, let C be the integral closure of A in E. By the general functorial properties of the Transfers (for instance Lemma 5.5) it is clear that if we may prove the theorem for $Spec(C) \to Spec(A)$ and for $Spec(B) \to Spec(C)$ then we may prove the theorem for $Spec(B) \to Spec(A)$.

To check the property, one may reduce first to the case A is an henselian essentially k-smooth d.v.r. In that case B is automatically also an henselian essentially k-smooth d.v.r.

It follows from all what we said that we may always further reduce to proving the theorem in the case $K \subset L$ is monogenous (choose a filtration of $K \subset L$ by monogenous fields extensions).

Let κ be the residue field of A and λ be that of B. Let $x \in B$ be a generator of $L|K$ contained in B. Then $B_0 := A[x] \subset B$ is the image of the monomorphism $A[X]/P(X) \subset B$, where $P(X) \in A[X]$ is the minimal polynomial of x over K. Consider the closed immersion $Y_0 := Spec(B_0) \subset \mathbb{P}_S^1$ defined by x, where $S = Spec(A)$. Let y_0 be the closed point of Y_i (note that

B_0 is henselian) and λ_0 be its residue field. Let $\mathbb{X}_1 \to \mathbb{P}^1_S$ be the blow-up of $\mathbb{P}^1_S$ at y_0 and $Y_1 \subset \mathbb{X}_1$ be the proper transform of Y_0; $Y_1 \to Y_0$ is a proper rational morphism between schemes of dimension 1, it is thus finite. Y_1 is thus local henselian, and we may go on and blow-up its closed point y_1 to get $\mathbb{X}_2 \to \mathbb{X}_1$. We let Y_2 be the proper transform, etc.

We get in this way a sequence of blow-up at closed points

$$f : \mathbb{X}_r \to \cdots \to \mathbb{X}_2 \to \mathbb{X}_1 \to \mathbb{P}^1_S$$

and we let $Y_i \subset \mathbb{X}_i$ be the proper transform of Y_{i-1} in $\mathbb{X}_i$. For r big enough it is well-known that Y_r is essentially k-smooth (see Remark 5.27 below), thus equals $Spec(B)$. Now applying Corollary 5.25 we see that

$$f_* : C^*(\mathbb{X}_r; \nu(f); M) \to C^*(\mathbb{P}^1_S; M)$$

is a morphism of Gersten complexes. But as $Y_r = Spec(B)$ is essentially k-smooth, the Gersten differential $\partial_z^{y_r}$ is equal to the Rost–Schmid differential. The theorem follows. □

Remark 5.27. The fact that we used in the proof, that one may eventually resolve the singularity of a local k-scheme of dimension 1 embedded into an essentially k-smooth scheme of dimension 2 by finitely many blow-ups at closed points is well-known, however we hardly found a reference in this generality. In our case keeping the previous notations and setting $Y_i = Spec(B_i)$, we see an increasing sequence $A[x] = B_0 \subset B_1 \subset \cdots \subset B_i \subset B$ of intermediary finite (thus henselian) local rings and the claim is that for i big enough $B_i = B$. This follows indeed from the fact that by the universal property of Blow-ups the ideal $\mathcal{M}_{i-1}.B_i \subset B_i$ generated in B_i by the maximal ideal $\mathcal{M}_{i-1}$ of B_{i-1} is a free B_i-module of rank one, contained in $\mathcal{M}_i$. Thus if we set $B' := \cup_i B_i$, this is a integral domain, local ring of dimension 1, contained in B, whose maximal ideal is a free B'-module of rank one, that is to say B' is a d.v.r. Thus by maximality of d.v.r.'s it follows that $B' = B$ and this holds for i big enough because B is a finite type A-module. □

Remark 5.28. 1) The previous proof would be a bit simpler if one could prove that for any finite extension $K \subset L$, where K is the fraction field of an henselian essentially k-smooth d.v.r A, there exists a finite filtration $K \subset L_1 \subset \ldots L_r = L$ such that letting $B_i \subset L_i$ be the integral closure (also an henselian d.v.r.) each extension $B_{i-1} \subset B_i$ is monogenous. One could then apply Theorem 5.19 at each step to be done. In characteristic 0 it is true, see the next point below, however it is unclear to us whether this is true in finite characteristic.

2) Let $K \subset L$ be a finite extension, where K is the fraction field of an henselian essentially k-smooth d.v.r. $A \subset K$. Let $L_{unr} \subset L$ be the maximal unramified intermediary extension of K. If we let $B_{unr} \subset L_{unr}$ be the integral

closure of A the morphism $Spec(B_{unr}) \to Spec(A)$ is étale and finite. It is well-known that in that case the extension $A \subset B_{unr}$ is monogenous (see [71, Chap. I, Prop. 16]). Thus we might reduce the proof of the preceding theorem to the case where the A is equal to its maximal unramified extension by Theorem 5.19.

If the characteristic of k is 0, this means exactly that the residue field extension induced by $A \to B$ is an isomorphism, that is to say that the extension is totally ramified. By [71, Chap. I, Prop. 18] this is again a monogenous extension. The theorem again follows from Theorem 5.19. $\square$

Corollary 5.29. *Let X be an essentially smooth k-scheme of dimension 2. Then the isomorphism $\Theta : C^*_{RS}(X; M) \to C^*(X; M)$ of graded abelian groups of Remark 5.14 is an isomorphism of complexes. In particular $\partial_{RS} \circ \partial_{RS} = 0$.*

Proof. Given a closed point $z \in X$ and $Y \subset X$ a closed integral subscheme of codimension 1, we have to prove that the Gersten differential $\partial_z^{\tilde{y}}$ equals the Rost–Schmid differential. By the Remark 5.27 above there is a finite succession of blow-up at closed points $\tilde{X} \to X$ such that the proper transform $\tilde{Y}$ of Y is essentially k-smooth (over z). Let $z_1, \ldots, z_r$ be the closed points in $\tilde{Y}$ lying over z. On $\tilde{X}$ we know the equalities $\partial_{z_i}^{\tilde{y}} = \partial_{RS,z_i}^{\tilde{y}}$ because $\tilde{Y}$ is essentially k-smooth. By Corollary 5.25 and by the very definition of the Rost–Schmid differential we get the claim. $\square$

We may now prove more generally:

Corollary 5.30. *Assume $f : X' \to X$ is a finite morphism. Then the morphism $f_* : C^*_{RS}(X; \nu(f); M) \to C^*_{RS}(Y; M)$ is a morphism of quasi-complexes.*

Proof. Let y' be a point of X' of codimension $n-1$, with image $y = f(y')$ in X, and let z be a point of codimension n in X in $\overline{y}$. We have to prove that the obvious square

$$M_{-n+1}(\kappa(y'); \lambda_{y'}^{X'} \otimes \nu(f)) \xrightarrow{\partial_{RS}^{X'}} \oplus_{z_i \in X'^{(n)}} M_{-n}(\kappa(z_i); \lambda_{z_i}^{X'} \otimes \nu(f))$$
$$\downarrow \qquad\qquad\qquad\qquad\qquad\qquad \downarrow$$
$$M_{-n+1}(\kappa(y); \lambda_y^X) \xrightarrow{\partial_{RS}^X} \qquad\qquad M_{-n}(\kappa(z); \lambda_z^X)$$

is commutative, where the z_i's are the finitely many points lying over z. We may localize at z and assume that X is local with closed point z and X' semi-local with closed points the z_i's. We may also replace X by its henselization at z; as the fiber product $X' \times_X X^h$ is the disjoint unions of the henselizations of X' at each of the z_i, we see moreover that we may reduce to the case both X and X' are both local and henselian. We now write z' the only closed point of X' lying over z.

Let $\tilde{Y} \to Y$ be the normalization of $Y = \overline{y}$ and $\tilde{Y}' \to Y'$ that of $Y' = \overline{y'}$. We let $\tilde{z}$ and $\tilde{z}'$ be the corresponding closed points. Observe that both $\tilde{Y}$ and

$\tilde{Y}'$ are essentially k-smooth henselian local schemes of dimension 1. We may apply the Theorem 5.26 to the induced morphism $\tilde{f} : \tilde{Y}' \to \tilde{Y}$, and we get that the following diagram commutes (with obvious notations for the residue fields):

$$
\begin{array}{ccc}
M_{-n+1}(\kappa(y'); \lambda_{\tilde{z}'}^{\tilde{Y}'} \otimes \nu(\tilde{f})) & \overset{\partial_{RS}^{\tilde{Y}'}}{\longrightarrow} & M_{-n}(\tilde{\kappa}'; \lambda_{\tilde{z}'}^{\tilde{Y}'} \otimes \nu(\tilde{f})) \\
\downarrow & & \downarrow \\
M_{-n+1}(\kappa(y); \lambda_{y}^{\tilde{Y}}) & \overset{\partial_{RS}^{\tilde{Y}}}{\longrightarrow} & M_{-n}(\tilde{\kappa}; \lambda_{\tilde{z}}^{\tilde{Y}})
\end{array}
$$

The result now follows from this diagram conveniently twisted (by $\omega(\tilde{Y}) \otimes \omega(X)^{-1}$ and $\omega(\tilde{Y}') \otimes \omega(X')^{-1}$), the definition of the Rost–Schmid differential and the commutativity of the diagram of absolute transfers:

$$
\begin{array}{ccc}
M_{-n}(\tilde{\kappa}'; \omega(\tilde{\kappa}')) & \to & M_{-n}(\kappa'; \omega(\kappa')) \\
\downarrow & & \downarrow \\
M_{-n}(\tilde{\kappa}; \omega(\tilde{\kappa})) & \to & M_{-n}(\kappa; \omega(\kappa))
\end{array}
$$

conveniently twisted. The corollary is proven.   □

Finally we get:

Theorem 5.31. *Let X be an essentially smooth k-scheme, then the Rost–Schmid complex $C_{RS}^*(X; M)$ is a complex.*

Proof. Let $z \in X$ be a point of codimension n and let Y be an integral closed subscheme of codimension $n-2$ with generic point y. We want to prove that the composent of $\partial \circ \partial$ starting from the summand $M_{-n+2}(\kappa(y); \lambda_y^X)$ and arriving at the summand $M_{-n}(\kappa(z); \lambda_z^X)$ is zero. We may reduce to the case X is of finite type, affine and smooth over $\kappa = \kappa(z)$ and z is a closed point of codimension n in X. By the Normalization lemma, [72, Théorème 2 p. 57], there exists a finite morphism $X \to \mathbb{A}_\kappa^n$ such that z maps to 0 (with same residue field) and such that the image of Y is a linear $\mathbb{A}_\kappa^2 \subset \mathbb{A}_\kappa^n$. Using the Corollary 5.30, we may reduce in this way to the case $X = \mathbb{A}_\kappa^n$, $Y = \mathbb{A}_\kappa^2 \subset \mathbb{A}_\kappa^n$ and $z = 0$. This case follows now from Corollary 5.29 for $\mathbb{A}_\kappa^2$ (conveniently twisted).   □

Remark 5.32. It follows from Corollary 5.29 that for any X the Rost–Schmid complex coincide in degree ≤ 2 with the Gersten complex of M. Using the results of Sect. 2.2 this implies that the H^i of the Rost–Schmid complex is equal for $i \leq 1$ to $H^i(X; M)$.   □

We will also need later the following:

Corollary 5.33. *Assume f is the projection morphism $\mathbb{P}_X^1 \to X$ for some essentially k-smooth X with the canonical orientation $\omega = \omega(\mathbb{P}_X^1|X)$ of the above remark.*

Then the morphism $f_*^\omega : C_{RS}^*(\mathbb{P}_X^1; M) \to C_{RS}^{*-1}(X; M_{-1})$ *is a morphism of complexes.*

Proof. One proceeds in the same way as in the proof of Lemma 5.23, using the previous results. We leave the details to the reader.   □

Remark 5.34. More generally, following [68], one may establish the statement of the for any proper morphism $f : X' \to X$ the morphism

$$f_* : C_{RS}^*(X'; \nu(f); M) \to C_{RS}^{*-d(f)}(X; M_{-d(f)})$$

commutes with the differential. We won't need the general case. We treat the case of a closed immersion in Lemma 5.35 below.   □

We finish this section by the following simple observation. given a closed immersion $i : Y \subset X$ between smooth k-schemes the pull-back morphism

$$j^* : C_{RS}^*(X; M) \to C_{RS}^*(U; M)$$

of Lemma 5.16 is onto. Its kernel is denoted by $C_{RS,Y}^*(X; M)$. Assume that the codimension of Y is everywhere d. The proof of the following lemma is a straightforward checking that we leave to the reader:

Lemma 5.35. *The push-forward morphism*

$$i_* : C_{RS}^*(Y; \nu(i); M_{-d}) \to C_{RS}^{*+d}(X; M_*)$$

is a morphism of complexes. It induces a canonical isomorphism of complexes:

$$C_{RS}^*(Y; \nu(i); M_{-d})[d] \cong C_{RS,Y}^*(X; M)$$

5.3 Gersten Complex *Versus* Rost–Schmid Complex

In this section M still denotes a fixed strongly $\mathbb{A}^1$-invariant sheaf of abelian groups.

Boundary morphism of a smooth divisor and multiplication by a unit. Following [68], given a closed immersion[1] $i : Y \subset X$ of codimension 1 between smooth k-schemes, we introduce a "boundary morphism" of the form

$$\partial_Y : C_{RS}^*(U; M) \to C_{RS}^*(Y; \nu(i); M_{-1})$$

---

[1] We will only treat the case of codimension 1, which is the one that we use below to check the homotopy invariance property.

with $U := X - Y$ as follows. Let $w \in U^{(n)}$ be a point of codimension n, $W \subset U$ be the corresponding integral closed subscheme and $\overline{W} \subset X$ its closure in X. Let $z_1, \ldots, z_r$ be the irreducible components of the closed complement $\overline{W} - W$, a closed subset of Y. These points also have codimension n in Y. On the summand $M_{-n}(\kappa(w); \lambda_w^U) = M_{-n}(\kappa(w); \lambda_w^X)$ we define ∂_Y to be the sum

$$\sum_i \partial_{z_i}^w : M_{-n}(\kappa(w); \lambda_y^X) \to \oplus_i M_{-n-1}(\kappa(z_i); \lambda_{z_i}^X) \cong \oplus_i M_{-n-1}(\kappa(z_i); \nu(i) \otimes \lambda_{z_i}^Y)$$

where we used the canonical isomorphism: $\lambda_{z_i}^X \cong \nu(i) \otimes \lambda_{z_i}^Y$.

Assume given an element $U \in \underline{\mathbf{K}}_1^{MW}(X)$, for instance an invertible function $u \in \mathcal{O}(X)^\times$ on X, we denote by

$$U. : C_{RS}^*(X; M) \to C_{RS}^*(X; M_{+1})$$

the morphism of graded abelian groups defined for $* > 0$ as follows. On each summand $M_{-*}(\kappa(y); \lambda_y^X)$ with $* \geq 1$ and y a point of codimension $*$, one takes the multiplication

$$U : M_{-*}(\kappa(y); \lambda_y^X) \to M_{-*+1}(\kappa(y); \lambda_y^X)$$

by the symbol $U \in \underline{\mathbf{K}}_1^{MW}$; this is not a morphism of complexes by Lemmas 5.10 and 5.5, we have instead the formula in each degree $* > 0$:

$$\partial(U.m) - \epsilon U.\partial(m) \tag{5.16}$$

We may now state the analogue of [68, Lemma 4.5]:

Lemma 5.36. *Let $g : X \to Y$ be a smooth morphism in Sm_k of relative dimension 1 and assume that $\sigma : Y \to X$ is a section of g and that $t \in \mathcal{O}(X)$ is a function defining $\sigma(Y)$ (that is to say t generate the sheaf of ideal $\mathcal{I}(Y)$ defining $\sigma(Y) \subset X$). Write $U := X - \sigma(Y)$ and let $\tilde{g} : U \to Y$ be the restriction of g to Y and let ∂_Y be the boundary morphism associated to $\sigma : Y \subset X$. Then:*

$$\partial_Y \circ \tilde{g}^* = 0 \quad and \quad \partial_Y \circ [t] \circ \tilde{g}^* = (Id_Y)_*$$

(the last equality should be considered only when it makes sense, that is for $* > 0$.)

Proof. The proof follows the line of *loc. cit.*. One reduces to checking the analogues of Axioms R3c and R3d, that is to say when Y is the spectrum of a field κ. One proceeds like in the proof of Lemma 5.10, by choosing an étale morphism $X \to \mathbb{A}_\kappa^1$ which defines y and induces t, and then using the morphism of cofibration sequences from $(X - y)_+ \to X_+ \to X/(X - y)$ to

$\mathbb{G}_m \wedge (X_+) \to \mathbb{A}^1 \wedge (X_+) \to T \wedge (X_+)$ induced by t. One observes to conclude that the composition $X/(X - y) \to T \wedge (X_+) \to T \wedge (Spec(\kappa)_+)$ is the canonical isomorphism induced by t. We leave the details to the reader. □

Remark 5.37. As it follows from Sect. 2.2 it is possible to adapt precisely the axioms for cycle modules of [68] to our context and to describe the category of strongly $\mathbb{A}^1$-invariant sheaves in terms of explicit data on fields in $\mathcal{F}_k$ like in *loc. cit.*. However, to write this down would make this work much longer, and in some sense we do not need this formulation at all, because given such a strongly $\mathbb{A}^1$-invariant sheaf M, we may check by hands the given axiom/property when we need it, each time we need it! This is what we did in the proof of the previous lemma and also what we will do in the proof of the homotopy invariance property below. □

The canonical homotopy of the Rost–Schmid complex. We are going to prove the following result:

Theorem 5.38. *Let $X \in Sm_k$ be a smooth k-scheme. Then the morphism of complexes defined by the projection $\pi : \mathbb{A}^1_X \to X$*

$$\pi^* : C^*_{RS}(X; M) \to C^*_{RS}(\mathbb{A}^1_X; M)$$

induces an isomorphism on cohomology groups.

The morphism π^* is the one from Lemma 5.16.

Remark 5.39. From Corollary 5.29 and Remark 5.32 we already know the result in degree ≤ 1. □

Proof of Theorem 5.38. Our method of proof is to use the formulas of [68, Sect. 9 (9.1)], conveniently adapted, to define an explicit homotopy of complexes. We proceed as follows.

We first define a morphism of complexes $r : C^*_{RS}(\mathbb{A}^1_X; M) \to C^*_{RS}(X; M)$ in degrees ≥ 1 as follows. Let $y \in (\mathbb{A}^1_X)^{(n)}$ be a point of codimension n, $n > 0$. Set $Y := \overline{y} \subset \mathbb{A}^1_X$ and $\overline{Y} \subset \mathbb{P}^1_X$ its closure in $\mathbb{P}^1_X$. We let Y_∞ be $Y_\infty := \overline{Y} \cap X_\infty$ the intersection of $\overline{Y}$ with the section at ∞. Let $z_1, \ldots, z_r$ be the generic points of Y_∞; the codimension of the z_i's in $X = X_\infty$ is n.

If $Y := \overline{y}$ is contained in the 0 section $X_0 \to \mathbb{A}^1_X$ the component of r starting from the summand $M_{-n}(\kappa(y); \lambda_y^{\mathbb{A}^1_X})$ (using the notations of Definition 5.7) is set to be 0.

If $Y := \overline{y}$ is not contained in the 0 section $X \to \mathbb{A}^1_X$, then the component

$$r^y_{z_i} : M_{-n}(\kappa(y); \lambda_y^{\mathbb{A}^1_X}) \to M_{-n}(\kappa(z_i); \lambda_{z_i}^X)$$

or r is defined as follows. One first takes the product (in the sense of Lemma 3.48) by the symbol $< -1 >^n \left[-\frac{1}{T} \right]$

$$M_{-n}(\kappa(y); \lambda_y^{\mathbb{A}^1_X}) \to M_{-n+1}(\kappa(y); \lambda_y^{\mathbb{A}^1_X})$$

(where T is as usual the section of $\mathcal{O}(1)$ which defines the divisor X_0. Observe that $-\frac{1}{T}$ then defines X_∞) and then composes with the component ∂^y_{RS,z_i}

$$M_{-n+1}(\kappa(y); \lambda_y^{\mathbb{A}^1_X}) = M_{-n+1}(\kappa(y); \lambda_y^{\mathbb{P}^1_X}) \xrightarrow{\partial^y_{RS,z_i}} M_{-n}(\kappa(z_i); \lambda_{z_i}^X)$$

Some comments. The explanation of the term $< -1>^n$ above (which doesn't exist of course in [68]) will appear below, in the computations of the homotopy. For the cup product with $-\frac{1}{T}$ in the case $n = 1$ observe that we reach $M(\kappa(y))$ which has no action of $\kappa(y)$. However in that case, Y is an irreducible hypersurface in $\mathbb{A}^1_X$ and the irreducible polynomial over the function field of X, F, of the function T on Y defines a canonical trivialization of $\lambda_y^{\mathbb{A}^1_X}$. Thus one may this way define the cup product. In general for $n \geq 2$ we use the fact that the pairing in Lemma 3.48 preserves the action of $\mathbf{GW}$. Finally in the last step, taking the differential morphism ∂^y_{RS,z_i}, we used the isomorphism

$$\lambda_y^{\mathbb{P}^1_X} \cong \omega(\kappa(y)) \otimes \omega(\mathbb{P}^1_X)^{-1} \cong \omega(\kappa(y)) \otimes \omega(X)^{-1} \otimes \mathcal{O}(-2)^{-1}$$

which allows (as $\mathcal{O}(-2)$ is a square) to reach the correct group.

We may also define r in degree 0, $r : M(F(T)) \to M(F)$ by observing that the uniformizing element at ∞ splits the short exact sequence

$$0 \to M(\mathcal{O}_{\mathbb{P}^1_X, X_\infty}) \subset M(F(T)) \xrightarrow{\partial^{\frac{1}{T}}_\infty} M_{-1}(F)$$

and then by taking $M(F(T)) \twoheadrightarrow M(\mathcal{O}_{\mathbb{P}^1_X, X_\infty}) \xrightarrow{s_\infty} M(K)$ (where s_∞ is just the restriction map). But we don't need this because the theorem is already known in degree 0 and 1.

We claim that r so defined commutes with the differentials in the Rost–Schmid complexes and that moreover is a section $\pi^* : C^*_{RS}(X; M) \to C^*_{RS}(\mathbb{A}^1_X; M)$. The last statement is easy. The first statement is checked by using the same formulas as in [68] conveniently adapted. But as π^* is injective, it also follows from the formula (5.17) below.

Now we define an explicit homotopy to the identity for the endomorphism $\pi^* \circ r$ of $C^*_{RS}(\mathbb{A}^1_X; M)$, by following precisely *loc. cit.*. This means that we are going to define for each $n \geq 2$ a group homomorphism $h_n : C^n_{RS}(\mathbb{A}^1_X; M) \to C^{n-1}_{RS}(\mathbb{A}^1_X; M)$ satisfying:

$$\partial^{n-1}_{RS} \circ h_n + h_{n+1} \circ \partial^{n+1}_{RS} = Id - \pi^* \circ r \tag{5.17}$$

This implies as usual the statement of the theorem in degree ≥ 2. We note that by the Remark 5.39 we already know the statement of the theorem in degree ≤ 1. The reader might check the definition below, and convince himself that there is no way to define the homotopy h_1 in general, as one would need some kind of transfers defined on M (see below).

Let y be a point of codimension $n \geq 2$ in $\mathbb{A}^1_X$. We write again $Y \subset \mathbb{A}^1_X$ for the corresponding closed integral subscheme whose generic point is y. There are two cases that may occur. Let y' be the image of y in X. Either y' has codimension n or $n-1$. In the second case, the induced residue field extension $\kappa(y') \subset \kappa(y)$ is finite.

If we are in the first case, then the homomorphism h_n on the summand $M_{-n}(\kappa(y); \lambda_y^{\mathbb{A}^1_X})$ of $C_{RS}^n(X; M)$ is defined to be zero. Now assume that we are in the second case. We write T for the canonical function on $\mathbb{A}^1_X$ and (S, T) for the two functions on $\mathbb{A}^1 \times \mathbb{A}^1_X$. We denote $y[S]$ the generic point in $\mathbb{A}^1 \times \mathbb{A}^1_X$ corresponding to the integral closed subscheme $\mathbb{A}^1 \times Y$, pull-back through the projection $p_2 : \mathbb{A}^1 \times \mathbb{A}^1_X \to \mathbb{A}^1_X$ which forget the variable S. Observe that the image of $y[S] \in \mathbb{A}^1 \times \mathbb{A}^1_X$ in $\mathbb{A}^1_X$ through the projection $p_1 : \mathbb{A}^1 \times \mathbb{A}^1_X \to \mathbb{A}^1_X$ which forgets T is the generic point $y'[S]$ of the integral closed subscheme $\mathbb{A}^1_{\overline{y'}} \subset \mathbb{A}^1_X$.

The component of h_n on $M_{-n}(\kappa(y); \lambda_y^{\mathbb{A}^1_X})$ is the morphism

$$h_{n,y} : M_{-n}(\kappa(y); \lambda_y^{\mathbb{A}^1_X}) \to M_{-n+1}(\kappa(y')(S); \lambda_{y'[S]}^{\mathbb{A}^1_X})$$

arriving into the summand of $C_{RS}^{n-1}(\mathbb{A}^1_X; M)$ corresponding to $y'[S]$ obtained as follows. First we take the pull-back map along the second projection $p_2 : \mathbb{A}^1 \times \mathbb{A}^1_X \to \mathbb{A}^1_X$ (forgetting S)

$$M_{-n}(\kappa(y); \lambda_y^{\mathbb{A}^1_X}) \to M_{-n}(\kappa(y)(S); \lambda_{\mathbb{A}^1[y]}^{\mathbb{A}^1 \times \mathbb{A}^1_X})$$

followed by the product with the element $(< -1 >)^n [S - T]$ of $\mathbf{K}_1^{MW}$

$$< -1 > [S - T]. : M_{-n}(\kappa(y)(S); \lambda_{\mathbb{A}^1[y]}^{\mathbb{A}^1 \times \mathbb{A}^1_X}) \to M_{-n+1}(\kappa(y)(S); \lambda_{\mathbb{A}^1[y]}^{\mathbb{A}^1 \times \mathbb{A}^1_X})$$

and then we finally compose with the transfer morphism:

$$M_{-n+1}(\kappa(y)(S); \lambda_{\mathbb{A}^1[y]}^{\mathbb{A}^1 \times \mathbb{A}^1_X}) \to M_{-n+1}(\kappa(y'); \lambda_{y'[S]}^{\mathbb{A}^1_X})$$

Observe that as $n \geq 2$, $n - 1 \geq 1$ and that as the field extension $\kappa(y')(S) \subset \kappa(y)(S)$ is finite and monogenous one may also define the Transfer morphism in the limit case $n = 2$ as the geometric transfer conveniently twisted.

Note that the previous formulas are exactly the same formulas as in [68, Sect. 9.1, p. 371]. In general there is no way to define the homotopy h_1, as already observe, because there is no transfers morphism in general on M.

The proof that these morphisms $\{h_n\}_{n \geq 2}$ define a homotopy as claimed above has the same form as in [68]. Write $\delta(h)$ for $\partial \circ h_* + h_{*+1} \circ \partial$. Here $*$ has to be at least 2. We want to prove (5.17) that is to say the equality $\delta(h) = Id - \pi^* \circ r$. We first observe that the previous homomorphisms h_* can be decomposed in three steps, first as:

$$C_{RS}^*(\mathbb{A}_X^1; M) \xrightarrow{p_2^*} C_{RS}^*((\mathbb{A}^1 \times \mathbb{A}^1 - \Delta)_X; M)$$

(where Δ means the diagonal subscheme) this is a morphism of complexes by Lemma 5.16, followed by the morphism of graded abelian groups

$$[[S - T]]. : C_{RS}^*((\mathbb{A}^1 \times \mathbb{A}^1 - \Delta)_X; M) \to C_{RS}^*((\mathbb{A}^1 \times \mathbb{A}^1 - \Delta)_X; M_{+1})$$

which is in degree $*$ the multiplication by the element $< -1 >^* [S - T] \in \mathbf{K}_1^{MW}((\mathbb{A}^1 \times \mathbb{A}^1 - \Delta)_X)$. This is not a morphism of complexes but almost. We have the formula (5.16) which implies that $\partial([[S - T]].m) = -[[S - T]].\partial(m)$. (Observe also that it is only defined for $* \geq 2$; one may define it also for $* = 1$ as for the morphism r above, but we don't need it). Finally we compose with the push-forward morphism of Sect. 5.2 with respect to $p_1 : (\mathbb{A}^1 \times \mathbb{A}^1 - \Delta)_X \to \mathbb{A}_X^1$ (forgetting T):

$$p_{1*} : C_{RS}^*((\mathbb{A}^1 \times \mathbb{A}^1 - \Delta)_X; M_{+1}) \to C^{*-1}(\mathbb{A}_X^1; M)$$

which is not a morphism of complexes. As in [68] we compute $\delta(h)$ and find $\delta(h) = \tilde{\delta}(p_{1*}) \circ [S - T] \circ p_2^*$, where we set $\tilde{\delta}\phi = \partial\phi - \phi\partial$.

The term $\tilde{\delta}(p_{1*})$ is analyzed as in *loc. cit.* by writing p_{1*} as the composition

$$(\mathbb{A}^1 \times \mathbb{A}^1 - \Delta)_X \xrightarrow{q_*} \mathbb{A}^1 \times \mathbb{P}_X^1 \xrightarrow{\overline{p}_{1*}} \mathbb{A}_X^1$$

where $q : (\mathbb{A}^1 \times \mathbb{A}^1 - \Delta)_X \subset \mathbb{A}^1 \times \mathbb{P}_X^1$ is the obvious open immersion and $\overline{p}_1 : \mathbb{A}^1 \times \mathbb{P}_X^1 \to \mathbb{A}_X^1$ is the obvious projection. We use here the fact that the normal line bundle of $\mathbb{A}^1 \times \mathbb{P}_X^1 \to \mathbb{A}_X^1$ is a square and thus no twist appears in the push-forward morphism. By Corollary 5.33 $\overline{p}_{1*}$ is a morphism of complexes so that we get $\tilde{\delta}(p_{1*}) = \overline{p}_{1*} \circ \tilde{\delta}(q_*)$ and

$$\delta(h) = \overline{p}_{1*} \circ \tilde{\delta}(q_*) \circ [S - T] \circ p_2^*$$

Once this is done, the end of the proof goes exactly as in Rost's argument on page 372. The term $\tilde{\delta}(q_*)$ is the sum of two terms like in *loc. cit.*

$$\tilde{\delta}(q_*) = (i_\Delta)_* \circ \partial_\Delta + (i_\infty)_* \circ \partial_\infty$$

one coming from the divisor Δ and the other one from $\mathbb{A}^1 \times \infty$. One concludes using the analogue of Lemma 5.5 of *loc. cit.*, that is to say Lemma 5.36 above. The theorem is proven. □

Remark 5.40. There is another way to prove the Theorem 5.38 by induction on the dimension of X and by using the ideas of the next paragraph. That is to say by proving also Theorem 5.41 by induction at the same time. One then ends up analysing $C_{RS}^*(\mathbb{A}_X^1; M)$ using the short exact sequence (where U is he complement of the closed point z of X):

$$0 \to C_{RS,z}^*(\mathbb{A}_X^1; M) \to C_{RS}^*(\mathbb{A}_X^1; M) \twoheadrightarrow C_{RS}^*(\mathbb{A}_U^1; M) \to 0$$

and by proceeding as in the proof of Theorem 4.13. □

Acyclicity for local essentially smooth k-schemes. This refers to the following:

Theorem 5.41. *For any integer n, for any localization X of a point of codimension $\leq n$ in a smooth k-scheme, the diagram of abelian groups*

$$0 \to M(X) \to C_{RS}^0(X; M) \to \cdots \to C_{RS}^{n-1}(X; M) \to C_{RS}^n(X; M) \to 0$$

is an exact sequence.

Proof. We proceed by induction on n. The result holds for $n \leq 2$ by Corollary 5.29. We also know from Remark 5.32 the exactness of the diagram of abelian groups

$$0 \to M(X) \to C_{RS}^0(X; M) \to C_{RS}^1(X; M) \to C_{RS}^2(X; M)$$

for any n. Assume the result proven for localizations at each point of codimension $\leq (n-1)$ in a smooth k-scheme. Let X be the localization at point z of codimension n in a smooth k-scheme. Let $2 \leq i \leq n$ and let $\gamma \in C_{RS}^i(X; M)$ be an i-cycle of the Rost–Schmid complex for X. We have to prove that γ is a boundary.

Write $\gamma = \alpha_1 + \cdots + \alpha_r$ as a finite sum with $\alpha_j \in M_{-i}(\kappa(y_j); \lambda_{y_j}^X)$, where the y_j's are points of X of codimension i.

By Gabber's presentation Lemma 6.6 one may choose an étale morphism $f : X \to \mathbb{A}_S^1$, such that S is the localization at a point of codimension $n-1$ in an affine space and such that the compositions $\overline{y_j} \subset X \to \mathbb{A}_S^1$ are closed immersions $\overline{y_j} \subset \mathbb{A}_S^1$ as well. As automatically the composition $Spec(\kappa(z)) \subset \mathbb{A}_S^1$ is a closed immersion this means that $X \to \mathbb{A}_S^1$ is a Nisnevich neighborhood of z.

We consider the morphism of complexes $f^* : C*_{RS}(\mathbb{A}_S^1; M) \to C^*(X; M)$ induced by f and Lemma 5.16. Now clearly, by the very construction, f^* is an isomorphism on each summand of the Rost–Schmid complex corresponding to one of the point y_j (and their images y'_j in $\mathbb{A}^1_S$), so that γ is equal to $f^*(\gamma')$ where $\gamma' = \alpha'_1 + \cdots + \alpha'_r$ is the "same" sum but viewed in the corresponding summand of $C^*_{RS}(\mathbb{A}^1_S; M)$. Moreover as f^* is a morphism of complexes, one sees that γ' is also a cycle in $C^*_{RS}(\mathbb{A}^1_S; M)$ (as the $\overline{y_j} \subset X \to \mathbb{A}^1_S$ are closed immersions.

By Theorem 5.38, $\pi^* : C^*_{RS}(S; M) \to C^*_{RS}(\mathbb{A}^1_S; M)$ is a quasi-isomorphism. By inductive assumption, $C^*_{RS}(S; M)$ is a resolution of $M(S)$. Thus $C^*_{RS}(\mathbb{A}^1_S; M)$ is acyclic in degree ≥ 1. Then γ' is a boundary in $C^*_{RS}(\mathbb{A}^1_S; M)$, and using the morphism of complexes f^*, we see that its image γ is a boundary. The theorem is proven. $\square$

To use our previous result, we will need the following classical result, see [19] for instance:

Lemma 5.42. *Given $X \in Sm_k$, the presheaves on $(X)_{Nis}$ of the form $U \mapsto C^n_{RS}(U; M)$, where $(X)_{Nis}$ is the small Nisnevich site of X, that is to say the category of étale morphism $U \to X$ with the Nisnevich topology, are sheaves in the Nisnevich topology. These sheaves are moreover acyclic in the Zariski and the Nisnevich topology, that is to say they satisfy for any U:*

$$H^*_{Zar}(U; C^n_{RS}(U; M)) = H^*_{Nis}(U; C^n_{RS}(U; M)) = 0$$

for $* > 0$.

Proof. The statement concerning the Zariski topology is clear: these are clearly sheaves which are flasque in the usual sense [26]. Thus their cohomology on any U is trivial in the Zariski topology by *loc. cit.*.

Now let us treat the case of the Nisnevich topology. We follow [19, Proof of Theorem 8.3.1]. Let $y_0 \in X$ be a point of codimension n in X, let κ_0 be its residue field, and let $M : (FS/\kappa_0)^{op} \to \mathcal{A}b$ be a presheaf of abelian groups on the category FS/κ_0 of finite separable field extensions of κ_0. Observe that M is the same thing as to give a sheaf in the Nisnevich topology on the small Nisnevich site $(Spec(\kappa_0))_{Nis}$. The correspondence:

$$(y_0)_* M :$$

$$(X)^{op}_{Nis} \ni U \mapsto \oplus_{y \in U^{(n)}, y \mapsto y_0} M(\kappa(y))$$

where $(X)_{Nis}$ is the small Nisnevich site of X that is to say the category of étale X-schemes $U \to X$, and the y's run over the set of points in U of codimension n is easily checked to be the presheaf $(y_0)_* M$, direct image of M through the morphism of schemes $y_0 : Spec(\kappa_0) \to X$, this presheaf is thus a sheaf in the Nisnevich topology.

The presheaves of the form $U \mapsto C^n_{RS}(U; M)$ are, by definition, the direct sum over the set of points y of X of codimension n, of the presheaves of the form $(y)_*(M_{-n}(-; \lambda^X_y)$ (with obvious notations). Thus $U \mapsto C^n_{RS}(U; M)$ is a sheaf in the Nisnevich topology and to prove the acyclicity property we have now to prove that $H^*(X; (y_0)_*M) = 0$ for $* > 0$, with our previous notations. This follows from the use of the spectral sequence of Grothendieck

$$H^p(X; R^q((y_0)_*M)) \Rightarrow H^{p+q}(X; R(y_0)_*M) \cong H^{p+q}(Spec(\kappa_0); M)$$

and the fact that the functor $M \mapsto (y_0)_*M$ is clearly exact.   □

Corollary 5.43. *For any essentially smooth k-scheme X, the Rost–Schmid complex is an acyclic resolution on X_{Nis} of M both in the Zariski and the Nisnevich topology. Consequently for any strongly $\mathbb{A}^1$-invariant sheaf of abelian groups M, one gets canonical isomorphisms*

$$H^*(C_{RS}^*(X; M) \cong H_{Zar}^*(X; M) \cong H_{Nis}^*(X; M)$$

This is a clear consequence of the preceding lemma and of Theorem 5.41. The following result claims that the Gersten complex and the Rost–Schmid complex are the "same":

Corollary 5.44. *For any essentially smooth k-scheme X, there is a canonical isomorphism of complexes*

$$\Theta : C_{RS}^*(X; M) \cong C^*(X; M)$$

which is natural (in the obvious way) with pull-back morphisms through smooth morphisms.

Proof. Let X be a smooth k-scheme. Recall from Sect. 4.1 and from [12], that the Gersten complex $C^*(X; M)$ is the line $q = 0$ of the E_1-term of the coniveau spectral sequence for X with coefficients in M

$$E_1^{p,q} = \oplus_{x \in X^{(p)}} H_x^{p+q}(X; M) \Rightarrow H_{Nis}^{p+q}(X; M)$$

Now looking at the definition of the spectral sequence in [12, 19], we may construct it by choosing a resolution $M \to \mathcal{I}^*$ of M by acyclic sheaves on X_{Nis}. We take of course the Rost–Schmid complex $M(-) \to C_{RS}^*(-; M)$. A careful checking of the definition shows that the Term $H_x^{p+q}(X; M)$, for x of codimension p, computed using the Rost–Schmid complex is canonically isomorphic to $M_{-p}(\kappa(x); \lambda_x^X)$ for $q = 0$ and 0 for $q \neq 0$. This comes from the fact that the if X is local of dimension n with closed point z, the kernel of the epimorphism $C_{RS}^*(X; M) \twoheadrightarrow C_{RS}^*(X - z; M)$ is canonically $M_{-n}(\kappa(z); \lambda_z^X)$.

A moment of reflection identifies the differential of the Gersten complex to exactly that of the Rost–Schmid complex.   □

The following is one of our most important results in the present work:

Corollary 5.45. *Any strongly $\mathbb{A}^1$-invariant sheaf of abelian groups M is strictly $\mathbb{A}^1$-invariant.*

Proof. This follows directly from Corollary 5.43 and Theorem 5.38.   □

We may then reformulate all of this in the following way, stated as Theorem 1.16 in he introduction:

Theorem 5.46. *Let M be a sheaf of abelian groups on Sm_k. Then the following conditions are equivalent:*

1) M is strongly $\mathbb{A}^1$-invariant;
2) M is strictly $\mathbb{A}^1$-invariant.

Proof. The implication 1) $\Rightarrow$ 2) is Corollary 5.45 and the converse implication is trivial.   □

Cohomological interpretation of the Chow group of oriented cycles.

Theorem 5.47. *Let X be a smooth k-scheme and let $\tilde{CH}^*(X)$ be the graded Chow group of oriented cycles defined in [7]. Let $n \geq 0$. Then there exist canonical isomorphisms*

$$H^n(X; \underline{\mathbf{K}}_n^{MW}) \cong H^n(X; \underline{\mathbf{I}}^n \times_{\underline{\mathbf{i}}^n} \underline{\mathbf{K}}_n^M) \cong \tilde{CH}^*(X)$$

Here we used the notation of [53] and [54]. The cohomology groups can be computed using either Zariski or Nisnevich topology.

Proof. One may write down explicitly the Rost–Schmid complex for both sheaves $\underline{\mathbf{K}}_n^{MW}$ and $\underline{\mathbf{I}}^n \times_{\underline{\mathbf{i}}^n} \underline{\mathbf{K}}_n^M$. Now one has $\underline{\mathbf{K}}_{n-1}^{MW} \cong (\underline{\mathbf{K}}_n^{MW})_{-1}$ and $\underline{\mathbf{I}}^{n-1} \times_{\underline{\mathbf{i}}^{n-1}} \underline{\mathbf{K}}_{n-1}^M \cong (\underline{\mathbf{I}}^n \times_{\underline{\mathbf{i}}^n} \underline{\mathbf{K}}_n^M)_{-1}$. Now the canonical morphism of sheaves $\underline{\mathbf{K}}_n^{MW} \twoheadrightarrow \underline{\mathbf{I}}^n \times_{\underline{\mathbf{i}}^n} \underline{\mathbf{K}}_n^M$, and the previous formulas, show that this morphism induces an isomorphism by taking the construction $(-)_{-i}$ for $i \geq n$ and an epimorphism for $i = n-1$ (because $\underline{\mathbf{K}}_1^{MW} = (\underline{\mathbf{K}}_n^{MW})_{-n+1}$ is generated by the symbols $[u]$, for u a unit). We conclude that the morphism of chain complexes

$$C_{RS}^*(X; \underline{\mathbf{K}}_n^{MW}) \to C_{RS}^*(X; \underline{\mathbf{I}}^n \times_{\underline{\mathbf{i}}^n} \underline{\mathbf{K}}_n^M)$$

induces an isomorphism in $* \geq n$ and an epimorphism in $* = n - 1$. This implies, by Corollary 5.43 that $H^n(X; \underline{\mathbf{K}}_n^{MW}) \cong H^n(X; \underline{\mathbf{I}}^n \times_{\underline{\mathbf{i}}^n} \underline{\mathbf{K}}_n^M)$ is an isomorphism, and that these groups maybe computed using the Rost–Schmid complex. Moreover, clearly the n-th cohomology group of the complex $C_{RS}^*(X; \underline{\mathbf{I}}^n \times_{\underline{\mathbf{i}}^n} \underline{\mathbf{K}}_n^M)$ is by definition the oriented n-th Chow group defined in [7].   □

Remark 5.48. 1) Using the product structure on Milnor–Witt K-theory

$$\underline{\mathbf{K}}_n^{MW} \times \underline{\mathbf{K}}_m^{MW} \to \underline{\mathbf{K}}_{n+m}^{MW}$$

one may deduce from the previous interpretation a graded commutative ring structure on $\tilde{CH}^*(X)$; this is most probably the one constructed in [24].

2) By [54] we know that $\underline{\mathbf{K}}_n^{MW} \twoheadrightarrow \mathbf{I}^n \times_{\underline{\mathbf{i}}^n} \underline{\mathbf{K}}_n^M$ is in fact an isomorphism of sheaves (because it is a morphism of strictly $\mathbb{A}^1$-invariant sheaves which induces an isomorphism on fields), but this is non trivial as it uses the Milnor conjectures. Our proof of Theorem 5.47 doesn't use these, and is completely elementary. □

Chapter 6
$\mathbb{A}^1$-Homotopy Sheaves and $\mathbb{A}^1$-Homology Sheaves

In this section we assume that the reader is (very) comfortable with [59]. We will freely use the basic notions and some of the results.

6.1 Strong $\mathbb{A}^1$-Invariance of the Sheaves $\pi_n^{\mathbb{A}^1}$, $n \geq 1$

Our aim in this section is to prove:

Theorem 6.1. *For any pointed space $\mathcal{B}$, its $\mathbb{A}^1$-fundamental sheaf of groups $\pi_1^{\mathbb{A}^1}(\mathcal{B})$ is strongly $\mathbb{A}^1$-invariant.*

To prove this theorem, we will prove by hand that the sheaf $\mathcal{G} := \pi_1^{\mathbb{A}^1}(\mathcal{B})$ is unramified and satisfies the assumption of Theorem 2.27.

We get the following important corollary of Theorems 6.1 and 5.46, which was stated as Theorem 1.9 in the introduction:

Corollary 6.2. *For any pointed space $\mathcal{B}$, and any integer $n \geq 1$, the sheaf of groups $\pi_1^{\mathbb{A}^1}(\mathcal{B})$ is strongly $\mathbb{A}^1$-invariant and for $n \geq 2$ the sheaf of abelian groups $\pi_n^{\mathbb{A}^1}(\mathcal{B})$ is strictly $\mathbb{A}^1$-invariant.*

Proof. Apply the theorem to the $(n-1)$-th iterated simplicial loop space $\Omega_s^{(n-1)}(\mathcal{B})$ of $\mathcal{B}$, which is still $\mathbb{A}^1$-local. $\qquad\square$

We also deduce the following characterization of connected pointed $\mathbb{A}^1$-local spaces:

Corollary 6.3. *For any pointed simplicially connected space $\mathcal{B}$, the following conditions are equivalent:*

1) *The space $\mathcal{B}$ is $\mathbb{A}^1$-local.*
2) *The sheaf of groups $\pi_1(\mathcal{B})$ is strongly $\mathbb{A}^1$-invariant and for any integer $n \geq 2$, the n-th simplicial homotopy sheaf of groups $\pi_n(\mathcal{B})$ is strictly $\mathbb{A}^1$-invariant.*

F. Morel, $\mathbb{A}^1$-*Algebraic Topology over a Field*, Lecture Notes in Mathematics 2052,   149
DOI 10.1007/978-3-642-29514-0_6, © Springer-Verlag Berlin Heidelberg 2012

Proof. The implication 1) $\Rightarrow$ 2) has just been proven. The other implication follows from the use of the Postnikov tower of $\mathcal{B}$ and the fact that $K(M, n)$ is $\mathbb{A}^1$-local if M is strictly $\mathbb{A}^1$-invariant. $\square$

Remark 6.4. The Theorem 6.1 holds over any field, however as Theorem 5.46 only holds at the moment over a perfect field, we can only conclude that Corollaries 6.2 and 6.3 hold over a perfect field. We believe they hold over any field. By refining a bit the method of Sect. 2.2 it is possible to prove over any field that for $n \geq 2$ the sheaf of abelian groups $\pi_n^{\mathbb{A}^1}(\mathcal{B})$ is n-strongly $\mathbb{A}^1$-invariant in the following sense: a sheaf of abelian groups M is said to be n-strongly $\mathbb{A}^1$-invariant if the presheaves $X \mapsto H^i(X; M)$ are $\mathbb{A}^1$-invariant for $i \leq n$. $\square$

We now start the proof of Theorem 6.1 with some remarks and preliminaries. We observe first that we may assume $\mathcal{B}$ is $\mathbb{A}^1$-local and, by the following lemma, we may assume further that $\mathcal{B}$ is 0-connected:

Lemma 6.5. *Given a pointed $\mathbb{A}^1$-local space $\mathcal{B}$, the connected component of the base point $\mathcal{B}^{(0)}$ is also $\mathbb{A}^1$-local and the morphism*

$$\pi_1^{\mathbb{A}^1}(\mathcal{B}^{(0)}) \to \pi_1^{\mathbb{A}^1}(\mathcal{B})$$

is an isomorphism.

Proof. Indeed, by [59] the $\mathbb{A}^1$-localization of a 0-connected space is still 0-connected; thus the morphism $L_{\mathbb{A}^1}(\mathcal{B}^{(0)}) \to \mathcal{B}$ induced by $\mathcal{B}^{(0)} \to \mathcal{B}$ and the fact that $L_{\mathbb{A}^1}(\mathcal{B}^{(0)})$ is $\mathbb{A}^1$-connected, induce $L_{\mathbb{A}^1}(\mathcal{B}^{(0)}) \to \mathcal{B}^{(0)}$, providing a left inverse to $\mathcal{B}^{(0)} \to L_{\mathbb{A}^1}(\mathcal{B}^{(0)})$. Thus $\mathcal{B}^{(0)}$ is a retract in $\mathcal{H}(k)$ of the $\mathbb{A}^1$-local space $L_{\mathbb{A}^1}(\mathcal{B}^{(0)})$ so is also $\mathbb{A}^1$-local. $\square$

From now on, $\mathcal{B}$ is a fixed $\mathbb{A}^1$-connected and $\mathbb{A}^1$-local space. For an open immersion $U \subset X$ and any $n \geq 0$ we set

$$\Pi_n(X, U) := [S^n \wedge (X/U), \mathcal{B}]_{\mathcal{H}_\bullet(k)} = \pi_n(\mathcal{B}(X/U))$$

where S^n denotes the simplicial n-sphere. For $n = 0$ these are just pointed sets, for $n = 1$ these are groups and for $i \geq 2$ these are abelian groups. In fact in the proof below we will only use the case $n = 0$ and $n = 1$. We may extend these definitions to an open immersion $U \subset X$ between essentially smooth k-schemes, by passing to the (co)limit.

The following is one of the main technical lemmas, and will be proven following the lines of [19, Key Lemma], using Gabber's presentation lemma:

Lemma 6.6. *Let X be a smooth k-scheme, $S \subset X$ be a finite set of points and $Z \subset X$ be a closed subscheme of codimension $d > 0$. Then there exists an open subscheme $\Omega \subset X$ containing S and a closed subscheme $Z' \subset \Omega$, of codimension $d - 1$, containing $Z_\Omega := Z \cap \Omega$ and such that the map of pointed sheaves*

$$\Omega/(\Omega - Z') \to \Omega/(\Omega - Z_\Omega)$$

is the trivial map in $\mathcal{H}_\bullet(k)$.

Proof. By Gabber's geometric presentation Lemma 1.15 there exists an open neighborhood Ω of S, and an étale morphism $\phi : \Omega \to \mathbb{A}_V^1$ with V some open subset in some affine space over k such that $Z_\Omega := Z \cap \Omega \to \mathbb{A}_V^1$ is a closed immersion, $\phi^{-1}(Z_\Omega) = Z_\Omega$ and $Z_\Omega \to V$ is a finite morphism. Let F denote the image of Z_Ω in V. Then set $Z' := \phi^{-1}(\mathbb{A}_F^1)$. Observe that $dim(F) = dim(Z)$ thus $codim(Z') = d - 1$. Because we work in the Nisnevich topology, the morphism of sheaves

$$\Omega/(\Omega - Z_\Omega) \to \mathbb{A}_V^1/(\mathbb{A}_V^1 - Z_\Omega)$$

is an isomorphism. The commutative square

$$\begin{array}{ccc} \Omega/(\Omega - Z') & \to & \Omega/(\Omega - Z_\Omega) \\ \downarrow & & \downarrow \wr \\ \mathbb{A}_V^1/(\mathbb{A}_V^1 - \mathbb{A}_F^1) & \to & \mathbb{A}_V^1/(\mathbb{A}_V^1 - Z_\Omega) \end{array}$$

implies that it suffices to show that the map of pointed sheaves

$$\mathbb{A}_V^1/(\mathbb{A}_V^1 - \mathbb{A}_F^1) \to \mathbb{A}_V^1/(\mathbb{A}_V^1 - Z_\Omega)$$

is the trivial map in $\mathcal{H}_\bullet(k)$. Now because $Z \to F$ is finite, the composition $Z \to \mathbb{A}_F^1 \subset \mathbb{P}_F^1$ is still a closed immersion, which has thus empty intersection with the section at infinity $s_\infty : V \to \mathbb{P}_V^1$. By the Mayer–Vietoris property the morphism $\mathbb{A}_V^1/(\mathbb{A}_V^1 - Z_\Omega) \to \mathbb{P}_V^1/(\mathbb{P}_V^1 - Z_\Omega)$ is an isomorphism of pointed sheaves. It suffices thus to check that

$$\mathbb{A}_V^1/(\mathbb{A}_V^1 - \mathbb{A}_F^1) \to \mathbb{P}_V^1/(\mathbb{P}_V^1 - Z_\Omega)$$

is the trivial map in $\mathcal{H}_\bullet(k)$. But the morphism $s_0 : V/(V - F) \to \mathbb{A}_V^1/(\mathbb{A}_V^1 - \mathbb{A}_F^1)$ induced by the zero section is an $\mathbb{A}^1$-weak equivalence. As the composition $s_0 : V/(V - F) \to \mathbb{A}_V^1/(\mathbb{A}_V^1 - \mathbb{A}_F^1) \to \mathbb{P}_V^1/(\mathbb{P}_V^1 - Z_\Omega)$ is $\mathbb{A}^1$-homotopic (by the obvious $\mathbb{A}^1$-homotopy which relates the zero section to the section at infinity) to the section at infinity $s_\infty : V/(V - F) \to \mathbb{P}_V^1/(\mathbb{P}_V^1 - Z_\Omega)$ we get the result because as noted previously s_∞ is disjoint from Z_Ω and thus $s_\infty : V/(V - F) \to \mathbb{P}_V^1/(\mathbb{P}_V^1 - Z_\Omega)$ is equal to the point. $\qquad\square$

Corollary 6.7. *Let X be a smooth (or essentially smooth) k-scheme, $s \in X$ be a point and $Z \subset X$ be a closed subscheme of codimension $d > 0$. Then there exists an open subscheme $\Omega \subset X$ containing s and a closed subscheme $Z' \subset \Omega$, of codimension $d - 1$, containing $Z_\Omega := Z \cap \Omega$ and such that for any $n \in \mathbb{N}$ the map*

$$\Pi_n(\Omega, \Omega - Z_\Omega) \to \Pi_n(\Omega, \Omega - Z')$$

is the trivial map.

In particular, observe that if Z has codimension 1 *and X is irreducible, Z'*
must be Ω. *Thus for any* $n \in \mathbb{N}$ *the map*

$$\Pi_n(\Omega, \Omega - Z_\Omega) \to \Pi_n(\Omega)$$

is the trivial map.

Proof. For X smooth this is an immediate consequence of the lemma. In
case X is an essentially smooth k-scheme, we get the result by an obvious
passage to the colimit, using standard results on limit of schemes [29].   □

Fix an essentially smooth k-scheme X. For any flag of open subschemes of
the form $V \subset U \subset X$ one has the following homotopy exact sequence (which
could be continued on the left):

$$\cdots \to \Pi_1(X, U) \to \Pi_1(X, V) \to \Pi_1(U, V) \to$$
$$\Pi_0(X, U) \to \Pi_0(X, V) \to \Pi_0(U, V) \tag{6.1}$$

where the exactness at $\Pi_0(X, V)$ is the exactness in the sense of pointed sets,
and at $\Pi_0(X, U)$ we observe that there is an action of the group $\Pi_1(X, U)$
on the set $\Pi_0(X, U)$ and the exactness is in the usual sense. The exactness
everywhere else is as diagram of groups.

We now assume that X is the localization of a smooth k-scheme at a
point x. We still denote by x the close point in X. For any flag $\mathcal{F}: Z^2 \subset Z^1 \subset X$ of closed reduced subschemes, with Z^i of codimension at least i,
we set $U_i = X - Z^i$ so that we get a corresponding flag of open subschemes
$U_1 \subset U_2 \subset X$. The set $\mathcal{F}$ of such flags is ordered by increasing inclusion (of
closed subschemes). Given a flag as above and applying the above observation
with $U = U_1$ and $V = \emptyset$ we get an exact sequence:

$$\cdots \to \Pi_1(X, U_1) \to \Pi_1(X) \to \Pi_1(U_1) \to \Pi_0(X, U_1) \to \Pi_0(X) \to \Pi_0(U_1)$$

By the corollary above, applied to X, to its closed point and to the closed
subset Z^1, we see that Ω must be X itself and thus that the maps (for any n)

$$\Pi_n(X, U_1) \to \Pi_n(X)$$

are trivial. We thus get a short exact sequence

$$1 \to \Pi_1(X) \to \Pi_1(U_1) \to \Pi_0(X, U_1) \to * \tag{6.2}$$

and a map of pointed sets $\Pi_0(X) \to \Pi_0(U_1)$ which has trivial kernel.

Passing to the right filtering colimit on flags we get a short exact sequence

$$1 \to \Pi_1(X) \to \Pi_1(F) \to colim_\mathcal{F}\Pi_0(X, U_1) \to * \tag{6.3}$$

and a pointed map with trivial kernel $\Pi_0(X) \to \Pi_0(F)$, where we denote by F the field of functions of X. But now we observe that $\mathcal{B}$ being 0-connected we have $\Pi_0(F) = *$, and thus $\Pi_0(X) = *$.

To understand a bit further the short exact sequence (6.3) we now consider for each flag $\mathcal{F}$ as above the part of the exact sequence obtained above for the flag of open subschemes $U_1 \subset U_2 \subset X$:

$$\to \Pi_0(X, U_2) \to \Pi_0(X, U_1) \to \Pi_0(U_2, U_1) \tag{6.4}$$

By the Corollary 6.7 applied to X, $S = \{x\}$ and to the closed subset $Z^2 \subset X$, we see that Ω must be X and that there exists $Z' \subset X$ of codimension 1, containing Z such that

$$\Pi_0(X, U_2) \to \Pi_0(X, X - Z')$$

is the trivial map. Define the flag $\mathcal{F}' : Z'^2 \subset Z'^1 \subset X$ by setting $Z'^2 = Z^2$ and $Z'^1 = Z1 \cup Z'$ we see that the map

$$colim_{\mathcal{F}}\Pi_0(X, U_2) \to colim_{\mathcal{F}}\Pi_0(X, U_1)$$

is trivial. Thus we conclude that

$$colim_{\mathcal{F}}\Pi_0(X, U_1) \to colim_{\mathcal{F}}\Pi_0(U_2, U_1) \tag{6.5}$$

has trivial kernel. However using now the exact sequence involving the flags of open subsets of the form $\emptyset \subset U_1 \subset U_2$ we see that there is a natural action of $\Pi_1(F)$ on $colim_{\mathcal{F}}\Pi_0(U_2, U_1)$ which makes the map (6.5) $\Pi_1(F)$-oquivariant. As the source $colim_{\mathcal{F}}\Pi_0(X, U_1)$ is one orbit under $\Pi_1(F)$ by (6.3), the equivariant map (6.5) which has trivial kernel must be injective. We thus have proven that if k is an infinite field and X is a smooth local k-scheme with function field F, the natural sequence:

$$1 \to \Pi_1(X) \to \Pi_1(F) \Rightarrow colim_{\mathcal{F}}\Pi_0(U_2, U_1)$$

(the double arrow referring to an action) is exact.

An interesting example is the case where X is the localization at a point x of codimension 1. The set $colim_{\mathcal{F}}\Pi_0(U_2, U_1)$ reduces to the $\Pi_1(F)$-set $\Pi_0(X, X - \{x\})$ because there is only one non-empty closed subset of codimension > 0, the closed point itself. Moreover by the exact sequence (6.2) shows that the action of $\Pi_1(F)$ on $\Pi_0(Y, U - \{y\})$ is transitive and the latter set can be identified with the quotient $\Pi_1(F)/\Pi_1(X)$; in that case we simply denote this set by $H_y^1(X; \Pi_1)$.

We observe that any étale morphism $X' \to X$ between smooth local k-schemes induces a morphism of corresponding associated exact sequences

$$1 \to \Pi_1(X) \to \Pi_1(F) \Rightarrow colim_{\mathcal{F}} \Pi_0(U_2, U_1)$$
$$\downarrow \qquad\qquad \downarrow \qquad\qquad\qquad \downarrow$$
$$1 \to \Pi_1(X') \to \Pi_1(F') \Rightarrow colim_{\mathcal{F}'} \Pi_0(U_2', U_1')$$

When $X' \to X$ runs over the set of localizations at points of codimension one in X we get a $\Pi_1(F)$-equivariant map

$$colim_{\mathcal{F}} \Pi_0(U_2, U_1) \to \Pi_{y \in X^{(1)}} H_y^1(\Pi_1)$$

Lemma 6.8. *(compare [19, Lemma 1.2.1]) The above map is injective and its image is the weak product, yielding a bijection:*

$$colim_{\mathcal{F}} \Pi_0(U_2, U_1) \cong \Pi'_{y \in X^{(1)}} H_y^1(\Pi_1)$$

Corollary 6.9. *1) Let X be a smooth local k-scheme with function field F. Then the natural sequence:*

$$1 \to \Pi_1(X) \to \Pi_1(F) \Rightarrow \Pi'_{y \in X^{(1)}} H_y^1(X; \Pi_1)$$

is exact.

2) The Zariski sheaf associated with $X \mapsto \Pi_1(X)$ is a sheaf in the Nisnevich topology and coincides with $\pi_1^{\mathbb{A}^1}(\mathcal{B})(F)$, which is thus unramified.

Proof. 1) is clear. Let's prove 2). Let's denote by $\mathcal{G}$ the sheaf $(\Pi_1)_{Zar}$. Observe that for X local $\mathcal{G}(X) = \Pi_1(X)$. 1) implies that for any k-smooth X irreducible with function field F the natural sequence:

$$1 \to \mathcal{G}(X) \to \mathcal{G}(F) \Rightarrow \Pi'_{y \in X^{(1)}} H_y^1(X; \mathcal{G})$$

is exact.

For X of dimension 1 with closed point y, the exact sequence (6.3) yields a bijection $H_y^1(X; \Pi_1) = H_y^1(X; \mathcal{G}) = H_{Nis}^1(X, X - \{y\}; \pi_1(\mathcal{B}))$.

If $V \to X$ is an étale morphism between local k-smooth schemes of dimension 1, with closed points y' and y respectively, and with same residue fields $\kappa(y) = \kappa(y')$, the map

$$H_{Nis}^1(X, X - \{y\}; \pi_1(\mathcal{B})) \to H_{Nis}^1(V, V - \{y'\}; \pi_1(\mathcal{B})) \qquad (6.6)$$

is thus bijective.

It follows that the correspondence $X \mapsto \Pi'_{y \in X^{(1)}} H_y^1(X; \mathcal{G})$ is a sheaf in the Nisnevich topology on $\widetilde{Sm}_k$.

Using our above exact sequence this implies easily that $X \mapsto \mathcal{G}(X)$ is a sheaf in the Nisnevich topology. The same exact sequence applied to the henselization X of a k-smooth local scheme implies that the obvious morphism $\mathcal{G}(X) \to \pi_1^{\mathbb{A}^1}(\mathcal{B})(X)$ is a bijection. Thus the morphism $\mathcal{G} \to \pi_1^{\mathbb{A}^1}(\mathcal{B})$ is an isomorphism of sheaves of groups in the Nisnevich topology. $\square$

Remark 6.10. The previous corollary applied to $\mathcal{B}$ being the $\mathbb{A}^1$-local space BG itself implies that a strongly $\mathbb{A}^1$-invariant sheaf of groups G on Sm_k is always unramified. This was used in Remark 2.28. $\square$

We now want to use the results of Sect. 1.2 to prove that $\mathcal{G} = \pi_1(\mathcal{B})$ is strongly $\mathbb{A}^1$-invariant.

We still denote by $\mathcal{G}$ the Nisnevich sheaf $\pi_1(\mathcal{B})$. By the previous corollary, for any smooth local k-scheme X, one has $\mathcal{G}(X) = \pi_1^{\mathbb{A}^1}(\mathcal{B})(X) = \Pi_1(X)$.

In view of Theorem 2.27 the next result implies Theorem 6.1.

Theorem 6.11. *The unramified sheaf of groups $\mathcal{G}$ satisfies the Axioms* **(A2')**, **(A5)** *and* **(A6)** *of Sect. 2.2.*

Proof. We first prove Axiom **(A5)**. Axiom **(A5)** (i) follows at once from the fact proven above that (6.6) is a bijection. From that fact we see that

$$1 \to \mathcal{G}(X) \to \mathcal{G}(F) \Rightarrow \Pi'_{y \in X^{(1)}} H^1_y(X; \mathcal{G})$$

defines on the category of smooth k-schemes of dimension ≤ 1 a short exact sequence of Zariski and Nisnevich sheaves. As the right hand side is flasque in the Nisnevich topology, we get for any smooth k-scheme V of dimension ≤ 1 a bijection

$$H^1_{Zar}(V; \mathcal{G}) = H^1_{Nis}(V; \mathcal{G}) = {}_{\mathcal{G}(F)} \backslash \Pi'_{y \in X^{(1)}} H^1_y(X; \mathcal{G})$$

For X a smooth local k-scheme of dimension 2 with closed point z and $V = X - \{z\}$ (which is of dimension 1), we get $H^1_{Nis}(V; \mathcal{G}) = H^2_z(X; \mathcal{G})$. Proceeding as in the proof of Lemma 2.24 we get Axiom **(A5)** (ii).

Now we prove Axiom **(A2')**. We recall from Lemma 6.8 that the map

$$colim_{\mathcal{F}} \Pi_0(U_2, U_1) \cong \Pi'_{y \in X^{(1)}} H^1_y(X; \mathcal{G})$$

is a bijection for any smooth k-scheme X.

Let $z \in X^{(2)}$. Denote by X_z the localization of X at z and by $V_z = X_z - \{z\}$. We have just proven that $H^1_{Zar}(V_z; \mathcal{G}) = H^1_{Nis}(V_z; \mathcal{G}) = H^2_z(X; \mathcal{G})$. The middle term is also equal to $\Pi_0(V_z) = [(V_z)_+, \mathcal{B}]_{\mathcal{H}_\bullet(k)}$ because $\mathcal{B}$ is connected with $\pi_1(\mathcal{B}) = \mathcal{G}$ and V_z is smooth of dimension 1.

Now for a fixed flag $\mathcal{F}$ in X, by definition, the composition $\Pi_0(U_2, U_1) \to H^2_z(X; \mathcal{G})$ is trivial if $z \in U_2$ and is the composition of the map $\Pi_0(U_2, U_1) \to \Pi_0(U_2)$ and of the map $\Pi_0(U_2) \to \Pi_0(V_z) = H^2_z(X; \mathcal{G})$. Thus given an element of $\Pi'_{y \in X^{(1)}} H^1_y(X; \mathcal{G})$ which comes from $\Pi_0(U_2, U_1)$, its boundary to $H^2_z(X; \mathcal{G})$ at points z of codimension 2 are trivial except maybe for those z not in U_2: but there are only finitely many of those, which establishes Axiom **(A2')**.

We now prove Axiom **(A6)**. Using the Lemma 6.12 below, we see by that for any field $F \in \mathcal{F}_k$, the map $[\Sigma((\mathbb{A}_F^1)_+), \mathcal{B}]_{\mathcal{H}_\bullet(k)} \to [\Sigma((\mathbb{A}_F^1)_+),$ $B(\mathcal{G}))]_{\mathcal{H}_\bullet(k)} = \mathcal{G}(\mathbb{A}_F^1)$ is onto. As $\mathcal{B}$ is $\mathbb{A}^1$-local, $[\Sigma((\mathbb{A}_F^1)_+), \mathcal{B}]_{\mathcal{H}_\bullet(k)} = [\Sigma(SpecF_+), \mathcal{B}]_{\mathcal{H}_\bullet(k)} = \mathcal{G}(F)$ and this shows that the map $\mathcal{G}(F) \to \mathcal{G}(\mathbb{A}_F^1)$ is onto. Thus it is an isomorphism as any F-rational point of $\mathbb{A}_F^1$ provides a left inverse. By part 2) of Lemma 2.16 this implies that $\mathcal{G}$ is $\mathbb{A}^1$-invariant.

By 2) of Lemma 6.12 we see that for any (essentially) smooth k-scheme X of dimension ≤ 1, the map $[(\mathbb{A}_X^1)_+, \mathcal{B}]_{\mathcal{H}_\bullet(k)} \to [(\mathbb{A}_X^1)_+, B(\mathcal{G})]_{\mathcal{H}_\bullet(k)} = H_{Nis}^1(\mathbb{A}_X^1; \mathcal{G})$ is onto. As $\mathcal{B}$ is 0-connected and $\mathbb{A}^1$-local, this shows that if moreover X is a local scheme $H_{Nis}^1(\mathbb{A}_X^1; \mathcal{G}) = *$.

As we know that $\mathcal{G}$ satisfies **(A5)**, Lemma 2.24 implies that $H_{Zar}^1(\mathbb{A}_X^1; \mathcal{G}) = *$. By Remark 2.22 we conclude that $C^*(\mathbb{A}_X^1; \mathcal{G})$ is exact, the axiom **(A6)** is proven, and the theorem as well. $\square$

Lemma 6.12. *1) For any smooth k-scheme X of dimension ≤ 1 the map*

$$Hom_{\mathcal{H}_{s,\bullet}(k)}(\Sigma(X_+), \mathcal{B}) \to Hom_{\mathcal{H}_{s,\bullet}(k)}(\Sigma(X_+), B(\mathcal{G})) = \mathcal{G}(X)$$

is surjective.

2) For any smooth k-scheme X of dimension ≤ 2 the map

$$Hom_{\mathcal{H}_{s,\bullet}(k)}(X_+, \mathcal{B}) \to Hom_{\mathcal{H}_{s,\bullet}(k)}(X_+, B(\mathcal{G})) = H_{Nis}^1(X; \mathcal{G})$$

is surjective and injective if $dim(X) \leq 1$.

Proof. This is proven using the Postnikov tower $\{P^n(\mathcal{B})\}_{n \in \mathbb{N}}$ of $\mathcal{B}$, see [59] for instance, together with standard obstruction theory, see Appendix B. $\square$

$\mathbb{A}^1$-Homotopy Sheaves of $\mathbb{G}_m$-Loop Spaces

Theorem 6.13. *For any pointed $\mathbb{A}^1$-local space $\mathcal{B}$ which is 0-connected, so is the function space $R\mathbf{Hom}_\bullet(\mathbb{G}_m, \mathcal{B})$ and for any integer $n > 0$, the canonical morphism*

$$\pi_n^{\mathbb{A}^1}(R\mathbf{Hom}_\bullet(\mathbb{G}_m, \mathcal{B})) \to (\pi_n^{\mathbb{A}^1}(\mathcal{B}))_{-1}$$

is an isomorphism.

In particular, by induction on $i \geq 0$, one gets an isomorphism for any $n > 0$

$$[S^n \wedge (\mathbb{G}_m)^{\wedge i}, \mathcal{B}]_{\mathcal{H}_\bullet(k)} \cong \pi_n^{\mathbb{A}^1}(\mathcal{B})_{-i}(k)$$

Proof. We first prove that $R\mathbf{Hom}_\bullet(\mathbb{G}_m, \mathcal{B})$ is $\mathbb{A}^1$-connected. From [52] that to show that a space $\mathcal{Z}$ is $\mathbb{A}^1$-connected, it suffices to show that the sets $[(Spec(F)_+; \mathcal{Z})$ are trivial for any $F \in \mathcal{F}_k$. By base change, we may reduce to $F = k$. $\mathbb{G}_m$ having dimension one, we conclude from the Lemma 6.14 below and an obstruction theory argument using Lemma 6.12.

Now we prove the second statement. The morphism is induced by the natural transformation of presheaves of groups "evaluation on the n-th homotopy sheaves"

$$[S^n \wedge \mathbb{G}_m \wedge (U_+), \mathcal{B}]_{\mathcal{H}_\bullet(k)} \to \pi_n(\mathcal{B})_{-1}(U)$$

Observe that the associated sheaf to the presheaf on the left is exactly $\pi_n^{\mathbb{A}^1}(R\mathbf{Hom}_\bullet(\mathbb{G}_m, \mathcal{B}))$.

Now by Lemma 2.32 and Corollary 6.2 both sheaves involved in the morphism are strongly $\mathbb{A}^1$-invariant. To check it is an isomorphism it is sufficient to check that it is an isomorphism on each $F \in \mathcal{F}_k$.

As the morphism in question in degree $n+1$ corresponding to $\mathcal{B}$ is the morphism in degree n applied to $R\Omega_s^1(\mathcal{B})$, by induction, it is sufficient to treat the case $n = 1$.

By a base change argument we may finally assume $F = k$ is the base field. Using Lemma 6.12 we get the result again from Lemma 6.14 below.   □

Lemma 6.14. *Let $\mathcal{G}$ be a strongly $\mathbb{A}^1$-invariant sheaf of groups. Then $H^1(\mathbb{G}_m; \mathcal{G})$ is trivial.*

Proof. For k infinite, we use the results of Sect. 2.2. For k finite we use the results of the appendix. We know from there that H^1 is always computed using the explicit complex $C^*(-; \mathcal{G})$. Thus we reduce to proving the fact that the action of $\mathcal{G}(k(T))$ on $\Pi'_{y \in (\mathbb{G}_m)^{(1)}} H_y^1(\mathbb{G}_m; \mathcal{G})$ is transitive. But this follows at once from the fact that the action of $\mathcal{G}(k(T))$ on $\Pi'_{y \in (\mathbb{A}^1)^{(1)}} H_y^1(\mathbb{A}^1; \mathcal{G})$ is transitive (because $H^1(\mathbb{A}^1; \mathcal{G})$ is trivial) and the fact that the epimorphism $\Pi'_{y \in (\mathbb{G}_m)^{(1)}} H_y^1(\mathbb{G}_m; \mathcal{G})$ is an obvious quotient of $\Pi'_{y \in (\mathbb{A}^1)^{(1)}} H_y^1(\mathbb{A}^1; \mathcal{G})$ as a $\mathcal{G}(k(T))$-set.   □

6.2 $\mathbb{A}^1$-Derived Category and Eilenberg–MacLane Spaces

The derived category. Let us denote by $\mathcal{A}b(k)$ the abelian category of sheaves of abelian groups on Sm_S in the Nisnevich topology. Let $\mathcal{C}_*(\mathcal{A}b(k))$ be the category of chain complexes[1] in $\mathcal{A}b(k)$.

The derived category of $\mathcal{A}b(k)$ is the category $D(\mathcal{A}b(k))$ obtained from $\mathcal{C}_*(\mathcal{A}b(k))$ by inverting the class Qis of quasi-isomorphisms between chain complexes. There are several ways to describe this category. The closest to the intuition coming from standard homological algebra [27] is the following.

---

[1] with differential of degree -1

Definition 6.15. 1) A morphism of chain complexes $C_* \to D_*$ in $\mathcal{C}_*(\mathcal{A}b(k))$
is said to be a *cofibration* if it is a monomorphism. It is called a *trivial
cofibration* if it is furthermore a quasi-isomorphism.

2) A chain complex K_* is said to be *fibrant* if for any trivial cofibration
$i : C_* \to D_*$ and any morphism $f : C_* \to K_*$, there exists a morphism
$g : D_* \to K_*$ such that $g \circ i = f$.

The following "fundamental lemma of homological algebra" seems to be
due to Joyal [35] in the more general context of chain complexes in a
Grothendieck abelian category [27]. One can find a proof in the case of abelian
categories of sheaves in [32]. In fact in both cases one endows the category
$\mathcal{C}_*(\mathcal{A}b(k))$ with a structure of a model category and apply the homotopical
algebra of Quillen [65].

Lemma 6.16. *1) For any chain complex $D_* \in \mathcal{C}_*(\mathcal{A}b(k))$ there exists a
functorial trivial cofibration $D_* \to D_*^f$ to a fibrant complex.*

2) A quasi-isomorphism between fibrant complexes is a homotopy equivalence.

*3) If D_* is a fibrant chain complex, then for any chain complex C_* the natural
map*

$$\pi(C_*, D_*) \to Hom_{D(\mathcal{A}b(k))}(C_*, D_*)$$

is an isomorphism.

Here we denote by $\pi(C_*, D_*)$ the group of homotopy classes of morphisms of chain complexes in the usual sense. Thus to compute the group
$Hom_{D(\mathcal{A}b(k))}(C_*, D_*)$ for any chain complexes C_* and D_*, one just chooses
a quasi-isomorphism $D_* \to D_*^f$ to a fibrant complex (also called a *fibrant
resolution*) and then one uses the chain of isomorphisms

$$\pi(C_*, D_*^f) \cong Hom_{D(\mathcal{A}b(k))}(C_*, D_*^f) \cong Hom_{D(\mathcal{A}b(k))}(C_*, D_*)$$

The main use we will make of this property is a "concrete" description
of internal derived Hom-complex $R\underline{Hom}(C_*, D_*)$: it is given by the naive
internal Hom-complex $\underline{Hom}(C_*, D_*^f)$, for C_* a chain complex which sections
on any smooth k-scheme are torsion free abelian groups (to simplify). Indeed,
it is clear that $\underline{Hom}(C_*, D_*^f)$ is fibrant; using part 2 of the above lemma and
obvious adjunction formula for homotopies of morphisms of chain complexes
we get that this functor $D(\mathcal{A}b(k)) \to D(\mathcal{A}b(k))$, $D_* \mapsto \underline{Hom}(C_*, D_*^f)$ is the
right adjoint to the functor $D(\mathcal{A}b(k)) \to D(\mathcal{A}b(k))$, $B_* \mapsto B_* \otimes C_*$.

The $\mathbb{A}^1$-derived category. The following definition was mentioned in [52,
Remark 9] and is directly inspired from [59, 80]:

Definition 6.17. 1) A chain complex $D_* \in \mathcal{C}_*(\mathcal{A}b(k))$ is called $\mathbb{A}^1$-*local* if
and only if for any $C_* \in \mathcal{C}_*(\mathcal{A}b(k))$, the projection $C_* \otimes \mathbb{Z}(\mathbb{A}^1) \to C_*$
induces a bijection:

$$Hom_{D(\mathcal{A}b(k))}(C_*, D_*) \to Hom_{D(\mathcal{A}b(k))}(C_* \otimes \mathbb{Z}(\mathbb{A}^1), D_*)$$

We will denote by $D_{\mathbb{A}^1-loc}(\mathcal{A}b(k)) \subset D(\mathcal{A}b(k))$ the full subcategory consisting of $\mathbb{A}^1$-local complexes.

2) A morphism $f : C_* \to D_*$ in $C_*(\mathcal{A}b(k))$ is called an $\mathbb{A}^1$-*quasi isomorphism* if and only if for any $\mathbb{A}^1$-local chain complex E_*, the morphism:

$$Hom_{D(\mathcal{A}b(k))}(D_*, E_*) \to Hom_{D(\mathcal{A}b(k))}(C_*, E_*)$$

is bijective. We will denote by $\mathbb{A}^1$-Qis the class of $\mathbb{A}^1$-quasi isomorphisms.

3) The $\mathbb{A}^1$-derived category $D_{\mathbb{A}^1}(\mathcal{A}b(k))$ is the category obtained by inverting the all the $\mathbb{A}^1$-quasi isomorphisms.

All the relevant properties we need are consequences of the following:

Lemma 6.18. *[52, 59] There exists a functor* $L_{\mathbb{A}^1}^{ab} : C_*(\mathcal{A}b(k)) \to C_*$ $(\mathcal{A}b(k))$, *called the (abelian)* $\mathbb{A}^1$-*localization functor, together with a natural transformation*

$$\theta : Id \to L_{\mathbb{A}^1}^{ab}$$

such that for any chain complex C_*, $\theta_{C_*} : C_* \to L_{\mathbb{A}^1}^{ab}(C_*)$ *is an* $\mathbb{A}^1$-*quasi isomorphism whose target is an* $\mathbb{A}^1$-*local fibrant chain complex.*

It is standard to deduce:

Corollary 6.19. *The functor* $L_{\mathbb{A}^1}^{ab} : C_*(\mathcal{A}b(k)) \to C_*(\mathcal{A}b(k))$ *induces a functor*

$$D(\mathcal{A}b(k)) \to D_{\mathbb{A}^1-loc}(\mathcal{A}b(k))$$

which is left adjoint to the inclusion $D_{\mathbb{A}^1-loc}(\mathcal{A}b(k)) \subset D(\mathcal{A}b(k))$, *and which induces an equivalence of categories*

$$D_{\mathbb{A}^1}(\mathcal{A}b(k)) \to D_{\mathbb{A}^1-loc}(\mathcal{A}b(k))$$

Proof of Lemma 6.18. We proceed as in [52]. We fix once for all a functorial fibrant resolution $C_* \to C_*^f$. Let C_* be a chain complex. We let $L_{\mathbb{A}^1}^{(1)}(C_*)$ be the cone in $C_*(\mathcal{A}b(k))$ of the obvious morphism

$$ev_1 : \underline{Hom}(\tilde{\mathbb{Z}}(\mathbb{A}^1), C_*^f) \to C_*^f$$

We let $C_* \to L_{\mathbb{A}^1}^{(1)}(C_*)$ denote the obvious morphism. Define by induction on $n \geq 0$, $L_{\mathbb{A}^1}^{(n)} := L_{\mathbb{A}^1}^{(1)} \circ L_{\mathbb{A}^1}^{(n-1)}$. We have natural morphisms, for any chain complex C_*, $L_{\mathbb{A}^1}^{(n-1)}(C_*) \to L_{\mathbb{A}^1}^{(n)}(C_*)$ and we set

$$L_{\mathbb{A}^1}^\infty(C_*) = colim_{n \in \mathbb{N}} L_{\mathbb{A}^1}^{(n)}(C_*)$$

As in [52, Theorem 4.2.1] we have:

Proposition 6.20. *For any chain complex C_* the complex $L_{\mathbb{A}^1}^\infty(C_*)$ is $\mathbb{A}^1$-local and the morphism*

$$C_* \to L_{\mathbb{A}^1}^\infty(C_*)$$

is an $\mathbb{A}^1$-quasi isomorphism.

This proves Lemma 6.18.

In the sequel we set $L_{\mathbb{A}^1}^{ab}(C_*) := L_{\mathbb{A}^1}^\infty(C_*)^f$: this is the $\mathbb{A}^1$-localization of C_*.

Remark 6.21. It should be noted that we have used implicitly the fact that we are working with the Nisnevich topology, as well as the B.G.-property from [59]: for a general topology on a site together with an interval in the sense of [59], the analogue localization functor would require more "iterations", indexed by some well chosen big enough ordinal number.   □

The (analogue of the) stable $\mathbb{A}^1$-connectivity theorem of [52] in $D(\mathcal{A}b(k))$ is the following:

Theorem 6.22. *Let C_* be a (-1)-connected chain complex. Then its $\mathbb{A}^1$-localization $L_{\mathbb{A}^1}^{ab}(C_*)$ is still (-1)-connected.*

The proof is exactly the same as the case of S^1-spectra treated in [52]. Following the same procedure as in *loc. cit.*, this implies that for an $\mathbb{A}^1$-local chain complex C_* each of its truncations $\tau_{\geq n}(C_*)$ is still $\mathbb{A}^1$-local and thus each of its homology sheaves are automatically strictly $\mathbb{A}^1$-invariant. Thus it is a characterization of $\mathbb{A}^1$-local chain complexes:

Corollary 6.23. *Let C_* be an arbitrary chain complex. The following conditions are equivalent:*

1) C_* is $\mathbb{A}^1$-local.
2) Each homology sheaf $H_n(C_*)$, $n \in \mathbb{Z}$, is strictly $\mathbb{A}^1$-invariant.

This fact endows the triangulated category $D(\mathcal{A}b(k))$ with a natural non degenerated t-structure [10] analogous to the homotopy t-structure of Voevodsky on $DM(k)$. The heart of that t-structure on $D(\mathcal{A}b(k))$ is precisely the category $\mathcal{A}b_{\mathbb{A}^1}(k)$ of strictly $\mathbb{A}^1$-invariant sheaves.

An easy consequence is:

Corollary 6.24. *The category $\mathcal{A}b_{\mathbb{A}^1}(k)$ of strictly $\mathbb{A}^1$-invariant sheaves is abelian, and the inclusion functor $\mathcal{A}b_{\mathbb{A}^1}(k) \subset \mathcal{A}b(k)$ is exact.*

Chain complexes and Eilenberg–MacLane spaces. Recall from [59], that for any simplicial sheaf of sets $\mathcal{X}$ we denote by $C_*(\mathcal{X})$ the (normalized) chain complex in $C_*(\mathcal{A}b(k))$ associated to the free simplicial sheaf of abelian groups $\mathbb{Z}(\mathcal{X})$ on $\mathcal{X}$. This construction defines a functor

$$C_* : \Delta^{op} Shv_{Nis}(Sm_k) \to C_*(\mathcal{A}b(k))$$

which is well known (see [44,59] for instance) to have a right adjoint

$$K : C_*(\mathcal{A}b(k)) \to \Delta^{op} Shv_{Nis}(Sm_k)$$

called the Eilenberg–MacLane space functor.

For an abelian sheaf $M \in \mathcal{A}b(k)$ and an integer n we define the pointed simplicial sheaf $K(M, n)$ (see [59, page 56]) by applying K to the shifted complex $M[n]$, of the complex M placed in degree 0. If $n < 0$, the space $K(M, n)$ is a point. If $n \geq 0$ then $K(M, n)$ has only one non-trivial homotopy sheaf which is the n-th and which is canonically isomorphic to M. More generally, for a chain complex C_*, the space KC_* has for n-th homotopy sheaf 0 for $n < 0$, and the n-th homology sheaf $H_n(C_*)$ for $n \geq 0$.

It is clear that $C_* : \Delta^{op} Shv_{Nis}(Sm_k) \to C_*(\mathcal{A}b(k))$ sends simplicial weak equivalences to quasi-isomorphisms and $K : C_*(\mathcal{A}b(k)) \to \Delta^{op} Shv_{Nis}(Sm_k)$ maps quasi-isomorphisms to simplicial weak equivalences. If C_* is fibrant, it follows that $K(C_*)$ is simplicially fibrant. Thus the two functors induce a pair of adjoint functors

$$C_* : \mathcal{H}_s(k) \to D(\mathcal{A}b(k))$$

and

$$K : D(\mathcal{A}b(k)) \to \mathcal{H}_s(k)$$

As a consequence it is clear that if C_* is an $\mathbb{A}^1$-local complex, the space $K(C_*)$ is an $\mathbb{A}^1$-local space. Thus $C_* : \mathcal{H}_s(k) \to D(\mathcal{A}b(k))$ maps $\mathbb{A}^1$-weak equivalences to $\mathbb{A}^1$-quasi isomorphisms and induces a functor

$$C_*^{\mathbb{A}^1} : \mathcal{H}(k) \to D_{\mathbb{A}^1}(\mathcal{A}b(k))$$

which in concrete terms, maps a space $\mathcal{X}$ to the $\mathbb{A}^1$-localization of $C_*(\mathcal{X})$. We denote the latter by $C_*^{\mathbb{A}^1}(\mathcal{X})$ and call it the $\mathbb{A}^1$-chain complex of $\mathcal{X}$. The functor $C_*^{\mathbb{A}^1} : \mathcal{H}(k) \to D_{\mathbb{A}^1}(\mathcal{A}b(k))$ admits as right adjoint the functor $K^{\mathbb{A}^1} : D_{\mathbb{A}^1}(\mathcal{A}b(k)) \to \mathcal{H}(k)$ induced by $C_* \mapsto K(L_{\mathbb{A}^1}(C_*))$. We observe that for an $\mathbb{A}^1$-local complex C_*, the space $K(C_*)$ is automatically $\mathbb{A}^1$-local and thus simplicially equivalent to the space $K^{\mathbb{A}^1}(C_*)$.

Proposition 6.25. *Let C_* be a 0-connected chain complex in $C_*(\mathcal{A}b(k))$. Then the following conditions are equivalent:*

(i) The space $K(C_*)$ is $\mathbb{A}^1$-local.
(ii) The chain complex C_* is $\mathbb{A}^1$-local.

Proof. It follows immediately from Corollaries 6.3 and 6.23. $\square$

For each complex C_* we simply denote by $(C_*)^{(\mathbb{A}^1)}$ the function complex $\underline{Hom}(\mathbb{Z}_\bullet(\mathbb{A}^1), C_*^f)$. And we let $(C_*)_{\geq 0}^{(\mathbb{A}^1)}$ denote the non negative part of $(C_*)^{(\mathbb{A}^1)}$. It is clear that the tautological $\mathbb{A}^1$-homotopy $(C_*)^{(\mathbb{A}^1)} \otimes \mathbb{Z}(\mathbb{A}^1) \to$

$(C_*)^{(\mathbb{A}^1)}$ between the Identity and the 0-morphism, induces an $\mathbb{A}^1$-homotopy $(C_*)^{(\mathbb{A}^1)}_{\geq 0} \otimes \mathbb{Z}(\mathbb{A}^1) \to (C_*)^{(\mathbb{A}^1)}_{\geq 0}$ as well. Thus $(C_*)^{(\mathbb{A}^1)}_{\geq 0}$ is $\mathbb{A}^1$-contractible.

We consider the morphism of "evaluation at one" $(C_*)^{(\mathbb{A}^1)}_{\geq 0} \to C_*$. And we set $U_{\mathbb{A}^1}(C_*) := cone((C_*)^{(\mathbb{A}^1)}_{\geq 0} \to C_*)$. By construction thus, $C_* \to U_{\mathbb{A}^1}(C_*)$ is an $\mathbb{A}^1$-quasi-isomorphism.

For each $n > 0$ we let $U^{(n)}_{\mathbb{A}^1}$ denote the n-iteration of that functor. We then denote by $U^{\infty}_{\mathbb{A}^1}(C_*)$ the colimit of the following diagram

$$C_* \to U_{\mathbb{A}^1}(C_*) \to \cdots \to U^{(n)}_{\mathbb{A}^1}(C_*) \to \cdots$$

in which each morphism is an $\mathbb{A}^1$-quasi-isomorphism. Thus so is

$$C_* \to U^{\infty}_{\mathbb{A}^1}(C_*)$$

Lemma 6.26. *For any chain complex C_*, the morphism of simplicial sheaves $K(C_*) \to K(U^{\infty}_{\mathbb{A}^1}(C_*))$ induced by the previous one is an $\mathbb{A}^1$-weak equivalence of spaces.*

If moreover C_* is 0-connected the space $K(U^{\infty}_{\mathbb{A}^1}(C_*))$ is 0-connected and $\mathbb{A}^1$-local.

Consequently, in that case, the morphism of simplicial sheaves

$$K(C_*) \to K(U^{\infty}_{\mathbb{A}^1}(C_*)) \cong K(L^{ab}_{\mathbb{A}^1}(C_*))$$

is the $\mathbb{A}^1$-localization of the source.

Proof. It suffices to prove that each $K(U^{(n)}_{\mathbb{A}^1}(C_*)) \to K(U^{(n+1)}_{\mathbb{A}^1}(C_*))$ is an $\mathbb{A}^1$-weak equivalence of spaces, and it suffices to treat the case $n = 0$ and to prove that $K(C_*) \to K(U_{\mathbb{A}^1}(C_*))$ is an $\mathbb{A}^1$-weak equivalence of spaces.

Now the above morphism is a principal $K((C_*)^{(\mathbb{A}^1)}_{\geq 0})$-principal fibration by construction. Thus $K(U^{(n+1)}_{\mathbb{A}^1}(C_*))$ is simplicially weakly equivalent to the Borel construction of $K(U^{(n)}_{\mathbb{A}^1}(C_*))$ with respect to the action of the group $K((C_*)^{(\mathbb{A}^1)}_{\geq 0})$. But now the Borel construction

$$E(K((C_*)^{(\mathbb{A}^1)}_{\geq 0})) \times_{K((C_*)^{(\mathbb{A}^1)}_{\geq 0})} K(U^{(n)}_{\mathbb{A}^1}(C_*))$$

is filtered by the skeleton of $E(K((C_*)^{(\mathbb{A}^1)})$. The first filtration is $K(U^{(n)}_{\mathbb{A}^1}(C_*))$ and the others are of the form $(K((C_*)^{(\mathbb{A}^1)}_{\geq 0}))^{\wedge i} \wedge S^i \wedge (K(U^{(n)}_{\mathbb{A}^1}(C_*))_+)$ with $i > 0$ which is thus $\mathbb{A}^1$-weakly contractible as the chain complex $(C_*)^{(\mathbb{A}^1)}_{\geq 0}$ is $\mathbb{A}^1$-contractible as we observed above.

It remains to prove that the space $K(U_{\mathbb{A}^1}^{\infty}(C_*))$ is $\mathbb{A}^1$-local in case C_* is 0-connected. We observe that by construction, each of the $U_{\mathbb{A}^1}^{(n)}(C_*)$ is 0-connected as well. It follows that the pointed space $K(U_{\mathbb{A}^1}^{\infty}(C_*))$ is also 0-connected. To prove that it is $\mathbb{A}^1$-local, it suffices now to prove that for any $n \geq 0$, and any smooth k-scheme X, any pointed morphism $(\mathbb{A}^1) \wedge (X_+) \wedge S^n \to K(U_{\mathbb{A}^1}^{\infty}(C_*))$ is trivial in the pointed simplicial homotopy category [50, Lemma 3.2.1].

Any such morphism factors through $K(U_{\mathbb{A}^1}^{(n)}(C_*)) \to K(U_{\mathbb{A}^1}^{\infty}(C_*))$ for n big enough. Now using the product $\mathbb{A}^1 \wedge \mathbb{A}^1 \to \mathbb{A}^1$ (where 0 is the base point) we see that any morphism $(\mathbb{A}^1) \wedge (X_+) \wedge S^n \to K(U_{\mathbb{A}^1}^n(C_*))$ factors canonically through the morphism of "evaluation at one" $(K(U_{\mathbb{A}^1}^n(C_*))_{\geq n}^{(\mathbb{A}^1)}) \to K(U_{\mathbb{A}^1}^n(C_*))$, which shows that the composition

$$(\mathbb{A}^1) \wedge (X_+) \wedge S^n \to K(U_{\mathbb{A}^1}^{n+1}(C_*)) \to K(U_{\mathbb{A}^1}^n(C_*))$$

is trivial, and proves the claim. □

Corollary 6.27. *For any 0-connected C_*, the complex $U_{\mathbb{A}^1}^{\infty}(C_*)$ is $\mathbb{A}^1$-local and $C_* \to U_{\mathbb{A}^1}^{\infty}(C_*)$ is thus the $\mathbb{A}^1$-localization $C_* \to L_{\mathbb{A}^1}^{ab}(C_*)$ of C_*.*

Consequently the functor $K(-)$ preserves $\mathbb{A}^1$-weak equivalences between complexes which are 0-connected and, quickly speaking, K commutes with $\mathbb{A}^1$-localizations for 0-connected complexes. In more precise term for any 0-connected chain complex C_* the morphism

$$L_{\mathbb{A}^1}(KC_*) \to K(L_{\mathbb{A}^1}^{ab}(C_*))$$

is a weak equivalence.

Proof. The first claim follows from Proposition 6.25 and the second claim because by construction $C_* \to U_{\mathbb{A}^1}^{\infty}(C_*)$ is an $\mathbb{A}^1$-quasi-isomorphism. The last claim follows from that. □

Remark 6.28. Observe that the last statement of the corollary is not at all formal. □

6.3 The Hurewicz Theorem and some of Its Consequences

The following definition was made in [52]:

Definition 6.29. Let $\mathcal{X}$ be a space and $n \in \mathbb{Z}$ be an integer. We let $H_n^{\mathbb{A}^1}(\mathcal{X})$ denote the n-th homology sheaf of the $\mathbb{A}^1$-chain complex $C_*^{\mathbb{A}^1}(\mathcal{X})$ of $\mathcal{X}$, and call it the n-th $\mathbb{A}^1$-homology sheaf of $\mathcal{X}$.

If $\mathcal{X}$ is pointed, we set $\tilde{H}_n^{\mathbb{A}^1}(\mathcal{X}) = Ker(H_n^{\mathbb{A}^1}(\mathcal{X}) \to H_n^{\mathbb{A}^1}(Spec(k)))$ and call it the n-th reduced homology sheaf of $\mathcal{X}$. As $H_n^{\mathbb{A}^1}(Spec(k)) = 0$ for $n \neq 0$ and $\mathbb{Z}$ for $n = 0$, this means that as graded abelian sheaves

$$H_*^{\mathbb{A}^1}(\mathcal{X}) = \mathbb{Z} \oplus \tilde{H}_*^{\mathbb{A}^1}(\mathcal{X})$$

Remark 6.30. We observe that the $\mathbb{A}^1$-localization functor commutes with the suspension in $D(\mathcal{A}b(k))$. As an immediate consequence, we see that there exists a canonical suspension isomorphism for any pointed space $\mathcal{X}$ and any integer $n \in \mathbb{Z}$:

$$\tilde{H}_n^{\mathbb{A}^1}(\mathcal{X}) \cong \tilde{H}_{n+1}^{\mathbb{A}^1}(\Sigma(\mathcal{X})) \qquad \square$$

Using the $\mathbb{A}^1$-connectivity Theorem 6.22 and its consequences, we get

Corollary 6.31. *The $\mathbb{A}^1$-homology sheaves $H_n^{\mathbb{A}^1}(\mathcal{X})$ of a space $\mathcal{X}$ vanish for $n < 0$ and are strictly $\mathbb{A}^1$-invariant sheaves for $n \geq 0$.*

Remark 6.32. We conjectured in [52] that this result should still hold over a general base; J. Ayoub produced in [5] a counter-example over a base of dimension 2. The case of a base of dimension 1 is still open. $\qquad \square$

Remark 6.33. In classical topology, one easily computes the whole homology of the sphere S^n: $H_i(S^n) = 0$ for $i > n$. In the $\mathbb{A}^1$-homotopy world, the analogue of this vanishing in big dimensions is unfortunately highly non-trivial and unknown. It is natural to make the:

Conjecture 6.34. Let X be a smooth quasi-projective k-scheme of dimension d. Then $H_n^{\mathbb{A}^1}(X) = 0$ for $n > 2d$ and in fact if moreover X is affine then $H_n^{\mathbb{A}^1}(X) = 0$ for $n > d$.

That would imply that the $\mathbb{A}^1$-homology of $(\mathbb{P}^1)^{\wedge n}$ vanishes in degrees $> 2n$. This is in fact a stronger version of the vanishing conjecture of Beilinson–Soulé. It was also formulated in [52].

Computations of higher $\mathbb{A}^1$-homotopy or $\mathbb{A}^1$-homology sheaves seem rather difficult in general. In fact, given a space, we now "understand" its first non-trivial $\mathbb{A}^1$-homotopy sheaf, but we do not know at the moment any "non-trivial" example where one can compute the next non-trivial $\mathbb{A}^1$-homotopy sheaf without using deep results like Milnor or Bloch–Kato conjectures. $\qquad \square$

Using the adjunction between the functors C_* and K it is clear that for a fixed pointed space $\mathcal{X}$ the adjunction morphism

$$\mathcal{X} \to K(C_*(\mathcal{X}))$$

induces a morphism, for each $n \in \mathbb{Z}$

$$\pi_n^{\mathbb{A}^1}(\mathcal{X}) \to H_n^{\mathbb{A}^1}(\mathcal{X})$$

which we call the *Hurewicz morphism*. Here $\pi_n^{\mathbb{A}^1}(\mathcal{X})$ means the n-th homotopy sheaf of the pointed space $L_{\mathbb{A}^1}(\mathcal{X})$, its $\mathbb{A}^1$-localization, constructed in [59], and is also called the n-th $\mathbb{A}^1$-homotopy sheaf of $\mathcal{X}$.

Given a space $\mathcal{X}$ we say that $\mathcal{X}$ is 0-connected if its (simplicial) sheaf $\pi_0(\mathcal{X})$ is the point and we say $\mathcal{X}$ is $\mathbb{A}^1$-connected if its sheaf $\pi_0^{\mathbb{A}^1}(\mathcal{X})$ is the point. For a given $n \geq 1$ we say that a pointed space $\mathcal{X}$ is n-connected if its (simplicial) homotopy sheaves $\pi_i(\mathcal{X})$ are trivial for $i \leq n$, and we say that $\mathcal{X}$ is n-$\mathbb{A}^1$-connected if its $\mathbb{A}^1$-homotopy sheaves $\pi_i^{\mathbb{A}^1}(\mathcal{X})$ are trivial for $i \leq n$.

The following two theorems form the weak form of our Hurewicz theorem:

Theorem 6.35. *Let $\mathcal{X}$ be a pointed $\mathbb{A}^1$-connected space. Then the Hurewicz morphism*

$$\pi_1^{\mathbb{A}^1}(\mathcal{X}) \to H_1^{\mathbb{A}^1}(\mathcal{X})$$

is the initial morphism from the sheaf of groups $\pi_1^{\mathbb{A}^1}(\mathcal{X})$ to a strictly $\mathbb{A}^1$-invariant sheaf (of abelian groups). This means that given a strictly $\mathbb{A}^1$-invariant sheaf M (of abelian groups) and a morphism of sheaves of groups

$$\pi_1^{\mathbb{A}^1}(\mathcal{X}) \to M$$

it factors uniquely through $\pi_1^{\mathbb{A}^1}(\mathcal{X}) \to H_1^{\mathbb{A}^1}(\mathcal{X})$.

Proof. Observe that $\pi_1^{\mathbb{A}^1}(\mathcal{X})$ is strongly $\mathbb{A}^1$-invariant by Theorem 6.1. Let M be a strongly $\mathbb{A}^1$-invariant sheaf of abelian groups. The group of morphisms of sheaves $Hom_{\mathcal{G}r}(\pi_1^{\mathbb{A}^1}(\mathcal{X}), M)$ is equal to the group of simplicial homotopy classes $Hom_{\mathcal{H}_s(k)}(L_{\mathbb{A}^1}(\mathcal{X}), K(M, 1))$ which, because $K(M, 1)$ is $\mathbb{A}^1$-local, is also $Hom_{\mathcal{H}(k)}(\mathcal{X}, K(M, 1))$; by our above adjunction, this is also $Hom_{D_{\mathbb{A}^1}(Ab(k))}(C_*^{\mathbb{A}^1}(\mathcal{X}), K(M, 1))$, and the latter is exactly $Hom_{Ab(k)}(H_1^{\mathbb{A}^1}(\mathcal{X}), M)$ because $C_*^{\mathbb{A}^1}(\mathcal{X})$ is 0-connected. $\square$

Remark 6.36. It is not yet known, though expected, that the Hurewicz morphism is an epimorphism in degree one and that its kernel is always the commutator subgroup. $\square$

Theorem 6.37. *Let $n \geq 2$ be an integer and let $\mathcal{X}$ be a pointed $(n-1)$-$\mathbb{A}^1$-connected space. Then for each $i \in \{0, \ldots, n-1\}$*

$$\tilde{H}_i^{\mathbb{A}^1}(\mathcal{X}) = 0$$

and the Hurewicz morphism

$$\pi_n^{\mathbb{A}^1}(\mathcal{X}) \to H_n^{\mathbb{A}^1}(\mathcal{X})$$

is an isomorphism between strictly $\mathbb{A}^1$-invariant sheaves.

Proof. Apply the same argument as in the previous theorem, using $K(M,n)$, and the fact from Corollary 6.2 that the $\mathbb{A}^1$-homotopy sheaves $\pi_n^{\mathbb{A}^1}(\mathcal{X})$ are strictly $\mathbb{A}^1$-invariant for $n \geq 2$. $\square$

The following immediate consequence is the unstable $\mathbb{A}^1$-connectivity theorem:

Theorem 6.38. *Let $n > 0$ be an integer and let $\mathcal{X}$ be a pointed $(n-1)$-connected space. Then its $\mathbb{A}^1$-localization is still simplicially $(n-1)$-connected.*

For any sheaf of sets F on Sm_k, let us denote by $\mathbb{Z}_{\mathbb{A}^1}(F)$ the strictly $\mathbb{A}^1$-invariant sheaf

$$\mathbb{Z}_{\mathbb{A}^1}(F) := H_0^{\mathbb{A}^1}(F)$$

where F is considered as a space in the right hand side. This strictly $\mathbb{A}^1$-invariant sheaf is the free one generated by F in the following sense: for any strictly $\mathbb{A}^1$-invariant sheaf M the natural map

$$Hom_{Ab(k)}(\mathbb{Z}_{\mathbb{A}^1}(F), M) \to Hom_{Shv(Sm_k)}(F, M)$$

is a bijection.

If F is pointed, we denote by $\mathbb{Z}_{\mathbb{A}^1,\bullet}(F)$ the reduced homology sheaf $\tilde{H}_0^{\mathbb{A}^1}(F)$.

Our previous results and proofs immediately yield:

Corollary 6.39. *For any integer $n \geq 2$ and any pointed sheaf of sets F the canonical morphism*

$$\pi_n^{\mathbb{A}^1}(\Sigma^n(F)) \to H_n^{\mathbb{A}^1}(\Sigma^n(F)) \cong \mathbb{Z}_{\mathbb{A}^1,\bullet}(F)$$

is an isomorphism.

The last isomorphism is the suspension isomorphism from Remark 6.30.

Now by Theorem 5.46, the free strictly $\mathbb{A}^1$-invariant sheaf generated by a (pointed) sheaf F is the same sheaf as the free strongly $\mathbb{A}^1$-invariant sheaf of abelian groups generated by the same (pointed) sheaf. Our main computation in Theorem 3.37 thus yields the following theorem which was announced as Theorem 1.23 in the introduction:

Theorem 6.40. *For $n \geq 2$ one has canonical isomorphisms of strictly $\mathbb{A}^1$-invariant sheaves*

$$\pi_{n-1}^{\mathbb{A}^1}(\mathbb{A}^n - \{0\}) \cong \pi_n^{\mathbb{A}^1}((\mathbb{P}^1)^{\wedge n}) \cong \mathbb{Z}_{\mathbb{A}^1,\bullet}((\mathbb{G}_m)^{\wedge n}) \cong \underline{\mathbf{K}}_n^{MW}$$

Remark 6.41. Observe that the previous computation of $\pi_1^{\mathbb{A}^1}(\mathbb{A}^2 - \{0\})$ requires a slightly more subtle argument, as it concerns the $\mathbb{A}^1$-fundamental group. The morphism $SL_2 \to \mathbb{A}^2 - \{0\}$ being an $\mathbb{A}^1$-weak equivalence, we know *a priori* that $\pi_1^{\mathbb{A}^1}(\mathbb{A}^2 - \{0\})$ is a strongly $\mathbb{A}^1$-invariant sheaf of abelian groups, as is the $\mathbb{A}^1$-fundamental group of any group (or h-group) as usual. The free strongly $\mathbb{A}^1$-invariant sheaf of groups on $\mathbb{G}_m \wedge \mathbb{G}_m$ is commutative and it is thus $\underline{\mathbf{K}}_2^{MW}$.

We postpone the computation of $\pi_1^{\mathbb{A}^1}(\mathbb{P}^1)$ to Sect. 7.3. □

Remark 6.42. For any $n \geq 0$ we let S^n denote $(S^1)^{\wedge n}$. We observe that $\mathbb{A}^n - \{0\}$ is canonically isomorphic in $\mathcal{H}_\bullet(k)$ to $S^{n-1} \wedge (\mathbb{G}_m)^{\wedge n}$ and $(\mathbb{P}^1)^{\wedge n}$ is canonically isomorphic to $S^n \wedge (\mathbb{G}_m)^{\wedge n}$, see [59, Sections Spheres, suspensions and Thom spaces p. 110]. It is thus natural for any $n \geq 0$ and any $i \geq 0$ to study the "sphere" of the form $S^n \wedge (\mathbb{G}_m)^{\wedge i}$.

The Hurewicz theorem implies that it is at least $n - 1$ connected and if $n \geq 2$, provides a canonical isomorphism

$$\pi_n^{\mathbb{A}^1}(S^n \wedge (\mathbb{G}_m)^{\wedge i}) \cong \underline{\mathbf{K}}_i^{MW}$$

for $i \geq 1$ and $\pi_n^{\mathbb{A}^1}(S^n) = \mathbb{Z}$ for $i = 0$ (and $n \geq 1$).

In case $n = 0$ our sphere is just a smash-power of $\mathbb{G}_m$ which is itself $\mathbb{A}^1$-invariant.

For $n = 1$ the question is harder and we only get, by the Hurewicz theorem, a canonical epimorphism $\pi_1^{\mathbb{A}^1}(S^1 \wedge (\mathbb{G}_m)^{\wedge i}) \twoheadrightarrow \underline{\mathbf{K}}_i^{MW}$. This epimorphism has a non trivial kernel for $i = 1$ (see the computation of $\pi_1^{\mathbb{A}^1}(\mathbb{P}^1)$ in Sect. 7.3). We have just observed in the previous Remark that this epimorphism is an isomorphism for $i = 2$. We don't know $\pi_1^{\mathbb{A}^1}(S^1 \wedge (\mathbb{G}_m)^{\wedge i})$ for $i > 2$. □

Corollary 6.43. *Let $(n, i) \in \mathbb{N}^2$ and $(m, j) \subset \mathbb{N}^2$ be pairs of integers. For $n \geq 2$ we have a canonical isomorphism:*

$$Hom_{\mathcal{H}_\bullet(k)}(S^m \wedge (\mathbb{G}_m)^{\wedge j}, S^n \wedge (\mathbb{G}_m)^{\wedge i}) \cong \begin{cases} 0 \text{ if } m < n \\ K_{i-j}^{MW}(k) \text{ if } m = n \text{ and } i > 0 \\ 0 \text{ if } m = n, \ j > 0 \text{ and } i = 0 \\ \mathbb{Z} \text{ if } m = n \text{ and } j = i = 0 \end{cases}$$

Proof. This follows immediately from our previous computation, from Theorem 6.13 and the fact from Sect. 3.2 that the product induces isomorphisms of sheaves $(\underline{\mathbf{K}}_n^{MW})_{-1} \cong \underline{\mathbf{K}}_{n-1}^{MW}$. □

$\mathbb{A}^1$-fibration sequences and applications. In this paragraph we give some natural consequences of the (weak) Hurewicz theorem and of our structure result for $\mathbb{A}^1$-homotopy sheaves Corollary 6.2.

We first recall some terminology.

Definition 6.44. 1) A simplicial fibration sequence between spaces

$$\Gamma \to \mathcal{X} \to \mathcal{Y}$$

with $\mathcal{Y}$ pointed, is a diagram such that the composition of the two morphisms is the trivial one and such that the induced morphism from Γ to the simplicial homotopy fiber of $\mathcal{X} \to \mathcal{Y}$ is a simplicial weak equivalence.

2) An $\mathbb{A}^1$-fibration sequence between spaces

$$\Gamma \to \mathcal{X} \to \mathcal{Y}$$

with $\mathcal{Y}$ pointed, is a diagram such that the composition of the two morphisms is the trivial one and such that the induced diagram between $\mathbb{A}^1$-localizations

$$L_{\mathbb{A}^1}(\Gamma) \to L_{\mathbb{A}^1}(\mathcal{X}) \to L_{\mathbb{A}^1}(\mathcal{Y})$$

is a simplicial fibration sequence.

A basic problem is that it is not true in general that a simplicial fibration sequence is an $\mathbb{A}^1$-fibration sequence. For instance, let $\mathcal{X}$ be a fibrant pointed space, denote by $\mathcal{P}(\mathcal{X})$ the pointed space $\underline{Hom}_{\bullet}(\Delta^1, \mathcal{X})$ of pointed paths $\Delta^1 \to \mathcal{X}$ in $\mathcal{X}$ so that we have a simplicial fibration sequence

$$\Omega^1(\mathcal{X}) \to \mathcal{P}(\mathcal{X}) \to \mathcal{X}$$

whose fiber $\Omega^1(\mathcal{X}) := \underline{Hom}_{\bullet}(S^1, \mathcal{X})$ is the simplicial loop space of $\mathcal{X}$ (with $S^1 = \Delta^1/\partial\Delta^1$ is the simplicial circle). The following observation is an immediate consequence of our definitions, the fact that if $\mathcal{X}$ is $\mathbb{A}^1$-fibrant so is $\Omega^1(\mathcal{X})$, and the fact that an $\mathbb{A}^1$-weak equivalence between $\mathbb{A}^1$-local space is a simplicial weak equivalence:

Lemma 6.45. *Let $\mathcal{X}$ be a simplicially fibrant pointed space. The paths simplicial fibration sequence $\Omega^1(\mathcal{X}) \to \mathcal{P}(\mathcal{X}) \to \mathcal{X}$ above is an $\mathbb{A}^1$-fibration sequence if and only if the canonical morphism*

$$L_{\mathbb{A}^1}(\Omega^1(\mathcal{X})) \to \Omega^1(L_{\mathbb{A}^1}(\mathcal{X}))$$

is a simplicial weak equivalence.

We now observe:

Theorem 6.46. *Let $\mathcal{X}$ be a simplicially fibrant pointed connected space. Then the canonical morphism*

$$L_{\mathbb{A}^1}(\Omega^1(\mathcal{X})) \to \Omega^1(L_{\mathbb{A}^1}(\mathcal{X}))$$

is a simplicial weak equivalence if and only if the sheaf of groups $\pi_0^{\mathbb{A}^1}(\Omega^1(\mathcal{X})) = \pi_0(L_{\mathbb{A}^1}(\Omega^1(\mathcal{X})))$ is strongly $\mathbb{A}^1$-invariant.

Proof. From Theorem 6.1 the condition is necessary. To prove the converse we may assume $\mathcal{X}$ is 0-connected (and fibrant). In that case the inclusion of $\mathcal{X}^{(0)} \subset \mathcal{X}$ of the sub-space consisting of "simplices whose vertices are the base point" is a simplicial weak equivalence: use [44] and stalks to check it. Using the Kan model $G(\mathcal{X}^{(0)})$ for the simplicial loop space on a pointed 0-reduced Kan simplicial set (*loc. cit.* for instance) one obtain a canonical morphism $\mathcal{X}^{(0)} \to B(G(\mathcal{X}^{(0)}))$ which is also a simplicial weak equivalence (by checking on stalks). Thus this defines in the simplicial homotopy category $\mathcal{H}_{s,\bullet}(k)$ a canonical pointed isomorphism between $\mathcal{X}$ and $B(G(\mathcal{X}^{(0)}))$ and in particular a canonical pointed isomorphism between $\Omega^1(\mathcal{X})$ and $G(\mathcal{X}^{(0)})$. Now we observe that by Lemma 6.47 below, we may choose $L_{\mathbb{A}^1}$ so that $L_{\mathbb{A}^1}$ maps groups to groups. Thus $G(\mathcal{X}^{(0)}) \to L_{\mathbb{A}^1}(G(\mathcal{X}^{(0)}))$ is an $\mathbb{A}^1$-weak equivalence between simplicial sheaves of groups. By Lemma 6.48 we see that

$$\mathcal{X} \cong B(G(\mathcal{X}^{(0)})) \to B(L_{\mathbb{A}^1}(G(\mathcal{X}^{(0)})))$$

is always an $\mathbb{A}^1$-weak equivalence. Now assuming that $\pi_0(L_{\mathbb{A}^1}(\Omega^1(\mathcal{X}))) \cong \pi_1(B(L_{\mathbb{A}^1}(G(\mathcal{X}^{(0)}))))$ is strongly $\mathbb{A}^1$-invariant, and the higher homotopy sheaves of $B(L_{\mathbb{A}^1}(G(\mathcal{X}^{(0)})))$ are strictly $\mathbb{A}^1$-invariant, we see using Corollary 6.3 that the space $B(L_{\mathbb{A}^1}(G(\mathcal{X}^{(0)})))$ is $\mathbb{A}^1$-local. It is thus the $\mathbb{A}^1$-localization of $\mathcal{X}$. □

Recall from [59] that an $\mathbb{A}^1$-resolution functor is a pair (Ex, θ) consisting of a functor $Ex : \Delta^{op} Shv(Sm_k) \to \Delta^{op} Shv(Sm_k)$ and a natural transformation $\theta : Id \to Ex$ such that for any space $\mathcal{X}$, $Ex(\mathcal{X})$ is fibrant and $\mathbb{A}^1$-local, and $\theta(\mathcal{X}) : \mathcal{X} \to Ex(\mathcal{X})$ is an $\mathbb{A}^1$-weak equivalence.

Lemma 6.47. *[59] There exists an $\mathbb{A}^1$-resolution functor (Ex, θ) which commutes with any finite products.*

Proof. Combine [59, Theorem 1.66 page 69] with the construction of the explicit I-resolution functor given page 92 of *loc. cit.* □

Recall that a principal fibration $G - \mathcal{X} \to \mathcal{Y}$ with simplicial group G is the same thing as a G-torsor over $\mathcal{Y}$.

Lemma 6.48. *Let*

$$\begin{array}{ccc} G - \mathcal{X} & \to & \mathcal{Y} \\ \downarrow \quad \downarrow & & \downarrow \\ G' - \mathcal{X}' & \to & \mathcal{Y}' \end{array}$$

be a commutative diagram of spaces in which the horizontal lines are principal fibrations with simplicial groups G and G'. Assume the vertical morphism (of simplicial groups) $G \to G'$ and the morphism of spaces $\mathcal{X} \to \mathcal{X}'$ are both $\mathbb{A}^1$-weak equivalence. Then

$$\mathcal{Y} \to \mathcal{Y}'$$

is an $\mathbb{A}^1$-weak equivalence.

Proof. Given a simplicial sheaf of groups G we use the model $E(G)$ of simplicially contractible space on which G acts freely given by the diagonal of the simplicial space $n \mapsto E(G_n)$ where $E(G)$ for a simplicially constant sheaf of group is the usual model (see [59, page 128] for instance). We may as well consider it as the diagonal of the simplicial space $m \mapsto G^{m+1}$, the action of G being the diagonal one. For any G-space $\mathcal{X}$ we introduce the Borel construction

$$EG \times_G \mathcal{X}$$

where G acts diagonally on $E(G) \times \mathcal{X}$. If the action of G is free on $\mathcal{X}$, the morphism $EG \times_G \mathcal{X} \to {}_G\backslash\mathcal{X}$ is a simplicial weak equivalence. Thus in the statement we may replace $\mathcal{Y}$ by $EG \times_G \mathcal{X}$ and $\mathcal{Y}'$ by $EG' \times_{G'} \mathcal{X}'$ respectively. Now from our recollection above, $EG \times_G \mathcal{X}$ is the diagonal space of the simplicial space $m \mapsto G^{m+1} \times_G \mathcal{X}$; it thus simplicially equivalent to its homotopy colimit (see [13] and [59, page 54]). The lemma thus follows from Lemma 2.12 page 73 of *loc. cit.* and the fact that for any m the morphism

$$G^{m+1} \times_G \mathcal{X} \to (G')^{m+1} \times_{G'} \mathcal{X}'$$

are $\mathbb{A}^1$-weak-equivalences. This is easy to prove by observing that the G-space G^{m+1} is functorially G-isomorphic to $G \times (G^m)$ with action given on the left factor only. Thus the spaces $G^{m+1} \times_G \mathcal{X}$ are separately (not taking the simplicial structure into account) isomorphic to $G^m \times \mathcal{X}$. $\qquad\square$

Definition 6.49. 1) A *homotopy principal G-fibration*

$$G - \mathcal{X} \to \mathcal{Y}$$

with simplicial group G consists of a G-space $\mathcal{X}$ and a G-equivariant morphism $\mathcal{X} \to \mathcal{Y}$ (with trivial action on $\mathcal{Y}$) such that the obvious morphism

$$EG \times_G \mathcal{X} \to \mathcal{Y}$$

is a simplicial weak equivalence.

2) Let $G - \mathcal{X} \to \mathcal{Y}$ be a (homotopy) G-principal fibration with structure group G. We say that it is an $\mathbb{A}^1$-homotopy G-principal fibration if the diagram

$$L_{\mathbb{A}^1}(\mathcal{X}) \to L_{\mathbb{A}^1}(\mathcal{Y})$$

is a homotopy principal fibration with structure group $L_{\mathbb{A}^1}(G)$.

In the previous statement, we used an $\mathbb{A}^1$-localization functor which commutes to finite product (such a functor exists by Lemma 6.47).

Theorem 6.50. *Let $G - \mathcal{X} \to \mathcal{Y}$ be a (homotopy) principal fibration with structure group G such that $\pi_0^{\mathbb{A}^1}(G)$ is strongly $\mathbb{A}^1$-invariant. Then it is an $\mathbb{A}^1$-homotopy G-principal fibration.*

Proof. We contemplate the obvious commutative diagram of spaces:

$$
\begin{array}{ccc}
G \quad - \quad \mathcal{X} & \rightarrow & \mathcal{Y} \\
\| \qquad\quad \uparrow \wr & & \uparrow \wr \\
G \quad - \quad E(G) \times \mathcal{X} & \rightarrow & E(G) \times_G \mathcal{X} \\
\downarrow \qquad\qquad \downarrow & & \downarrow \\
L_{\mathbb{A}^1}(G) - E(L_{\mathbb{A}^1}(G)) \times L_{\mathbb{A}^1}(\mathcal{X}) \rightarrow E(L_{\mathbb{A}^1}(G)) \times_{L_{\mathbb{A}^1}(G)} L_{\mathbb{A}^1}(\mathcal{X})
\end{array}
$$

where the upper vertical arrows are simplicial weak equivalences. By Lemma 6.48 the right bottom vertical arrow is an $\mathbb{A}^1$-weak equivalence. By the very definition, to prove the claim we only have to show now that the obvious morphism $E(L_{\mathbb{A}^1}(G)) \times_{L_{\mathbb{A}^1}(G)} L_{\mathbb{A}^1}(\mathcal{X}) \rightarrow L_{\mathbb{A}^1}(\mathcal{Y})$ is a simplicial weak equivalence.

As $E(G) \times_G \mathcal{X} \rightarrow E(L_{\mathbb{A}^1}(G)) \times_{L_{\mathbb{A}^1}(G)} L_{\mathbb{A}^1}(\mathcal{X})$ is an $\mathbb{A}^1$-weak equivalence, we only have to show that the space $E(L_{\mathbb{A}^1}(G)) \times_{L_{\mathbb{A}^1}(G)} L_{\mathbb{A}^1}(\mathcal{X})$ is $\mathbb{A}^1$-local. But it fits, by construction, into a simplicial fibration sequence of the form

$$
L_{\mathbb{A}^1}(\mathcal{X}) \rightarrow E(L_{\mathbb{A}^1}(G)) \times_{L_{\mathbb{A}^1}(G)} L_{\mathbb{A}^1}(\mathcal{X}) \rightarrow B(L_{\mathbb{A}^1}(G))
$$

As $\pi_0^{\mathbb{A}^1}(G)$ is strongly $\mathbb{A}^1$-invariant the 0-connected space $B(L_{\mathbb{A}^1}(G))$ is $\mathbb{A}^1$-local by Corollary 6.2. This implies the claim using the Lemma 6.51 above. $\square$

Lemma 6.51. *Let $\Gamma \rightarrow \mathcal{X} \rightarrow \mathcal{Y}$ be a simplicial fibration sequence with $\mathcal{Y}$ pointed and 0-connected. If Γ and $\mathcal{Y}$ are $\mathbb{A}^1$-connected, then so is $\mathcal{X}$.*

Proof. We use the commutative diagram of spaces

$$
\begin{array}{ccc}
\Gamma \rightarrow \mathcal{X} \rightarrow \mathcal{Y} \\
\downarrow \qquad \downarrow \qquad \downarrow \\
\Gamma^{\mathbb{A}^1} \rightarrow \mathcal{X}^{\mathbb{A}^1} \rightarrow \mathcal{Y}^{\mathbb{A}^1}
\end{array}
$$

where the horizontal rows are both simplicial fibration sequences (we denote here by $\mathcal{Z}^{\mathbb{A}^1}$ the right simplicially derived functor $R\underline{Hom}(\mathbb{A}^1, \mathcal{Z})$, see [59]). We must prove that the middle vertical arrow is a simplicial weak equivalence knowing that both left and right vertical arrows are. But using stalks we reduce easily to the corresponding case for simplicial sets, which is well-known. $\square$

Example 6.52. 1) For instance any SL_n-torsor, $n \geq 2$, satisfy the property of the theorem because $\pi_0^{\mathbb{A}^1}(SL_n) = *$: this follows from the fact that over a field $F \in \mathcal{F}_k$, any element of $SL_n(F)$ is a product of elementary matrices, which shows that over $\pi_0^{\mathbb{A}^1}(SL_n)(F) = *$. From [50] this implies the claim.

2) Any GL_n-torsor, for $n \geq 1$, satisfy this condition as well as $\pi_0^{\mathbb{A}^1}(GL_n) = \mathbb{G}_m$ is strictly $\mathbb{A}^1$-invariant. This equality follows from the previous statement.

3) This is also the case for finite groups or abelian varieties: as these are flasque as sheaves, H^1_{Nis} is trivial.

4) In fact we do not know any example of smooth algebraic group G over k whose $\pi_0^{\mathbb{A}^1}$ is not strongly $\mathbb{A}^1$-invariant.   □

Theorem 6.53. *Let $\Gamma \to \mathcal{X} \to \mathcal{Y}$ be a simplicial fibration sequence with $\mathcal{Y}$ pointed and 0-connected. Assume that the sheaf of groups $\pi_0^{\mathbb{A}^1}(\Omega^1(\mathcal{Y})) = \pi_0(L_{\mathbb{A}^1}(\Omega^1(\mathcal{Y})))$ is strongly $\mathbb{A}^1$-invariant. Then $\Gamma \to \mathcal{X} \to \mathcal{Y}$ is also an $\mathbb{A}^1$-fibration sequence.*

Proof. This theorem is an easy reformulation of the previous one (using a little bit its proof) by considering a simplicial group G with a simplicial weak equivalence $\mathcal{Y} \cong B(G)$.   □

We observe that the assumptions of the theorem are fulfilled if $\mathcal{Y}$ is simplicially 1-connected, or if it is 0-connected and if $\pi_1(\mathcal{Y})$ itself is strongly $\mathbb{A}^1$-invariant. This follows from the following lemma applied to $\Omega^1(\mathcal{Y})$.

Lemma 6.54. *Let $\mathcal{X}$ be a space. Assume its sheaf $\pi_0(\mathcal{X})$ is $\mathbb{A}^1$-invariant. Then the morphism $\pi_0(\mathcal{X}) \to \pi_0^{\mathbb{A}^1}(\mathcal{X}) = \pi_0(L_{\mathbb{A}^1}(\mathcal{X}))$ is an isomorphism.*

Proof. This lemma follows from the fact that $\pi_0(\mathcal{X}) \to \pi_0^{\mathbb{A}^1}(\mathcal{X})$ is always an epimorphism [59, Corollary 3.22 page 94] and the fact that as a space the $\mathbb{A}^1$-invariant sheaf $\pi_0(\mathcal{X})$ is $\mathbb{A}^1$-local. This produces a factorization of the identity of $\pi_0(\mathcal{X})$ as $\pi_0(\mathcal{X}) \to \pi_0^{\mathbb{A}^1}(\mathcal{X}) = \pi_0(L_{\mathbb{A}^1}(\mathcal{X})) \to \pi_0(\mathcal{X})$ which implies the result.   □

The Relative $\mathbb{A}^1$-Connectivity Theorem

Definition 6.55. A morphism of spaces $\mathcal{X} \to \mathcal{Y}$ is said to be n-connected for some integer $n \geq 0$ if each stalk of that morphism (at any point of any smooth k-scheme) is n-connected in the usual sense.

When the spaces are pointed and $\mathcal{Y}$ is 0-connected this is equivalent to the fact that the simplicial homotopy fiber of the morphism is n-connected.

The relative $\mathbb{A}^1$-connectivity theorem refers to:

Theorem 6.56. *Let $f : \mathcal{X} \to \mathcal{Y}$ be a morphism with $\mathcal{Y}$ pointed and 0-connected. Assume that the sheaf of groups $\pi_0^{\mathbb{A}^1}(\Omega^1(\mathcal{Y})) = \pi_0(L_{\mathbb{A}^1}(\Omega^1(\mathcal{Y})))$ is strongly $\mathbb{A}^1$-invariant (for instance if $\mathcal{Y}$ is simplicially 1-connected, or if $\pi_1(\mathcal{Y})$ itself is strongly $\mathbb{A}^1$-invariant). Let $n \geq 1$ be an integer and assume f is $(n-1)$-connected, then so is the morphism*

$$L_{\mathbb{A}^1}(\mathcal{X}) \to L_{\mathbb{A}^1}(\mathcal{Y})$$

Proof. Let $\Gamma \to \mathcal{X}$ be the homotopy fiber. By Theorem 6.53 above the diagram $L_{\mathbb{A}^1}(\Gamma) \to L_{\mathbb{A}^1}(\mathcal{X}) \to L_{\mathbb{A}^1}(\mathcal{Y})$ is a simplicial fibration sequence. Our connectivity assumption is that $\pi_i(\Gamma) = 0$ for $i \in \{0, \ldots, n-1\}$. By the unstable $\mathbb{A}^1$-connectivity Theorem 6.38, the space $L_{\mathbb{A}^1}(\Gamma)$ is also $(n-1)$-connected. Thus so is $L_{\mathbb{A}^1}(\mathcal{X}) \to L_{\mathbb{A}^1}(\mathcal{Y})$.   □

The strong form of the Hurewicz theorem. This refers to the following classical improvement of the weak Hurewicz theorem:

Theorem 6.57. *Let $n > 1$ be an integer and let $\mathcal{X}$ be a pointed $(n-1)$-$\mathbb{A}^1$-connected space. Then $H_i^{\mathbb{A}^1}(\mathcal{X}) = 0$ for each $i \in \{0, \ldots, n-1\}$, the Hurewicz morphism $\pi_n^{\mathbb{A}^1}(\mathcal{X}) \to H_n^{\mathbb{A}^1}(\mathcal{X})$ is an isomorphism, and moreover the Hurewicz morphism*

$$\pi_{n+1}^{\mathbb{A}^1}(\mathcal{X}) \to H_{n+1}^{\mathbb{A}^1}(\mathcal{X})$$

is an epimorphism of sheaves.

Proof. We may assume $\mathcal{X}$ fibrant and $\mathbb{A}^1$-local. Consider the canonical morphism $\mathcal{X} \to K(C_*(\mathcal{X}))$ and let us denote by Γ its simplicial homotopy fiber. The classical Hurewicz theorem for simplicial homotopy tells us that Γ is simplicially n-connected (just compute on the stalks).

Now as $K(C_*(\mathcal{X}))$ is 1-connected the Theorem 6.56 above tells us that the morphism $\mathcal{X} = L_{\mathbb{A}^1}(\mathcal{X}) \to L_{\mathbb{A}^1}(K(C_*(\mathcal{X})))$ is still n-connected. But as $K(L_{\mathbb{A}^1}(C_*(\mathcal{X}))) \to L_{\mathbb{A}^1}(K(C_*(\mathcal{X})))$ is a simplicial weak equivalence by Corollary 6.27 we conclude that $\mathcal{X} \to K(C_*^{\mathbb{A}^1}(\mathcal{X}))$ is n-connected, which gives exactly the strong form of Hurewicz theorem.   □

Remark 6.58. For $n = 1$ if one assumes that $\pi_1^{\mathbb{A}^1}(\mathcal{X})$ is abelian (thus strictly $\mathbb{A}^1$-invariant) the theorem remains true.   □

A stability result. Recall that for a fibrant space $\mathcal{X}$ and an integer n the space $P^{(n)}(\mathcal{X})$ denotes the n-th stage of the Postnikov tower for $\mathcal{X}$ [59, page 55]. If $\mathcal{X}$ is pointed, we denote by $\mathcal{X}^{(n+1)} \to \mathcal{X}$ the homotopy fiber at the point of $\mathcal{X} \to P^{(n)}(\mathcal{X})$. The space $\mathcal{X}^{(n+1)}$ is of course n-connected. There exists by functoriality a canonical morphism $\mathcal{X}^{(n)} \to (L_{\mathbb{A}^1}(\mathcal{X}))^{(n)}$. As the target is $\mathbb{A}^1$-local, we thus get a canonical morphism of pointed spaces

$$L_{\mathbb{A}^1}(\mathcal{X}^{(n)}) \to (L_{\mathbb{A}^1}(\mathcal{X}))^{(n)}$$

Theorem 6.59. *Let $\mathcal{X}$ be a pointed connected space. Assume $n > 0$ is an integer such that the sheaf $\pi_1(\mathcal{X})$ is strongly $\mathbb{A}^1$-invariant and for each $1 < i \leq n$, the sheaf $\pi_i(\mathcal{X})$ is strictly $\mathbb{A}^1$-invariant. Then for each $i \leq n+1$ the above morphism $L_{\mathbb{A}^1}(\mathcal{X}^{(i)}) \to (L_{\mathbb{A}^1}(\mathcal{X}))^{(i)}$ is a simplicial weak equivalence.*

We obtain immediately the following:

Corollary 6.60. *Let $\mathcal{X}$ be a pointed connected space. Assume $n > 0$ is an integer such that the sheaf $\pi_1(\mathcal{X})$ is strongly $\mathbb{A}^1$-invariant and for each $1 < i \leq n$, the sheaf $\pi_i(\mathcal{X})$ is strictly $\mathbb{A}^1$-invariant. Then for $i \leq n$ the morphism*

$$\pi_i(\mathcal{X}) \to \pi_i^{\mathbb{A}^1}(\mathcal{X}) = \pi_i(L_{\mathbb{A}^1}(\mathcal{X}))$$

is an isomorphism and the morphism

$$\pi_{n+1}(\mathcal{X}) \to \pi_{n+1}^{\mathbb{A}^1}(\mathcal{X}) = \pi_{n+1}(L_{\mathbb{A}^1}(\mathcal{X}))$$

is the universal morphism from $\pi_{n+1}(\mathcal{X})$ to a strictly $\mathbb{A}^1$-invariant sheaf.

Proof. We proceed by induction on n. Assume the statement of the theorem is proven for $n-1$. We apply Theorem 6.53 to the simplicial fibration sequence $\mathcal{X}^{(n+1)} \to \mathcal{X} \to P^{(n)}(\mathcal{X})$; $P^{(n)}(\mathcal{X})$ satisfies indeed the assumptions. Thus we get a simplicial fibration sequence

$$L_{\mathbb{A}^1}(\mathcal{X}^{(n+1)}) \to L_{\mathbb{A}^1}(\mathcal{X}) \to L_{\mathbb{A}^1}(P^{(n)}(\mathcal{X}))$$

Then we observe that by induction and the Corollary 6.60 above that the morphism $P^{(n)}(\mathcal{X}) \to P^{(n)}(L_{\mathbb{A}^1}(\mathcal{X}))$ is a simplicial weak equivalence. Thus $P^{(n)}(\mathcal{X}) \cong L_{\mathbb{A}^1}(P^{(n)}(\mathcal{X})) \cong P^{(n)}(L_{\mathbb{A}^1}(\mathcal{X}))$. These two facts imply the claim. $\qquad\square$

The $\mathbb{A}^1$-Simplicial Suspension Theorem

Theorem 6.61. *Let $\mathcal{X}$ be a pointed space and let $n \geq 2$ be an integer. If $\mathcal{X}$ is $(n-1)$-$\mathbb{A}^1$-connected space the canonical morphism*

$$L_{\mathbb{A}^1}(\mathcal{X}) \to \Omega^1(L_{\mathbb{A}^1}(\Sigma^1(\mathcal{X})))$$

is $2(n-1)$-$(\mathbb{A}^1)$-connected.

Proof. We first observe that the classical suspension theorem implies that for any simplicially $(n-1)$-connected space $\mathcal{Y}$ the canonical morphism

$$\mathcal{Y} \to \Omega^1(\Sigma^1(\mathcal{Y}))$$

is simplicially $2(n-1)$-connected. Thus the theorem follows from: We apply this to the space $\mathcal{Y} = L_{\mathbb{A}^1}(\mathcal{X})$ itself, which is simplicially $(n-1)$-connected. Thus the morphism $L_{\mathbb{A}^1}(\mathcal{X}) \to \Omega^1(\Sigma^1(L_{\mathbb{A}^1}(\mathcal{X})))$ is simplicially $2(n-1)$-connected. This implies in particular that the suspension morphisms $\pi_i^{\mathbb{A}^1}(\mathcal{X}) \to \pi_{i+1}(\Sigma_s^1(L_{\mathbb{A}^1}(\mathcal{X})))$ are isomorphisms for $i \leq 2(n-1)$ and an epimorphism for $i = 2n - 1$.

From Theorem 6.59 and its corollary, this implies that $\Sigma^1(\mathcal{X}) \to L_{\mathbb{A}^1}(\Sigma^1(\mathcal{X}))$ induces an isomorphism on π_i for $i \leq 2n - 1$ and that the morphism $\pi_{2n}(\Sigma^1(\mathcal{X})) \to \pi_{2n}(L_{\mathbb{A}^1}(\Sigma^1(\mathcal{X})))$ is the universal morphism to a strictly $\mathbb{A}^1$-invariant sheaf. Thus it follows formally that $\pi_{2n-1}^{\mathbb{A}^1}(\mathcal{X}) \to$

$\pi_{2n}(L_{\mathbb{A}^1}(\Sigma_s^1(\mathcal{X})))$ is a categorical epimorphism in the category of strictly $\mathbb{A}^1$-invariant sheaves. As by Corollary 6.24 this category is an abelian category for which the inclusion into $\mathcal{A}b(k)$ is exact, it follows that the morphism $\pi_{2n-1}^{\mathbb{A}^1}(\mathcal{X}) \to \pi_{2n}(L_{\mathbb{A}^1}(\Sigma_s^1(\mathcal{X})))$ is actually an epimorphism of sheaves. Thus the morphism $\Sigma^1(\mathcal{X}) \to L_{\mathbb{A}^1}(\Sigma^1(\mathcal{X}))$ is a $(2n-1)$-simplicial weak-equivalence. The morphism

$$\Omega^1(\Sigma^1(\mathcal{X})) \to \Omega^1(L_{\mathbb{A}^1}(\Sigma(\mathcal{X})))$$

is thus a $2(n-1)$-simplicial weak-equivalence. The composition

$$L_{\mathbb{A}^1}(\mathcal{X}) \to \Omega^1(\Sigma^1(L_{\mathbb{A}^1}(\mathcal{X}))) \to \Omega^1(L_{\mathbb{A}^1}(\Sigma(\mathcal{X})))$$

is thus also simplicially $2(n-1)$-connected. $\qquad\qquad\square$

Chapter 7
$\mathbb{A}^1$-Coverings, $\pi_1^{\mathbb{A}^1}(\mathbb{P}^n)$ and $\pi_1^{\mathbb{A}^1}(SL_n)$

7.1 $\mathbb{A}^1$-Coverings, Universal $\mathbb{A}^1$-Covering and $\pi_1^{\mathbb{A}^1}$

Definition 7.1. 1) A *simplicial covering* $\mathcal{Y} \to \mathcal{X}$ is a morphism of spaces
which has the unique right lifting property with respect to simplicially
trivial cofibrations. This means that given any commutative square of
spaces

$$
\begin{array}{ccc}
\mathcal{A} & \to & \mathcal{Y} \\
\downarrow & & \downarrow \\
\mathcal{B} & \to & \mathcal{X}
\end{array}
$$

in which $\mathcal{A} \to \mathcal{B}$ is an simplicially trivial cofibration, there exists one and
exactly one morphism $\mathcal{B} \to \mathcal{Y}$ which lets the whole diagram commutative.
2) An $\mathbb{A}^1$-*covering* $\mathcal{Y} \to \mathcal{X}$ is a morphism of spaces which has the unique
right lifting property with respect to $\mathbb{A}^1$-trivial cofibrations.[1]

Lemma 7.2. *A morphism* $\mathcal{Y} \to \mathcal{X}$ *is a simplicial (resp $\mathbb{A}^1$-) covering if and
only if it has the unique right lifting property with respect to any simplicial
(resp $\mathbb{A}^1$-) weak equivalence.*

Proof. It suffices to prove that coverings have the unique lifting property
with respect to weak-equivalences (both in the simplicial and in the $\mathbb{A}^1$-
structure). Pick up a commutative square as in the definition with $\mathcal{A} \to \mathcal{B}$
a weak-equivalence. Factor it as a trivial cofibration $\mathcal{A} \to \mathcal{C}$ and a trivial
fibration $\mathcal{C} \to \mathcal{B}$. In this way we reduce to the case $\pi : \mathcal{A} \to \mathcal{B}$ is a
trivial fibration. Uniqueness is clear as trivial fibrations are epimorphisms
of spaces. Let's prove the existence statement. For both structures the spaces
are cofibrant thus one gets a section $i : \mathcal{B} \to \mathcal{A}$ which is of course a trivial
cofibration. Now we claim that $f : \mathcal{A} \to \mathcal{Y}$ composed with $i \circ \pi : \mathcal{A} \to \mathcal{A}$ is

---

[1]Remember [59] that this means both a monomorphism and an $\mathbb{A}^1$-weak equivalence.

F. Morel, $\mathbb{A}^1$-*Algebraic Topology over a Field*, Lecture Notes in Mathematics 2052,   177
DOI 10.1007/978-3-642-29514-0_7, © Springer-Verlag Berlin Heidelberg 2012

equal to f. This follows from the unique lifting property applied to i. Thus $f \circ i : \mathcal{B} \to \mathcal{Y}$ is a solution and we are done.   □

Remark 7.3. It would be interesting to have a geometric criterium for a morphism in Sm_k to be an $\mathbb{A}^1$-covering. Of course such a morphism has to have the right lifting property with respect to only the 0-sections morphisms of the form $U \to \mathbb{A}^1 \times U$, for $U \in Sm_k$. It is not clear whether this is sufficient.

□

The Simplicial Theory

Lemma 7.4. *If* $\mathcal{Y} \to \mathcal{X}$ *is a simplicial covering for each* $x \in X \in Sm_k$ *the morphism of simplicial sets* $\mathcal{Y}_x \to \mathcal{X}_x$ *is a covering of simplicial sets.*

Proof. For $i \in \{0, \ldots, n\}$ we let as usual $\Lambda^{n,i} \subset \Delta^n$ be the union of all the faces of Δ^n but the i-th. The inclusion $\Lambda^{n,i} \subset \Delta^n$ is then a simplicial equivalence (of simplicial sets). Now for any $U \in Sm_k$ and any inclusion $\Lambda^{n,i} \subset \Delta^n$ as above, we apply the definition of simplicial covering to $\Lambda^{n,i} \times U \subset \Delta^n \times U$. When U runs over the set of Nisnevich neighborhoods of $x \in X$, this easily implies that $\mathcal{Y}_x \to \mathcal{X}_x$ has the right lifting property with respect to the $\Lambda^{n,i} \subset \Delta^n$, proving our claim.   □

For any pointed simplicially connected space $\mathcal{Z}$ there exists a canonical morphism in $\mathcal{H}_{s,\bullet}(k)$ of the form $\mathcal{Z} \to BG$, where G is the fundamental group sheaf $\pi_1(\mathcal{Z})$; this relies on the Postnikov tower [59] for instance. Using now Prop. 1.15 p.130 of *loc. cit.* one gets a canonical isomorphism class $\tilde{\mathcal{Z}} \to \mathcal{Z}$ of G-torsor. Choosing one representative, we may point it by lifting the base point of $\mathcal{Z}$. Now this pointed G-torsor is canonical up to isomorphism. To prove this we first observe that $\tilde{\mathcal{Z}}$ is simplicially 1-connected. Now we claim that any pointed simplicially 1-connected covering $\mathcal{Z}' \to \mathcal{Z}$ over $\mathcal{Z}$ is canonically isomorphic to this one.

Indeed, one first observe that the composition $\mathcal{Z}' \to \mathcal{Z} \to BG \to \mathcal{B}G$ (where $\mathcal{B}G$ means a simplicially fibrant resolution of BG) is homotopically trivial. This follows from the fact that $\mathcal{Z}'$ is 1-connected.

Now let $\mathcal{E}G \to \mathcal{B}G$ be the universal covering of $\mathcal{B}G$ (given by Prop. 1.15 p.130 of *loc. cit.*). Clearly this is also a simplicial fibration, thus $\mathcal{E}G$ is simplicially fibrant. Thus we get the existence of a lifting $\mathcal{Z}' \to \mathcal{E}G$. Now the commutative square

$$\begin{array}{ccc} \mathcal{Z}' & \to & \mathcal{E}G \\ \downarrow & & \downarrow \\ \mathcal{Z} & \to & \mathcal{B}G \end{array}$$

Using the lemma above, we see that this square is cartesian on each stalk (by the classical theory), thus cartesian. This proves precisely that $\mathcal{Z}'$ as a covering is isomorphic to $\tilde{\mathcal{Z}}$. But then as a pointed covering, it is canonically isomorphic to $\tilde{\mathcal{Z}} \to \mathcal{Z}$ because the automorphism group of the pointed covering $\tilde{\mathcal{Z}} \to \mathcal{Z}$ is trivial.

Now given any pointed simplicial covering $\mathcal{Z}' \to \mathcal{Z}$ one may consider the connected component of the base point $\mathcal{Z}'^{(0)}$ of $\mathcal{Z}'$. Clearly $\mathcal{Z}'^{(0)} \to \mathcal{Z}$ is still a pointed (simplicial) covering. Now the universal covering (constructed above) of $\mathcal{Z}'^{(0)}$ is also the universal covering of $\mathcal{Z}$. One thus get a unique isomorphism from the pointed universal covering of $\mathcal{Z}$ to that of $\mathcal{Z}'^{(0)}$. The composition $\tilde{\mathcal{Z}} \to \mathcal{Z}'$ is the unique morphism of pointed coverings (use stalks). Thus $\tilde{\mathcal{Z}} \to \mathcal{Z}$ is the universal object in the category of pointed coverings of $\mathcal{Z}$.

$\square$

The $\mathbb{A}^1$-Theory

We want to prove the analogue statement in the case of $\mathbb{A}^1$-coverings. We observe that as any simplicially trivial cofibration is an $\mathbb{A}^1$-trivial cofibration, an $\mathbb{A}^1$-covering is in particular a simplicial covering.

Before that we first establish the following lemma which will provide us with our two basic examples of $\mathbb{A}^1$-coverings.

Lemma 7.5. *1) A G-torsor $\mathcal{Y} \to \mathcal{X}$ with G a strongly $\mathbb{A}^1$-invariant sheaf is an $\mathbb{A}^1$-covering.*

2) Any G-torsor $\mathcal{Y} \to \mathcal{X}$ in the étale topology, with G a finite étale k-group of order prime to the characteristic, is an $\mathbb{A}^1$-covering.

Proof. 1) Recall from [59, Prop. 1.15 p. 130] that the set $H^1(\mathcal{X}; G)$ of isomorphism classes (denoted by $P(\mathcal{X}; G)$ in *loc. cit.*) of G-torsors over a space $\mathcal{X}$ is in one-to-one correspondence with $[\mathcal{X}, BG]_{\mathcal{H}_s(k)}$ (observe we used the simplicial homotopy category). By the assumption on G, BG is $\mathbb{A}^1$-local. Thus we get now a one-to-one correspondence $H^1(\mathcal{X}; G) \cong [\mathcal{X}, BG]_{\mathcal{H}(k)}$. Now let us choose a commutative square like in the definition, with the right vertical morphism a G-torsor. This implies that the pull-back of this G-torsor to $\mathcal{B}$ is trivial when restricted to $\mathcal{A} \subset \mathcal{B}$. By the property just recalled, we get that $H^1(\mathcal{B}; G) \to H^1(\mathcal{A}; G)$ is a bijection, thus the G-torsor over $\mathcal{B}$ itself is trivial. This fact proves the existence of a section $s : \mathcal{B} \to Y$ of $Y \to X$.

The composition $s \circ (\mathcal{A} \subset \mathcal{B}) : \mathcal{A} \to \mathcal{Y}$ may not be equal to the given top morphism $s_0 : \mathcal{A} \to \mathcal{Y}$ in the square. But then there exists a morphism $g : \mathcal{A} \to G$ with $s = g.s_0$ (by one of the properties of torsors).

But as G is $\mathbb{A}^1$-invariant the restriction map $G(\mathcal{B}) \to G(\mathcal{A})$ is an isomorphism. Let $\tilde{g} : \mathcal{B} \to G$ be the extension of g. Then $\tilde{g}^{-1}.s : \mathcal{B} \to \mathcal{Y}$ is still a section of the torsor, but now moreover its restriction to $\mathcal{A} \subset \mathcal{B}$ is equal to s_0. We have proven the existence of an $s : \mathcal{B} \to \mathcal{Y}$ which makes the diagram commutative. The uniqueness follows from the previous reasoning as the restriction map $G(\mathcal{B}) \to G(\mathcal{A})$ is an isomorphism.

2) Recall from [59, Prop. 3.1 p. 137] that the étale classifying space $B_{et}(G) = R\pi_*(BG)$ is $\mathbb{A}^1$-local. Here $\pi : (Sm_k)_{et} \to (Sm_k)_{Nis}$ is the canonical morphism of sites. But then for any space $\mathcal{X}$, the set $[\mathcal{X}, B_{et}(G)]_{\mathcal{H}(k)} \cong$

$[\mathcal{X}, B_{et}(G)]_{\mathcal{H}_s(k)}$ is by adjunction (see *loc. cit.* Section Functoriality p. 61]) in natural bijection with $Hom_{\mathcal{H}_s(Sm_k)_{et}}(\pi^*(\mathcal{X}), BG) \cong H^1_{et}(\mathcal{X}; G)$.

This proves also in that case that the restriction map $H^1_{et}(\mathcal{B}; G) \to H^1_{et}(\mathcal{A}; G)$ is a bijection. We know moreover that G is $\mathbb{A}^1$-invariant as space, thus $G(\mathcal{B}) \to G(\mathcal{A})$ is also an isomorphism. The same reasoning as previously yields the result.   □

Example 7.6. 1) Any $\mathbb{G}_m$-torsor $\mathcal{Y} \to \mathcal{X}$ is an $\mathbb{A}^1$-covering. Thus any line bundle yields a $\mathbb{A}^1$-covering. In particular, a connected smooth projective k-variety of dimension ≥ 1 has always non trivial $\mathbb{A}^1$-coverings!

2) Any finite étale Galois covering $Y \to X$ between smooth k-varieties whose Galois group has order prime to $char(k)$ is an $\mathbb{A}^1$-covering. More generally, one could show that any finite étale covering between smooth k-varieties which can be covered by a surjective étale Galois covering $Z \to X$ with group a finite étale k-group G of order prime to $char(k)$ is an $\mathbb{A}^1$-covering. In characteristic 0, for instance, any finite étale covering is an $\mathbb{A}^1$-covering.   □

Lemma 7.7. *1) Any pull-back of an $\mathbb{A}^1$-covering is an $\mathbb{A}^1$-covering.*
2) The composition of two $\mathbb{A}^1$-coverings is a $\mathbb{A}^1$-covering.
3) Any $\mathbb{A}^1$-covering is an $\mathbb{A}^1$-fibration in the sense of [59].
4) A morphism $\mathcal{Y}_1 \to \mathcal{Y}_2$ of $\mathbb{A}^1$-coverings $\mathcal{Y}_i \to \mathcal{X}$ which is an $\mathbb{A}^1$-weak equivalence is an isomorphism.

Proof. Only the last statement requires an argument. It follows from Lemma 7.2: applying it to $\mathcal{Y}_1 \to \mathcal{Y}_2$ one first get a retraction $\mathcal{Y}_2 \to \mathcal{Y}_1$ and to check that this retraction composed with $\mathcal{Y}_1 \to \mathcal{Y}_2$ is the identity of $\mathcal{Y}_2$ one uses once more the Lemma 7.2.   □

We now come to the main result of this section:

Theorem 7.8. *Any pointed $\mathbb{A}^1$-connected space $\mathcal{X}$ admits a universal pointed $\mathbb{A}^1$-covering $\tilde{\mathcal{X}} \to \mathcal{X}$ in the category of pointed coverings of $\mathcal{X}$. It is (up to unique isomorphism) the unique pointed $\mathbb{A}^1$-covering whose source is $\mathbb{A}^1$-simply connected. It is a $\pi_1^{\mathbb{A}^1}(\mathcal{X})$-torsor over $\mathcal{X}$ and the canonical morphism*

$$\pi_1^{\mathbb{A}^1}(\mathcal{X}) \to Aut_{\mathcal{X}}(\tilde{\mathcal{X}})$$

is an isomorphism.

Proof. Let $\mathcal{X}$ be a pointed $\mathbb{A}^1$-connected space. Let $\mathcal{X} \to L_{\mathbb{A}^1}(\mathcal{X})$ be its $\mathbb{A}^1$-localization. Let $\tilde{\mathcal{X}}_{\mathbb{A}^1}$ be the universal covering of $L_{\mathbb{A}^1}(\mathcal{X})$ in the simplicial meaning. It is a $\pi_1^{\mathbb{A}^1}(\mathcal{X})$-torsor by construction. From Lemma 7.5 $\tilde{\mathcal{X}}_{\mathbb{A}^1} \to L_{\mathbb{A}^1}(\mathcal{X})$ is thus also an $\mathbb{A}^1$-covering (as $\pi_1^{\mathbb{A}^1}(\mathcal{X})$ is strongly $\mathbb{A}^1$-invariant. Let $\tilde{\mathcal{X}} \to \mathcal{X}$ be its pull back to $\mathcal{X}$. This is a pointed $\pi_1^{\mathbb{A}^1}(\mathcal{X})$-torsor and still a pointed $\mathbb{A}^1$-covering. We claim it is the universal pointed $\mathbb{A}^1$-covering of $\mathcal{X}$.

Next we observe that $\tilde{\mathcal{X}}$ is $\mathbb{A}^1$-simply connected. This follows from the left properness property of the $\mathbb{A}^1$-model category structure on the category of spaces [59] that $\tilde{\mathcal{X}} \to \tilde{\mathcal{X}}_{\mathbb{A}^1}$ is an $\mathbb{A}^1$-weak equivalence.

Now we prove the universal property. Let $\mathcal{Y} \to \mathcal{X}$ be a pointed $\mathbb{A}^1$-covering. Let $\mathcal{Y}^{(0)} \subset \mathcal{Y}$ the inverse image of (the image of) the base point in $\pi_0^{\mathbb{A}^1}(\mathcal{Y})$. We claim (like in the above simplicial case) that $\mathcal{Y}^{(0)} \to \mathcal{X}$ is still an $\mathbb{A}^1$-covering. It follows easily from the fact that an $\mathbb{A}^1$-trivial cofibration induces an isomorphism on $\pi_0^{\mathbb{A}^1}$. In this way we reduce to proving the universal property for pointed $\mathbb{A}^1$-coverings $\mathcal{Y} \to \mathcal{X}$ with $\mathcal{Y}$ also $\mathbb{A}^1$-connected.

By Lemma 7.9 below there exists a cartesian square of pointed spaces

$$
\begin{array}{ccc}
\mathcal{Y} & \to & \mathcal{Y}' \\
\downarrow & & \downarrow \\
\mathcal{X} & \to & L_{\mathbb{A}^1}(\mathcal{X})
\end{array}
$$

with $\mathcal{Y}' \to L_{\mathbb{A}^1}(\mathcal{X})$ a pointed $\mathbb{A}^1$-covering of $L_{\mathbb{A}^1}(\mathcal{X})$. By the above theory of simplicial coverings, there exists a unique morphism of pointed coverings

$$
\begin{array}{ccc}
\tilde{\mathcal{X}}_{\mathbb{A}^1} & \to & \mathcal{Y}' \\
\downarrow & & \downarrow \\
L_{\mathbb{A}^1}(\mathcal{X}) & = & L_{\mathbb{A}^1}(\mathcal{X})
\end{array}
$$

Pulling-back this morphism to $\mathcal{X}$ yields a pointed morphism of $\mathbb{A}^1$-coverings

$$
\begin{array}{ccc}
\tilde{\mathcal{X}} & \to & \mathcal{Y} \\
\downarrow & & \downarrow \\
\mathcal{X} & = & \mathcal{X}
\end{array}
$$

Now it suffices to check that there is only one such morphism. Let f_1 and f_2 be morphisms $\tilde{\mathcal{X}} \to \mathcal{Y}$ of pointed $\mathbb{A}^1$-coverings of $\mathcal{X}$. We want to prove they are equal. We again apply Lemma 7.9 below to each f_i and get a cartesian square of pointed spaces

$$
\begin{array}{ccc}
\tilde{\mathcal{X}} & \xrightarrow{\tilde{f}_i} & \tilde{\mathcal{X}}_i' \\
\downarrow & & \downarrow \\
\mathcal{Y} & \to & \mathcal{Y}'
\end{array}
$$

in which $\tilde{\mathcal{X}} \to \tilde{\mathcal{X}}_i$ is an $\mathbb{A}^1$-weak equivalence. As a consequence the pointed $\mathbb{A}^1$-coverings $\tilde{\mathcal{X}}_i' \to \mathcal{Y}'$ to the $\mathbb{A}^1$-local space $\mathcal{Y}'$ are simply $\mathbb{A}^1$-connected and are thus both the simplicial universal pointed covering of $\mathcal{Y}'$ (and of $L_{\mathbb{A}^1}(\mathcal{X})$): let $\phi : \tilde{\mathcal{X}}_1' \cong \tilde{\mathcal{X}}_2'$ be the canonical isomorphism of pointed coverings. To check $f_1 = f_2$, it suffices to check that $\phi \circ \tilde{f}_1 = \tilde{f}_2$. But there exists $\psi : \tilde{\mathcal{X}} \to \pi_1^{\mathbb{A}^1}(\mathcal{X})$ such that $\tilde{f}_2 = \psi.(\phi \circ \tilde{f}_1)$. But as $\tilde{\mathcal{X}}$ is $\mathbb{A}^1$-connected, ψ is constant, i.e. factor as $\tilde{\mathcal{X}} \to * \to \pi_1^{\mathbb{A}^1}(\mathcal{X})$. But as all the morphisms are pointed, that constant $* \to \pi_1^{\mathbb{A}^1}(\mathcal{X})$ must be the neutral element so that $\phi \circ \tilde{f}_1 = \tilde{f}_2$.

We observe that if $\mathcal{Y} \to \mathcal{X}$ is a pointed $\mathbb{A}^1$-covering with $\mathcal{Y}$ simply $\mathbb{A}^1$-connected, the unique morphism $\tilde{\mathcal{X}} \to \mathcal{Y}$ is an $\mathbb{A}^1$-weak equivalence and thus an isomorphism by Lemma 7.7 4).

Finally it only remains to prove the statement concerning the morphism

$$\pi_1^{\mathbb{A}^1}(\mathcal{X}) \to Aut_{\mathcal{X}}(\tilde{\mathcal{X}})$$

Here the right hand side means the sheaf of groups on Sm_k which to U associates the group of automorphisms $Aut_{\mathcal{X}}(\tilde{\mathcal{X}})(U)$ of the covering $\tilde{\mathcal{X}} \times U \to \mathcal{X} \times U$. We observe that if two automorphisms $\phi_i \in Aut_{\mathcal{X}}(\tilde{\mathcal{X}})(U)$, $i \in \{1,2\}$, coincide on the base-point section $U \to \tilde{\mathcal{X}} \times U$ then $\phi_1 = \phi_2$. Indeed as $\tilde{\mathcal{X}} \times U \to \mathcal{X} \times U$ is a $\pi_1^{\mathbb{A}^1}(\mathcal{X})$-torsor, there is $\alpha : \tilde{\mathcal{X}} \times U \to \pi_1^{\mathbb{A}^1}(\mathcal{X})$ with $\phi_2 = \alpha.\phi_1$. But $\pi_0^{\mathbb{A}^1}(\tilde{\mathcal{X}} \times U) = \pi_0^{\mathbb{A}^1}(U)$ and α factors through $\pi_0^{\mathbb{A}^1}(U) \to \pi_1^{\mathbb{A}^1}(\mathcal{X})$. As the composition of α with the base-point section $U \to \tilde{\mathcal{X}} \times U$ is the neutral element, we conclude that α is the neutral element and $\phi_1 = \phi_2$.

This first shows that the above morphism is a monomorphism. Let $\phi \in Aut_{\mathcal{X}}(\tilde{\mathcal{X}})(U)$. Composing ϕ with the base-point section $U \to \tilde{\mathcal{X}} \times U$ we get $\psi \in \pi_1^{\mathbb{A}^1}(\mathcal{X})(U)$. But the automorphisms ϕ and ψ coincide by construction on the base-point section. Thus they are equal and our morphism is also onto. The theorem is proven.   □

Lemma 7.9. *Let $\mathcal{Y} \to \mathcal{X}$ be a pointed $\mathbb{A}^1$-covering between pointed $\mathbb{A}^1$-connected spaces. Then for any $\mathbb{A}^1$-weak equivalence $\mathcal{X} \to \mathcal{X}'$ any there exists a cartesian square of spaces*

$$\begin{array}{ccc} \mathcal{Y} & \to & \mathcal{Y}' \\ \downarrow & & \downarrow \\ \mathcal{X} & \to & \mathcal{X}' \end{array}$$

in which the right vertical morphism is an $\mathbb{A}^1$-covering (and thus the top horizontal morphism an $\mathbb{A}^1$-weak equivalence).

Proof. Let $\mathcal{X}' \to L_{\mathbb{A}^1}(\mathcal{X}')$ be the $\mathbb{A}^1$-localization of $\mathcal{X}'$. As by construction, $L_{\mathbb{A}^1}(-)$ is a functor on spaces we get a commutative square

$$\begin{array}{ccc} \mathcal{Y} & \to & L_{\mathbb{A}^1}(\mathcal{Y}) \\ \downarrow & & \downarrow \\ \mathcal{X} & \to & L_{\mathbb{A}^1}(\mathcal{X}') \end{array}$$

in which the horizontal arrows are $\mathbb{A}^1$-weak equivalences. As the left vertical arrow is an $\mathbb{A}^1$-fibration (by Lemma 7.7) with $\mathbb{A}^1$-homotopy fiber equal to the fiber $\Gamma \subset \mathcal{Y}$, which is an $\mathbb{A}^1$-invariant sheaf, thus is $\mathbb{A}^1$-local, the $\mathbb{A}^1$-homotopy fiber of the pointed morphism $L_{\mathbb{A}^1}(\mathcal{Y}) \to L_{\mathbb{A}^1}(\mathcal{X}')$ is $\mathbb{A}^1$-

equivalent to the previous one (because the square is obviously $\mathbb{A}^1$-homotopy cartesian).

As both $L_{\mathbb{A}^1}(\mathcal{Y})$ and $L_{\mathbb{A}^1}(\mathcal{X}')$ are $\mathbb{A}^1$-fibrant and (simplicially) connected, this means (using the theory of simplicial coverings for $L_{\mathbb{A}^1}(\mathcal{X}')$) that there exists a commutative square

$$
\begin{array}{ccc}
L_{\mathbb{A}^1}(\mathcal{Y}) & \to & \mathcal{Y}' \\
\downarrow & & \downarrow \\
L_{\mathbb{A}^1}(\mathcal{X}') & = & L_{\mathbb{A}^1}(\mathcal{X}')
\end{array}
$$

in which $\mathcal{Y}' \to L_{\mathbb{A}^1}(\mathcal{X}')$ is an ($\mathbb{A}^1$-)covering and $L_{\mathbb{A}^1}(\mathcal{Y}) \to \mathcal{Y}'$ an ($\mathbb{A}^1$-)weak equivalence.

This $\mathbb{A}^1$-homotopy cartesian square induces a commutative square

$$
\begin{array}{ccc}
\mathcal{Y} & \to & \mathcal{Y}'' \\
\downarrow & & \downarrow \\
\mathcal{X} & = & \mathcal{X}
\end{array}
\tag{7.1}
$$

in which both vertical morphisms are $\mathbb{A}^1$-coverings and the top horizontal morphism is an $\mathbb{A}^1$-weak equivalence (by the properness of the $\mathbb{A}^1$-model structure [59]), where $\mathcal{Y}''$ is the fiber product $\mathcal{Y}' \times_{L_{\mathbb{A}^1}(\mathcal{X}')} \mathcal{X}$. By Lemma 7.7 $\mathcal{Y} \to \mathcal{Y}''$ is an isomorphism. This finishes our proof as $\mathcal{Y}''$ is the pull-back of an $\mathbb{A}^1$-covering of $\mathcal{X}'$ because $\mathcal{X} \to L_{\mathbb{A}^1}(\mathcal{X}')$ factor through $\mathcal{X} \to \mathcal{X}'$. $\square$

Remark 7.10. Let us denote by $Cov_{\mathbb{A}^1}(\mathcal{X})$ the category of $\mathbb{A}^1$-coverings of a fixed pointed $\mathbb{A}^1$-connected space $\mathcal{X}$. The fiber Γ_{x_0} of an $\mathbb{A}^1$-covering $\mathcal{Y} \to \mathcal{X}$ over the base point x_0 is an $\mathbb{A}^1$-invariant sheaf of sets. One may define a natural right action of $\pi_1^{\mathbb{A}^1}(\mathcal{X})$ on $\Gamma_{x_0}(\mathcal{Y} \to \mathcal{X})$ and it can be shown that the induced functor Γ_{x_0} from $Cov_{\mathbb{A}^1}(\mathcal{X})$ to the category of $\mathbb{A}^1$-invariant sheaves with a right action of $\pi_1^{\mathbb{A}^1}(\mathcal{X})$ is an equivalence of categories.

When $\mathcal{X}$ is an arbitrary space, this correspondence can be extended to an equivalence between the category $Cov_{\mathbb{A}^1}(\mathcal{X})$ and some category of "functor-sheaves" defined on the fundamental $\mathbb{A}^1$-groupoid of $\mathcal{X}$. $\square$

We end this section by mentioning the (easy version of the) Van-Kampen theorem.

Remark 7.11. The trick to deduce these kind of results is to observe that for any pointed connected space $\mathcal{X}$, the map $[\mathcal{X}, BG]_{\mathcal{H}_{s,\bullet}(k)} \to Hom_{\mathcal{G}r}(\pi_1(\mathcal{X}), G)$ is a bijection. This follows as usual by considering the functoriality of the Postnikov tower [59]. But then if G is a strongly $\mathbb{A}^1$-invariant sheaf, we get in the same way:

$$
[\mathcal{X}, BG]_{\mathcal{H}_\bullet(k)} \to Hom_{\mathcal{G}r_{\mathbb{A}^1}}(\pi_1(\mathcal{X}), G)
$$

where $\mathcal{G}r_{\mathbb{A}^1}$ denotes the category of strongly $\mathbb{A}^1$-invariant sheaves of groups. It follows at once that the inclusion $\mathcal{G}r_{\mathbb{A}^1} \subset \mathcal{G}r$ admits a left adjoint $G \mapsto G_{\mathbb{A}^1}$, with $G_{\mathbb{A}^1} := \pi_1^{\mathbb{A}^1}(BG) = \pi_1(L_{\mathbb{A}^1}(BG))$. As a consequence, $\mathcal{G}r_{\mathbb{A}^1}$ admits all colimits. For instance we get the existence of sums denoted by $*^{\mathbb{A}^1}$ in $\mathcal{G}r_{\mathbb{A}^1}$: if G_i is a family of strongly $\mathbb{A}^1$-invariant sheaves, their sum $*_i^{\mathbb{A}^1} G_i$ is $(*_i G_i)_{\mathbb{A}^1}$ where $*$ means the usual sum in $\mathcal{G}r$.   □

Theorem 7.12. *Let X be an $\mathbb{A}^1$-connected pointed smooth scheme. Let $\{U_i\}_i$ be a an open covering of X by $\mathbb{A}^1$-connected open subschemes which contains the base point. Assume furthermore that each intersection $U_i \cap U_j$ is still $\mathbb{A}^1$-connected. Then for any strongly $\mathbb{A}^1$-invariant sheaf of groups G, the following diagram*

$$*_{i,j}^{\mathbb{A}^1} \pi_1^{\mathbb{A}^1}(U_i \cap U_j) \rightrightarrows *_i^{\mathbb{A}^1} \pi_1^{\mathbb{A}^1}(U_i) \to \pi_1^{\mathbb{A}^1}(X) \to *$$

is right exact in $\mathcal{G}r_{\mathbb{A}^1}$.

Proof. We let $\check{C}(\mathcal{U})$ the simplicial space associated to the covering U_i of X (the Čech object of the covering). By definition, $\check{C}(\mathcal{U}) \to X$ is a simplicial weak equivalence. Thus from Remark 7.11, it follows that for any $G \in \mathcal{G}r_{\mathbb{A}^1}$

$$Hom_{\mathcal{G}r_{\mathbb{A}^1}}(\pi_1(X), G) = [\check{C}(\mathcal{U}), BG]_{\mathcal{H}_{s,\bullet}(k)}$$

Now the usual skeletal filtration of $\check{C}(\mathcal{U})$ easily yields the fact that the obvious diagram (of sets)

$$Hom_{\mathcal{G}r}(\pi_1^{\mathbb{A}^1}(\check{C}(\mathcal{U})), G) \to \Pi_i Hom_{\mathcal{G}r}(\pi_1^{\mathbb{A}^1}(U_i), G)$$

$$\rightrightarrows \Pi_{i,j} Hom_{\mathcal{G}r}(\pi_1^{\mathbb{A}^1}(U_i \cap U_j), G)$$

is exact. Putting all these together we obtain our claim.   □

7.2   Basic Computation: $\pi_1^{\mathbb{A}^1}(\mathbb{P}^n)$ and $\pi_1^{\mathbb{A}^1}(SL_n)$ for $n \geq 2$

The following is the easiest application of the preceding results:

Theorem 7.13. *For $n \geq 2$ the canonical $\mathbb{G}_m$-torsor*

$$\mathbb{G}_m - (\mathbb{A}^{n+1} - \{0\}) \to \mathbb{P}^n$$

is the universal $\mathbb{A}^1$-covering of $\mathbb{P}^n$. This defines a canonical isomorphism $\pi_1^{\mathbb{A}^1}(\mathbb{P}^n) \cong \mathbb{G}_m$.

Proof. For $n \geq 2$, the pointed space $\mathbb{A}^{n+1} - \{0\}$ is $\mathbb{A}^1$-simply connected by Theorem 6.38. We now conclude by Theorem 7.8. $\square$

For $n = 1$, $\mathbb{A}^2 - \{0\}$ is no longer 1-$\mathbb{A}^1$-connected. We now compute $\pi_1^{\mathbb{A}^1}(\mathbb{A}^2 - \{0\})$. As $SL_2 \to \mathbb{A}^2 - \{0\}$ is an $\mathbb{A}^1$-weak equivalence, $\pi_1^{\mathbb{A}^1}(\mathbb{A}^2 - \{0\}) \cong \pi_1^{\mathbb{A}^1}(SL_2)$. Now, the $\mathbb{A}^1$-fundamental sheaf of groups $\pi_1^{\mathbb{A}^1}(G)$ of a group-space G is always a sheaf of abelian groups by the classical argument. Here we mean by "group-space" a group object in the category of spaces, that is to say a simplicial sheaf of groups on Sm_k.

By the Hurewicz theorem and Theorem 6.40 we get canonical isomorphisms $\pi_1^{\mathbb{A}^1}(SL_2) = H_1^{\mathbb{A}^1}(SL_2) = H_1^{\mathbb{A}^1}(\mathbb{A}^2 - \{0\}) = \underline{\mathbf{K}}_2^{MW}$.

Finally the classical argument also yields:

Lemma 7.14. *Let G be a group-space which is $\mathbb{A}^1$-connected. Then there exists a unique group structure on the pointed space $\tilde{G}$ for which the $\mathbb{A}^1$-covering $\tilde{G} \to G$ is an (epi-)morphism of group-spaces. The kernel is central and canonically isomorphic to $\pi_1^{\mathbb{A}^1}(\tilde{G})$.*

Altogether we have obtained:

Theorem 7.15. *The universal $\mathbb{A}^1$-covering of SL_2 given by Theorem 7.8 admits a group structure and we get in this way a central extension of sheaves of groups*

$$0 \to \underline{\mathbf{K}}_2^{MW} \to \tilde{SL}_2 \to SL_2 \to 1$$

Remark 7.16. In fact this extension can also be shown to be a (central) extension in the Zariski topology by the theorem. $\square$

This central extension can be constructed in the following way:

Lemma 7.17. *Let $B(SL_2)$ denote the simplicial classifying space of SL_2. Then there exists a unique $\mathcal{H}_{s,\bullet}(k)$-morphism*

$$e_2 : B(SL_2) \to K(\underline{\mathbf{K}}_2^{MW}, 2)$$

which composed with $\Sigma(SL_2) \subset B(SL_2)$ gives the canonical cohomology class $\Sigma(SL_2) \cong \Sigma(\mathbb{A}^2 - \{0\}) \to K(\underline{\mathbf{K}}_2^{MW}, 2)$.

The central extension of SL_2 associated with this element of $H^2(SL_2; \underline{\mathbf{K}}_2^{MW})$ is canonically isomorphic to the central extension of Theorem 7.15.

Proof. We use the skeletal filtration F_s of the classifying space BG; it has the property that (simplicially) $F_s/F_{s-1} \cong \Sigma^s(G^{\wedge s})$. Clearly now, using the long exact sequences in cohomology with coefficients in $\underline{\mathbf{K}}_2^{MW}$ one sees that the restriction:

$$H^2(B(SL_2); \underline{\mathbf{K}}_2^{MW}) \to H^2(F_1; \underline{\mathbf{K}}_2^{MW}) = H^2(\Sigma(SL_2); \underline{\mathbf{K}}_2^{MW})$$

is an isomorphism.

Now it is well-known that such an element in $H^2(B(SL_2); \underline{\mathbf{K}}_2^{MW})$ corresponds to a central extension of sheaves as above: just take the pointed simplicial homotopy fiber Γ of (a representative of) the previous morphism $B(SL_2) \to K(\underline{\mathbf{K}}_2^{MW}, 2)$. Using the long exact homotopy sequence of simplicial homotopy sheaves of this fibration yields the required central extension:

$$0 \to \underline{\mathbf{K}}_2^{MW} \to \pi_1(\Gamma) \to SL_2 \to 0$$

To check it is the universal $\mathbb{A}^1$-covering for SL_2, just observe that the map $B(SL_2) \to K(\underline{\mathbf{K}}_2^{MW}, 2)$ is onto on $\pi_2^{\mathbb{A}^1}$ as the map $\Sigma(SL_2) \cong \Sigma(\mathbb{A}^2 - \{0\}) \to K(\underline{\mathbf{K}}_2^{MW}, 2)$ is already onto (actually an isomorphism) on $\pi_2^{\mathbb{A}^1}$. Now by Theorem 6.50 the simplicial homotopy fiber sequence is also an $\mathbb{A}^1$-fiber homotopy sequence. Then the long exact homotopy sequence in $\mathbb{A}^1$-homotopy sheaves this time shows that Γ is simply 2-$\mathbb{A}^1$-connected. Thus the group-object $\pi_1(\Gamma)$ is simply $\mathbb{A}^1$-connected thus is canonically isomorphic to $\tilde{SL}_2$.

$\square$

Remark 7.18. 1) As a $\underline{\mathbf{K}}_2^{MW}$-torsor (forgetting the group structure) $\tilde{SL}_2$ can easily be described as follows. We use the morphism $SL_2 \to \mathbb{A}^2 - \{0\}$. It is thus sufficient to describe a $\underline{\mathbf{K}}_2^{MW}$-torsor over $\mathbb{A}^2 - \{0\}$. We use the open covering of $\mathbb{A}^2 - \{0\}$ by the two obvious open subsets $\mathbb{G}_m \times \mathbb{A}^1$ and $\mathbb{A}^1 \times \mathbb{G}_m$. Their intersection is exactly $\mathbb{G}_m \times \mathbb{G}_m$. The tautological symbol $\mathbb{G}_m \times \mathbb{G}_m \to \underline{\mathbf{K}}_2^{MW}$ (see Sect. 3.3) defines a 1-cocycle on $\mathbb{A}^2 - \{0\}$ with values in $\underline{\mathbf{K}}_2^{MW}$ and thus an $\underline{\mathbf{K}}_2^{MW}$-torsor. The pull-back of this torsor to SL_2 is $\tilde{SL}_2$. It suffices to check that it is simply $\mathbb{A}^1$-connected. This follows in the same way as in the previous proof from the fact the $\mathcal{H}_{\bullet}(k)$-morphism $\mathbb{A}^2 - \{0\} \to K(\underline{\mathbf{K}}_2^{MW}, 1)$ induced by the previous 1-cocycle is an isomorphism on $\pi_1^{\mathbb{A}^1}$.

2) For any SL_2-torsor ξ over a smooth scheme X (or equivalently a rank two vector bundle ξ over X with a trivialization of $\Lambda^2(\xi)$) the composition of the $\mathcal{H}_{s,\bullet}(k)$-morphisms $X \to B(SL_2)$ classifying ξ and of $e_2 : B(SL_2) \to K(\underline{\mathbf{K}}_2^{MW}, 2)$ defines an element $e_2(\xi) \in H^2(X; \underline{\mathbf{K}}_2^{MW})$; this can be shown to coincide with the Euler class of ξ defined in [7].

$\square$

The Computation of $\pi_1^{\mathbb{A}^1}(SL_n)$, $n \geq 3$

Lemma 7.19. *1) For $n \geq 3$, the inclusion $SL_n \subset SL_{n+1}$ induces an isomorphism*

$$\pi_1^{\mathbb{A}^1}(SL_n) \cong \pi_1^{\mathbb{A}^1}(SL_{n+1})$$

2) The inclusion $SL_2 \subset SL_3$ induces an epimorphism

$$\pi_1^{\mathbb{A}^1}(SL_2) \twoheadrightarrow \pi_1^{\mathbb{A}^1}(SL_3)$$

Proof. We denote by $SL'_n \subset SL_{n+1}$ the subgroup formed by the matrices of the form

$$\begin{pmatrix} 1 & 0 \ldots 0 \\ ? & \\ \vdots & M \\ ? & \end{pmatrix}$$

with $M \in SL_n$. Observe that the group homomorphism $SL'_n \to SL_n$ is an $\mathbb{A}^1$-weak equivalence: indeed the inclusion $SL_n \subset SL'_n$ shows SL'_n is the semi-direct product of SL_n and $\mathbb{A}^n$ so that as a space SL'_n is the product $\mathbb{A}^n \times SL_n$.

The group SL'_n is the isotropy subgroup of $(1, 0, \ldots, 0)$ under the right action of SL_{n+1} on $\mathbb{A}^{n+1} - \{0\}$. The following diagram

$$SL'_n - SL_{n+1} \to \mathbb{A}^{n+1} - \{0\} \tag{7.2}$$

is thus an SL'_n-Zariski torsor over $\mathbb{A}^{n+1} - \{0\}$, where the map $SL_{n+1} \to \mathbb{A}^{n+1} - \{0\}$ assigns to a matrix its first horizontal line.

By Theorem 6.53, and our computations, the simplicial fibration sequence (7.2) is still an $\mathbb{A}^1$-fibration sequence. The associated long exact sequence of $\mathbb{A}^1$-homotopy sheaves, together with the fact that $\mathbb{A}^{n+1} - \{0\}$ is $(n-1)$-$\mathbb{A}^1$-connected and that $SL_n \subset SL'_n$ an $\mathbb{A}^1$-weak equivalence implies the claim. $\qquad\square$

Now we may state the following result which implies the point 2) of Theorem 1.27:

Theorem 7.20. *The canonical isomorphism $\pi_1^{\mathbb{A}^1}(SL_2) \cong \mathbf{K}_2^{MW}$ induces through the inclusions $SL_2 \to SL_n$, $n \geq 3$, an isomorphism*

$$\mathbf{K}_2^M = \mathbf{K}_2^{MW}/\eta \cong \pi_1^{\mathbb{A}^1}(SL_n) = \pi_1^{\mathbb{A}^1}(SL_\infty) = \pi_1^{\mathbb{A}^1}(GL_\infty)$$

Remark 7.21. Let $\mathbb{A}^3 - \{0\} \to B(SL'_2)$ be the morphism in $\mathcal{H}_{s,\bullet}(k)$ which classifies the SL'_2-torsor (7.2) over $\mathbb{A}^3 - \{0\}$. Applying $\pi_2^{\mathbb{A}^1}$ yields a morphism:

$$\mathbf{K}_3^{MW} = \pi_2^{\mathbb{A}^1}(\mathbb{A}^3 - \{0\}) \to \pi_2^{\mathbb{A}^1}(B(SL'_2)) \cong \pi_2^{\mathbb{A}^1}(B(SL_2)) = \pi_1^{\mathbb{A}^1}(SL_2) = \mathbf{K}_2^{MW}$$

This morphism can be shown in fact to be the Hopf morphism η in Milnor–Witt K-theory sheaves. The proof we give below only gives that this morphism is η up to multiplication by a unit in $W(k)$. $\qquad\square$

Remark 7.22. We will use in the proof the "second Chern class morphism", a canonical $\mathcal{H}_\bullet(k)$-morphism

$$c_2 : B(GL_\infty) \to K(\mathbf{K}_2^M, 2)$$

more generally the n-th Chern class morphism $c_n : B(GL_\infty) \to K(\underline{\mathbf{K}}_n^M, n)$ is defined as follows: in $\mathcal{H}_\bullet(k)$, $B(GL_\infty)$ is canonically isomorphic to the infinite Grassmannian $\mathbb{G}r_\infty$ [59]. This space is the filtering colimit of the finite Grassmannian $\mathbb{G}r_{m,i}$ [loc. cit. p. 138]. But, $[\mathbb{G}r_{m,i}, K(\underline{\mathbf{K}}_n^M, n)]_{\mathcal{H}_\bullet(k)}$ is the cohomology group $H^n(\mathbb{G}r_{m,i}; \underline{\mathbf{K}}_n^M)$. This group is isomorphic to the n-th Chow group $CH^n(\mathbb{G}r_{m,i})$ by Rost [68], and we let $c_n \in [\mathbb{G}r_{m,i}, K(K_n^M, n)]_{\mathcal{H}_\bullet(k)}$ denote the n-th Chern class of the tautological rank m vector over bundle on $\mathbb{G}r_{m,i}$ [28]. As the Chow groups of the Grassmanians stabilize loc. cit., the Milnor exact sequence gives a canonical element $c_n \in [B(GL_\infty), K(\underline{\mathbf{K}}_n^M, n)]_{\mathcal{H}_\bullet(k)}$.

Form this definition it is easy to check that c_2 is the unique morphism $B(GL_\infty) \to K(\underline{\mathbf{K}}_2^M, 2)$ whose composite with $\Sigma(GL_2) \to B(GL_2) \to B(GL_\infty) \to K(\underline{\mathbf{K}}_2^M, 2))$ is the canonical composition $\Sigma(GL_2) \to \Sigma(\mathbb{A}^2 - \{0\}) \to K(\underline{\mathbf{K}}_2^{MW}, 2)) \to K(\underline{\mathbf{K}}_2^M, 2))$. $\qquad\Box$

Proof of Theorem 7.20. Lemma 7.19 implies that we only have to show that the epimorphism

$$\pi : \underline{\mathbf{K}}_2^{MW} = \pi_1^{\mathbb{A}^1}(SL_2) \twoheadrightarrow \pi_1^{\mathbb{A}^1}(SL_3)$$

has exactly has kernel the image $\mathcal{I}(\eta) \subset \underline{\mathbf{K}}_2^{MW}$ of $\eta : \underline{\mathbf{K}}_3^{MW} \to \underline{\mathbf{K}}_2^{MW}$.

The long exact sequence of homotopy sheaves of the $\mathbb{A}^1$-fibration sequence (7.2): $SL_2' - SL_3 \to \mathbb{A}^3 - \{0\}$ and the $\mathbb{A}^1$-weak equivalence $SL_2' \to SL_2$ provides an exact sequence

$$\underline{\mathbf{K}}_3^{MW} = \pi_1^{\mathbb{A}^1}(\mathbb{A}^3 - \{0\}) \to \underline{\mathbf{K}}_2^{MW} = \pi_1^{\mathbb{A}^1}(SL_2) \twoheadrightarrow \pi_1^{\mathbb{A}^1}(SL_3) \to 0$$

But from the fact that $\underline{\mathbf{K}}_n^{MW}$ is the free strongly/strictly $\mathbb{A}^1$-invariant sheaf on $(\mathbb{G}_m)^{\wedge n}$ we get that the obvious morphism

$$Hom_{Ab(k)}(\underline{\mathbf{K}}_3^{MW}, \underline{\mathbf{K}}_2^{MW}) \to \underline{\mathbf{K}}_{-1}^{MW}(k) = W(k)$$

is an isomorphism. Thus this means that the connecting homomorphism $\underline{\mathbf{K}}_3^{MW} \to \underline{\mathbf{K}}_2^{MW}$ is a multiple of η. This proves the inclusion $Ker(\pi) \subset \mathcal{I}(\eta)$. Now the morphism $\pi_1^{\mathbb{A}^1}(SL_2) \twoheadrightarrow \pi_1^{\mathbb{A}^1}(SL_3) \to \underline{\mathbf{K}}_2^M$ induced by the second Chern class (cf. Remark 7.22) is the obvious projection $\underline{\mathbf{K}}_2^{MW} \to \underline{\mathbf{K}}_2^{MW}/\eta = \underline{\mathbf{K}}_2^M$. This shows the converse inclusion. $\qquad\Box$

7.3 The Computation of $\pi_1^{\mathbb{A}^1}(\mathbb{P}^1)$

We recall from [59] that there is a canonical $\mathcal{H}_\bullet(k)$-isomorphism $\mathbb{P}^1 \cong \Sigma(\mathbb{G}_m)$ induced by the covering of $\mathbb{P}^1$ by its two standard $\mathbb{A}^1$'s intersecting to $\mathbb{G}_m$. Thus to compute $\pi_1^{\mathbb{A}^1}(\mathbb{P}^1)$ is the same thing as to compute $\pi_1^{\mathbb{A}^1}(\Sigma(\mathbb{G}_m))$.

Let us denote by $Shv_\bullet$ the category of sheaves of pointed sets on Sm_k. For any $\mathcal{S} \in Shv_\bullet$, we denote by $\theta_\mathcal{S} : \mathcal{S} \to \pi_1^{\mathbb{A}^1}(\Sigma(\mathcal{S}))$ the canonical $Shv_\bullet$-morphism obtained by composing $\mathcal{S} \to \pi_1(\Sigma(\mathcal{S}))$ and $\pi_1(\Sigma(\mathcal{S})) \to \pi_1^{\mathbb{A}^1}(\Sigma(\mathcal{S}))$.

Lemma 7.23. *The morphism $\mathcal{S}$ induces for any strongly $\mathbb{A}^1$-invariant sheaf of groups G a bijection $Hom_{\mathcal{G}r}(\pi_1^{\mathbb{A}^1}(\Sigma(\mathcal{S})), G) \cong Hom_{Shv_\bullet}(\mathcal{S}, G)$.*

Proof. As the classifying space BG is $\mathbb{A}^1$-local the map $[\Sigma(\mathcal{S}), BG]_{\mathcal{H}_{s,\bullet}(k)} \to [\Sigma(\mathcal{S}), BG]_{\mathcal{H}_\bullet(k)}$ is a bijection.

Now the obvious map $[\Sigma(\mathcal{S}), BG]_{\mathcal{H}_\bullet(k)} \to Hom_{\mathcal{G}r}(\pi_1^{\mathbb{A}^1}(\Sigma(\mathcal{S})), G)$ given by the functor π_1 is bijective, see Remark 7.11.

The classical adjunction $[\Sigma(\mathcal{S}), BG]_{\mathcal{H}_{s,\bullet}(k)} \cong [\mathcal{S}, \Omega^1(BG)]_{\mathcal{H}_{s,\bullet}(k)}$ and the canonical $\mathcal{H}_{s,\bullet}(k)$-isomorphism $G \cong \Omega^1(BG)$ are checked to provide the required bijection. $\qquad\square$

The previous result can be expressed by saying that the sheaf of groups $F_{\mathbb{A}^1}(\mathcal{S}) := \pi_1^{\mathbb{A}^1}(\Sigma(\mathcal{S}))$ is the "free strongly $\mathbb{A}^1$-invariant" sheaf of groups on the pointed sheaf of sets $\mathcal{S}$. In the sequel we will simply denote, for $n \geq 1$, by $F_{\mathbb{A}^1}(n)$ the sheaf $\pi_1^{\mathbb{A}^1}(\Sigma((\mathbb{G}_m)^{\wedge n}))$.

We have proven in Sect. 7.2 that $F_{\mathbb{A}^1}(2)$ is abelian and (thus) isomorphic to $\mathbf{K}_2^{MW}$. Our aim is to describe $F_{\mathbb{A}^1}(1) = \pi_1^{\mathbb{A}^1}(\mathbb{P}^1)$.

The Hopf Map of a Sheaf of Group

Recall that for two pointed spaces $\mathcal{X}$ and $\mathcal{Y}$ we let $\mathcal{X} * \mathcal{Y}$ denote the *reduced join* of $\mathcal{X}$ and $\mathcal{Y}$, that is to say the quotient of $\Delta^1 \times \mathcal{X} \times \mathcal{Y}$ by the relations $(0, x, y) = (0, x, y')$, $(1, x, y) = (1, x', y)$ and $(t, x_0, y_0) - (t, x_0, y_0)$ where x_0 (resp. y_0) is the base point of $\mathcal{X}$ (resp. $\mathcal{Y}$). It is a homotopy colimit of the diagram of pointed spaces

$$\begin{array}{c} \mathcal{X} \\ \uparrow \\ \mathcal{X} \times \mathcal{Y} \to \mathcal{Y} \end{array}$$

Example 7.24. $\mathbb{A}^2 - \{0\}$ has canonically the $\mathbb{A}^1$-homotopy type of $\mathbb{G}_m * \mathbb{G}_m$: use the classical covering of $\mathbb{A}^2 - \{0\}$ by $\mathbb{G}_m \times \mathbb{A}^1$ and $\mathbb{A}^1 \times \mathbb{G}_m$ with intersection $\mathbb{G}_m \times \mathbb{G}_m$. $\qquad\square$

The join $\mathcal{X} * (point)$ of $\mathcal{X}$ and the point is called the cone of $\mathcal{X}$ and is denoted by $C(\mathcal{X})$. It is the smash product $\Delta^1 \wedge \mathcal{X}$ with Δ^1 pointed by 1. we let $\mathcal{X} \subset C(\mathcal{X})$ denote the canonical inclusion. The quotient is obviously $\Sigma(\mathcal{X})$. The "anticone" $C'(\mathcal{X})$ is the smash product $\Delta^1 \wedge \mathcal{X}$ with Δ^1 pointed by 0.

The join obviously contains the wedge $C(\mathcal{X}) \vee C'(\mathcal{Y})$. Clearly the quotient $(\mathcal{X} * \mathcal{Y})/(\mathcal{X} \vee \mathcal{Y})$ is $\Sigma(\mathcal{X} \times \mathcal{Y})$ and the quotient $(\mathcal{X} * \mathcal{Y})/(C(\mathcal{X}) \vee C'(\mathcal{Y}))$ is $\Sigma(\mathcal{X} \wedge \mathcal{Y})$.

The morphism of pointed spaces $\mathcal{X} * \mathcal{Y} \to \Sigma(\mathcal{X} \wedge \mathcal{Y})$ is thus a simplicial weak-equivalence. The diagram of pointed spaces

$$
\begin{array}{cc}
\Sigma(\mathcal{X} \times \mathcal{Y}) & \Sigma(\mathcal{X} \times \mathcal{Y}) \\
\uparrow & \downarrow \\
\mathcal{X} * \mathcal{Y} & \xrightarrow{\sim} \Sigma(\mathcal{X} \wedge \mathcal{Y})
\end{array}
$$

defines a canonical $\mathcal{H}_{s,\bullet}(k)$-morphism

$$
\omega_{\mathcal{X},\mathcal{Y}} : \Sigma(\mathcal{X} \times \mathcal{Y}) \to \Sigma(\mathcal{X} \times \mathcal{Y})
$$

The following result is classical:

Lemma 7.25. *The $\mathcal{H}_{s,\bullet}(k)$-morphism $\omega_{\mathcal{X},\mathcal{Y}}$ is (for the co-h-group structure on $\Sigma(\mathcal{X} \times \mathcal{Y})$ equal in $\mathcal{H}_{s,\bullet}(k)$ to $(\pi_1)^{-1}.Id_{\Sigma(\mathcal{X} \times \mathcal{Y})}.(\pi_2)^{-1}$, where π_1 is the obvious composition $\Sigma(\mathcal{X} \times \mathcal{Y}) \to \Sigma(\mathcal{X}) \to \Sigma(\mathcal{X} \times \mathcal{Y})$ and π_2 is defined the same way using $\mathcal{Y}$.*

Proof. To prove this, the idea is to construct an explicit model for the map $\Sigma(\mathcal{X} \times \mathcal{Y}) \to \mathcal{X} * \mathcal{Y}$. One may use as model for $\Sigma(\mathcal{X} \times \mathcal{Y})$ the amalgamate sum Σ of $C(\mathcal{X} \times \mathcal{Y})$, $\Delta^1 \mathcal{X} \times \mathcal{Y}$ and $C'(\mathcal{X} \times \mathcal{Y})$ obviously glued together. Collapsing $C(\mathcal{X} \times \mathcal{Y}) \vee C'(\mathcal{X} \times \mathcal{Y})$ in that space gives exactly $\Sigma(\mathcal{X} \times \mathcal{Y})$ thus $\Sigma \to \Sigma(\mathcal{X} \times \mathcal{Y})$ is a simplicial weak equivalence. Now there is an obvious map $\Sigma \to \mathcal{X} * \mathcal{Y}$ given by the obvious inclusions of the cones and the canonical projection on the middle. It then remains to understand the composition $\Sigma \to \mathcal{X} * \mathcal{Y} \to \Sigma(\mathcal{X} \times \mathcal{Y})$. This is easily analyzed and yields the result. □

Now let G be a sheaf of groups. We consider the pointed map

$$
\mu'_G : G \times G \to G, \quad (g, h) \mapsto g^{-1}.h
$$

This morphism induces a morphism $\Delta^1 \times G \times G \to \Delta \times G$ which is easily seen to induce a canonical morphism

$$
\eta_G : G * G \to \Sigma(G)
$$

which is called the (geometric) Hopf map of G.

We will still denote by $\eta_G : \Sigma(G \wedge G) \to \Sigma(G)$ the canonical $\mathcal{H}_\bullet(k)$-morphism obtained as the composition of the geometric Hopf map and the inverse to $G * G \to \Sigma(G \wedge G)$.

Example 7.26. Example 7.24 implies that the Hopf fibration $\mathbb{A}^2 - \{0\} \to \mathbb{P}^1$ is canonically $\mathbb{A}^1$-equivalent to the geometric Hopf map $\eta_{\mathbb{G}_m}$

$$
\Sigma(\mathbb{G}_m \wedge \mathbb{G}_m) \to \Sigma(\mathbb{G}_m) \qquad\qquad \square
$$

We observe that G acts diagonally on $G * G$ and that the geometric Hopf map $\eta_G : G * G \to \Sigma(G)$ is a G-torsor. It is well known that the classifying map $\Sigma(G) \to BG$ for this G torsor is the canonical one [46]. By Theorem 6.53 if $\pi_0^{\mathbb{A}^1}(G)$ is a strongly $\mathbb{A}^1$-invariant sheaf, the simplicial fibration

$$G * G \to \Sigma(G) \to BG \tag{7.3}$$

is also an $\mathbb{A}^1$-fibration sequence.

Remark 7.27. Examples are given by $G = SL_n$ and $G = GL_n$ for any $n \geq 1$. In fact we do not know any connected smooth algebraic k-group which doesn't satisfy this assumption. $\qquad\qquad\square$

The following result is an immediate consequence of Lemma 7.25:

Corollary 7.28. *For any sheaf of groups G, the composition*

$$\Sigma(G \times G) \to \Sigma(G \wedge G) \overset{\eta_G}{\to} \Sigma(G)$$

is equal in $[\Sigma(G \times G), \Sigma(G)]_{\mathcal{H}_{s,\bullet}(k)}$ (for the usual group structure) to

$$(\Sigma(\chi_1))^{-1}.\Sigma(\mu').(\Sigma(pr_2))^{-1}$$

where χ_1 is the obvious composition $G \times G \overset{pr_1}{\to} G \overset{g \mapsto g^{-1}}{\to} G \overset{Id_G \times *}{\to} G \times G$ and pr_2 is the composition $G \times G \overset{pr_2}{\to} G \overset{* \times Id_G}{\to} G \times G$.

We now specialize to $G = \mathbb{G}_m$. From what we have just done, the fibration sequence (7.3) $\mathbb{G}_m * \mathbb{G}_m \to \Sigma(\mathbb{G}_m) \to B\mathbb{G}_m$ is $\mathbb{A}^1$-equivalent to

$$\mathbb{A}^2 - \{0\} \to \mathbb{P}^1 \to \mathbb{P}^\infty$$

As the spaces $\Sigma(\mathbb{G}_m) \cong \mathbb{P}^1$ and $B(\mathbb{G}_m) \cong \mathbb{P}^\infty$ are $\mathbb{A}^1$-connected, the long exact sequence of homotopy sheaves induces at once a short exact sequence of sheaves of groups

$$1 \to \mathbf{K}_2^{MW} \to F_{\mathbb{A}^1}(1) \to \mathbb{G}_m \to 1 \tag{7.4}$$

We simply denote by $\theta : \mathbb{G}_m \to F_{\mathbb{A}^1}(1)$ the section $\theta_{\mathbb{G}_m}$. As the sheaf of pointed sets $F_{\mathbb{A}^1}(1)$ is the product $\mathbf{K}_2^{MW} \times \mathbb{G}_m$ (using θ), the following result entirely describes the group structure on $F_{\mathbb{A}^1}(1)$ and thus the sheaf of groups $F_{\mathbb{A}^1}(1)$:

Theorem 7.29. *1) The morphism of sheaves of sets*

$$\mathbb{G}_m \times \mathbb{G}_m \to \mathbf{K}_2^{MW}$$

induced by the morphism $(U,V) \mapsto \theta(U^{-1})^{-1}\theta(U^{-1}V)\theta(V)^{-1}$ *is equal to the symbol* $(U,V) \mapsto [U][V]$.

2) *The short exact sequence (7.4):*

$$1 \to \underline{\mathbf{K}}_2^{MW} \to F_{\mathbb{A}^1}(1) \to \mathbb{G}_m \to 1$$

is a central extension.

Proof. 1) Follows directly from the definitions and the Corollary 7.28.

2) For two units U and V in some $F \in \mathcal{F}_k$ the calculation in 1) easily yields $\theta(U)\theta(V)^{-1} = -[U][-V]\theta(U^{-1}V)^{-1}$ and $\theta(U)^{-1}\theta(V) = [U^{-1}][-V]\theta(U^{-1}V)$.

Now we want to check that the action by conjugation of $\mathbb{G}_m$ on $\underline{\mathbf{K}}_2^{MW}$ (through θ) is trivial. It suffices to check it on fields. For units U, V and W in some field $F \in \mathcal{F}_k$, we get (using the previous formulas):

$$\theta(W)([U][V])\theta(W)^{-1}$$

$$= (-[W][-U^{-1}] + [UW][-U^{-1}.V] - [WV][-V])\theta(W^{-1})^{-1}.\theta(W)^{-1}$$

Now applying 1) to $U = W = V$ yields (as θ is pointed) $\theta(W^{-1})^{-1}.\theta(W)^{-1} = [W][W]$.

Now the claim follows from the easily checked equality in $\underline{\mathbf{K}}_2^{MW}(F)$

$$-[W][-U^{-1}] + [UW][-U^{-1}.V] - [WV][-V] + [W][W] = [U][V]$$

which finally yields $\theta(W)([U][V])\theta(W)^{-1} = [U][V]$. $\square$

Remark 7.30. Though it is the more "geometric" way to describe $F_{\mathbb{A}^1}(1)$ it is not the most natural. We denote by $F(\mathcal{S})$ the free sheaf of groups on the pointed sheaf of sets $\mathcal{S}$. This is also the sheaf $\pi_1(\Sigma(\mathcal{S}))$. Its stalks are the free groups generated by the pointed stalks of $\mathcal{S}$.

For a sheaf of group G let $c_G : F(G) \twoheadrightarrow G$ be the canonical epimorphism induced by the identity of G, which admits θ_G as a section (in $Shv_\bullet$). Consider the $Shv_\bullet$-morphism $\theta^{(2)} : G^{\wedge 2} \to F(G)$ given by $(U,V) \mapsto \theta_G(U).\theta_G(U).\theta_G(UV)^{-1}$. This morphism induces a morphism $F(G^{\wedge 2}) \to F(G)$.

A classical result of group theory, a proof of which can be found in [20, Theorem 4.6] gives that the diagram

$$1 \to F(G^{\wedge 2}) \to F(G) \to G \to 1$$

is a short exact sequence of sheaves of groups. If G is strongly $\mathbb{A}^1$-invariant, we deduce the compatible short exact sequence of strongly $\mathbb{A}^1$-invariant

$$1 \to F_{\mathbb{A}^1}(G^{\wedge 2}) \to F_{\mathbb{A}^1}(G) \to G \to 1$$

But now $\theta_G(U).\theta_G(U).\theta_G(UV)^{-1}$ is the tautological symbol $G^2 \to F_{\mathbb{A}^1}(G^{\wedge 2})$. In the case of $\mathbb{G}_m$ this implies (in a easier way) that the extension

$$1 \to \mathbf{K}_2^{MW} = F_{\mathbb{A}^1}(\mathbb{G}_m^{\wedge 2}) \to F_{\mathbb{A}^1}(1) \to \mathbb{G}_m \to 1$$

is central. It is of course isomorphic to (7.4) but not equal as an extension! Indeed, as a consequence of the theorem, we get for the extension (7.4) the formula $\theta(U)\theta(V) =< -1 > [U][V]\theta(UV)$, but for the previous extension one has by construction $\theta(U)\theta(V) = [U][V]\theta(UV)$. $\qquad\square$

Remark 7.31. As a consequence we also see that the sheaf $F_{\mathbb{A}^1}(1)$ is never abelian. Indeed the formula $\theta(U)\theta(V) =< -1 > [U][V]\theta(UV)$ implies

$$\theta(U)\theta(V)\theta(U)^{-1} = h([U][V])\theta(V)$$

Now given any field k one can show that there always exists such an F and such units with $h([U][V]) \neq 0 \in \mathbf{K}_2^{MW}(F)$. Take $F = k(U,V)$ to be the field of rational fraction in U and V over k. The composition of the residues morphisms ∂^U and ∂^V: $K_2^{MW}(k(U,V) \to K_0^{MW}(k)$ commutes to multiplication by h. As the image of the symbol $[U][V]$ is one, the claim follows by observing that $h \neq 0 \in K_0^{MW}(k)$. $\qquad\square$

Endomorphisms of $F_{\mathbb{A}^1}(1) = \pi_1^{\mathbb{A}^1}(\mathbb{P}^1)$

We want to understand the monoid of endomorphisms $\mathcal{E}nd(F_{\mathbb{A}^1}(1))$ of the sheaf of groups $F_{\mathbb{A}^1}(1)$. As $F_{\mathbb{A}^1}(1)$ is the free strongly $\mathbb{A}^1$-invariant sheaf on the pointed sheaf $\mathbb{G}_m$, we see that as a set $\mathcal{E}nd(F_{\mathbb{A}^1}(1)) = Hom_{Shv_\bullet}(\mathbb{G}_m, F_{\mathbb{A}^1}(1))$. By definition the latter set is $F_{\mathbb{A}^1}(1)_{-1}(k)$. As a consequence we observe that there is a natural group structure on $\mathcal{E}nd(F_{\mathbb{A}^1}(1))$.

Remark 7.32. It follows from our results that the obvious map

$$[\mathbb{P}^1, \mathbb{P}^1]_{\mathcal{H}_\bullet(k)} \to \mathcal{E}nd(F_{\mathbb{A}^1}(1))$$

is a bijection. The above group structure comes of course from the natural group structure on $[\mathbb{P}^1, \mathbb{P}^1]_{\mathcal{H}_\bullet(k)} = [\Sigma(\mathbb{G}_m), \mathbb{P}^1]_{\mathcal{H}_\bullet(k)}$. $\qquad\square$

The functor $\mathcal{G} \mapsto (\mathcal{G})_{-1}$ is exact in the following sense:

Lemma 7.33. *For any short exact sequence* $1 \to \mathcal{K} \to \mathcal{G} \to \mathcal{H} \to 1$ *of strongly $\mathbb{A}^1$-invariant sheaves yields, the diagram of*

$$1 \to (\mathcal{K})_{-1} \to (\mathcal{G})_{-1} \to (\mathcal{H})_{-1} \to 1$$

is still a short exact sequence of strongly $\mathbb{A}^1$-invariant sheaves.

Proof. We already know from Lemma 2.32 that the sheaves are strongly $\mathbb{A}^1$-invariant sheaves. The only problem is in fact to check that the morphism $\mathcal{G}_{-1} \to \mathcal{H}_{-1}$ is still an epimorphism. For any $x \in X \in Sm_k$, let X_x be the localization of X at x. We claim that the morphism $\mathcal{G}(\mathbb{G}_m \times X_x) \to \mathcal{H}(\mathbb{G}_m \times X_x)$ is an epimorphism of groups. Using the very definition of the functor $\mathcal{G} \mapsto (\mathcal{G})_{-1}$ this claim easily implies the result.

Now a element in $\mathcal{H}(\mathbb{G}_m \times X_x)$ comes from an element $\alpha \in \mathcal{H}(\mathbb{G}_m \times U)$, for some open neighborhood U of x. Pulling back the short exact sequence to $\mathbb{G}_m \times U$ yields a $\mathcal{K}$-torsor on $\mathbb{G}_m \times U$. But $H^1(\mathbb{G}_m \times X_x; \mathcal{G})$ is trivial by our results of Sect. 2.2. This means that up to shrinking U a bit the $\mathcal{K}$ torsor is trivial. But this means exactly that there is a $\beta \in \mathcal{G}(\mathbb{G}_m \times U)$ lifting α. The lemma is proven. $\square$

Applying this to the short exact sequence (7.4) $1 \to \mathbf{K}_2^{MW} \to F_{\mathbb{A}^1}(1) \to \mathbb{G}_m \to 1$, which is a central extension by Theorem 7.29, obviously yields a central extension as well:

$$0 \to (\underline{\mathbf{K}}_2^{MW})_{-1} \to (F_{\mathbb{A}^1}(1))_{-1} \to (\mathbb{G}_m)_{-1} \to 1$$

But now observe that $\mathbb{Z} = (\mathbb{G}_m)_{-1}$ so that the epimorphism $(F_{\mathbb{A}^1}(1))_{-1} \to (\mathbb{G}_m)_{-1}$ admits a canonical section sending one to the identity.

Corollary 7.34. *The sheaf of groups $(F_{\mathbb{A}^1}(1))_{-1}$ is abelian and is canonically isomorphic to $\mathbb{Z} \oplus \underline{\mathbf{K}}_1^{MW}$.*

Proof. The only remaining point is to observe from Lemma 2.48 applied to unramified Milnor–Witt K-theory that the products induce an isomorphism $\underline{\mathbf{K}}_1^{MW} \cong (\underline{\mathbf{K}}_2^{MW})_{-1}$. $\square$

We let $\rho : \mathbb{G}_m \to (F_{\mathbb{A}^1}(1))_{-1} = \mathbb{Z} \oplus \underline{\mathbf{K}}_1^{MW}$ be the morphism of sheaves which maps U to $(1, [U])$. Observe it is not a morphism of sheaves of groups.

Theorem 7.35. *Endowed with the previous abelian group structure and the composition of morphisms $\mathcal{E}nd(F_{\mathbb{A}^1}(1)) \cong [\mathbb{P}^1, \mathbb{P}^1]_{\mathcal{H}_\bullet(k)}$ is an associative ring. $\rho(k)$ induces a group homomorphism $k^\times \to \mathcal{E}nd(F_{\mathbb{A}^1}(1))^\times$ to the group of units and the induced ring homomorphism*

$$\Theta(k) : \mathbb{Z}[k^\times] \to \mathcal{E}nd(F_{\mathbb{A}^1}(1))$$

is onto. As a consequence, $\mathcal{E}nd(F_{\mathbb{A}^1}(1))$ is a commutative ring.

Proof. Let $\mathbb{Z}(k^\times)$ be the free abelian group on $k^\times$ with the relation the symbol $1 \in k^\times$ equals 0. It is clear that $\mathbb{Z}[k^\times]$ splits as $\mathbb{Z} \oplus \mathbb{Z}(k^\times)$ in a compatible way to the splitting of Corollary 7.34 so that $\Theta(k)$ decomposes as the identity of $\mathbb{Z}$ plus the obvious symbol $\mathbb{Z}(k^\times) \to \underline{\mathbf{K}}_1^{MW}(k)$. But then we know from Lemma 3.6 that this is an epimorphism. $\square$

The canonical morphism

$$[\mathbb{P}^1, \mathbb{P}^1]_{\mathcal{H}_\bullet(k)} = \mathcal{E}nd(F_{\mathbb{A}^1}(1)) \to K_0^{MW}(k) = \mathcal{E}nd(\underline{\mathbf{K}}_1^{MW})$$

given by the "Brouwer degree" (which means evaluation of $\mathbb{A}^1$-homology in degree 1) is thus an epimorphism as $\mathbb{Z}[k^\times] \to K_0^{MW}(k)$ is onto.

To understand this a bit further, we use Theorem 3.46 and its corollary which show that $\underline{\mathbf{K}}_1^W \to \mathbf{I}$ is an isomorphism.

In fact $K_0^{MW}(k)$ splits canonically as $\mathbb{Z} \oplus I(k)$ as an abelian group, and moreover this decomposition is compatible through the above epimorphism to that of Corollary 7.34. This means that the kernel of

$$[\mathbb{P}^1, \mathbb{P}^1]_{\mathcal{H}_\bullet(k)} = [\Sigma(\mathbb{G}_m), \Sigma(\mathbb{G}_m)]_{\mathcal{H}_\bullet(k)} \twoheadrightarrow K_0^{MW}(k)$$

is isomorphic to the kernel of $K_1^{MW}(k) \to I(k)$. As $K_1^{MW}(k) \xrightarrow{h} K_1^{MW}(k) \to I(k) \to 0$ is always an exact sequence by Theorem 3.46 and its corollary, and as $K_1^{MW}(k) \xrightarrow{h} K_1^{MW}(k)$ factors through $K_1^{MW}(k) \twoheadrightarrow K_1^{MW}(k)/\eta = k^\times$ we get an exact sequence of the form

$$k^\times \to K_1^{MW}(k) \to I(k) \to 0$$

where the map $k^\times \to K_1^{MW}(k)$ arises from multiplication by h. Clearly -1 is mapped to zero in $K_1^{MW}(k)$ because $h.[-1] = [-1] + <-1>[-1] = [(-1)(-1)] = [1] = 0$. Moreover the composition $k^\times \to K_1^{MW}(k) \to k^\times$ is the squaring map. Thus $(k^\times)/_{(+1)} \to K_1^{MW}(k)$ is injective and equal to the kernel. We thus get altogether:

Theorem 7.36. *The diagram previously constructed*

$$0 \to (k^\times)/_{(\pm 1)} \to [\mathbb{P}^1, \mathbb{P}^1]_{\mathcal{H}_\bullet(k)} \to GW(k) \to 0$$

is a short exact sequence of abelian groups.

Remark 7.37. J. Lannes has observed that as a ring, $\tilde{GW}(k) := [\mathbb{P}^1, \mathbb{P}^1]_{\mathcal{H}_\bullet(k)} = \mathcal{E}nd(F_{\mathbb{A}^1}(1))$ is the Grothendieck ring of isomorphism classes of symmetric inner product spaces over k with a given diagonal basis, where an isomorphism between two such objects is a linear isomorphism preserving the inner product and with determinant one in the given basis. It fits in the following cartesian square of rings

$$\begin{array}{ccc} \tilde{GW}(k) & \to & \mathbb{Z} \oplus k^\times \\ \downarrow & & \downarrow \\ GW(k) & \to & \mathbb{Z} \oplus (k^\times/2) \end{array}$$

The bottom horizontal map is the rank plus the determinant. The group structure on the right hand side groups is the obvious one. The product structure is given by $(n, U).(m, V) = (nm, U^m.V^n)$.

C. Cazanave [16–18] has a different method to produce the invariant $[\mathbb{P}^1, \mathbb{P}^1]_{\mathcal{H}_\bullet(k)} \to \widetilde{GW}(k)$ using the approach of Barge and Lannes on Bott periodicity for orthogonal algebraic K-theory [6].   □

Remark 7.38. We may turn $\Theta(k)$ into a morphism of sheaves of abelian groups $\Theta : \mathbb{Z}[\mathbb{G}_m] \to (F_{\mathbb{A}^1}(1))_{-1} \cong \mathbb{Z} \oplus \underline{\mathbf{K}}_1^{MW}$ induced by ρ. Here $\mathbb{Z}[\mathcal{S}]$ means the free sheaf of abelian groups generated by the sheaf of sets $\mathcal{S}$. $\Theta : \mathbb{Z}[\mathbb{G}_m] \to (F_{\mathbb{A}^1}(1))_{-1} \cong \mathbb{Z} \oplus \underline{\mathbf{K}}_1^{MW}$ is then the universal morphism of sheaves of abelian groups to a strictly $\mathbb{A}^1$-invariant sheaf. As a consequence the target is also a sheaf of commutative rings: it is the $\mathbb{A}^1$-group ring on $\mathbb{G}_m$.

□

Free homotopy classes $[\mathbb{P}^1, \mathbb{P}^1]_{\mathcal{H}(k)}$. By Remark 3.44 to understand the set $[\mathbb{P}^1, \mathbb{P}^1]_{\mathcal{H}(k)}$ we have to understand the action of $\pi_1^{\mathbb{A}^1}(\mathbb{P}^1)(k)$ on $[\mathbb{P}^1, \mathbb{P}^1]_{\mathcal{H}_\bullet(k)}$ and to compute the quotient.

Clearly, as $[\mathbb{P}^1, \mathbb{P}^1]_{\mathcal{H}_\bullet(k)} \cong [\mathbb{P}^1, B(F_{\mathbb{A}^1}(1))]_{\mathcal{H}_\bullet(k)} \cong \mathcal{E}nd(F_{\mathbb{A}^1}(1))$, this action is given on the right hand side by the action by conjugation of $F_{\mathbb{A}^1}(1)$ on the target. Now the abelian group structure comes from the source and thus this action is an action of the group $F_{\mathbb{A}^1}(1)$ on the abelian group $\mathcal{E}nd(F_{\mathbb{A}^1}(1))$. As $K_2^{MW}(k) \subset F_{\mathbb{A}^1}(1)$ is central this action factors through an action of $k^\times$ on the abelian group $\mathcal{E}nd(F_{\mathbb{A}^1}(1))$.

Lemma 7.39. *The action of $k^\times$ on $\mathcal{E}nd(F_{\mathbb{A}^1}(1)) = K_1^{MW}(k) \oplus \mathbb{Z}$ is given as follows. For any $u \in k^\times$ and any $(v, n) \in k^\times \times \mathbb{Z}$ one has in $K_1^{MW}(k) \oplus \mathbb{Z}$*

$$c_u([v], n) = ([v] - nh[u], n)$$

Proof. To find the action of $k^\times$ by conjugation on $\mathcal{E}nd(F_{\mathbb{A}^1}(1)) = (F_{\mathbb{A}^1})_{-1}(k)$ we observe that by Remark 7.31 we understand this action on $F_{\mathbb{A}^1}(1)$.

We may explicit this action on $F_{\mathbb{A}^1}(k(T)$ and observe that the isomorphism $\mathcal{E}nd(F_{\mathbb{A}^1}(1)) = K_1^{MW}(k) \oplus \mathbb{Z} = (F_{\mathbb{A}^1})_{-1}(k)$ is obtained by cup-product by T on the left $[T] \cup : K_1^{MW}(k) \oplus \mathbb{Z} \to F_{\mathbb{A}^1}(k(T))$. Thus for $(v, n) \in k^\times \times \mathbb{Z}$ the corresponding element $[T] \cup ([v], n)$ in $F_{\mathbb{A}^1}(k(T))$ is $[T][v].\theta(T)^n$.

Now by the formula in Remark 7.31 we get for any $u \in k^x$, and any $(v, n) \in k^\times \times \mathbb{Z}$:

$$c_u([T][v].\theta(T)^n) = [T][v].(h[u][T].\theta(T))^n$$

$$= ([T][v] + nh[u][T])\theta(T)^n = [T]([v] - nh[u])\theta(T)^n$$

This implies the lemma.   □

Corollary 7.40. *Assume that for each $n \geq 1$ the map $k^\times \to k^\times$, $u \mapsto u^n$ is onto. Then the surjective map*

$$[\mathbb{P}^1, \mathbb{P}^1]_{\mathcal{H}(k)} \to [\mathbb{P}^1, \mathbb{P}^\infty]_{\mathcal{H}(k)} = \mathbb{Z}$$

has trivial fibers over any integer $n \neq 0$ and its fiber over 0 is exactly $K_1^{MW}(k) = [\mathbb{P}^1, \mathbb{A}^2 - \{0\}]_{\mathcal{H}(k)}$.

Proof. First if every unit is a square, by Proposition 3.13, we know that $K_1^{MW}(k) \to k^\times$ is an isomorphism. On the set of pairs $(v, n) \in K_1^{MW}(k) \times \mathbb{Z} = \mathcal{E}nd(F_{\mathbb{A}^1}(1))$ the action of $u \in k^\times$ is thus given as

$$c_u([v], n) = ([vu^{-2n}], n)$$

But for $n \neq 0$, any unit w can be written vu^{-2n} for some u and v by assumption on k. Thus the result. □

Corollary 7.4 ...

Chapter 8
$\mathbb{A}^1$-Homotopy and Algebraic Vector Bundles

8.1 $\mathbb{A}^1$-Homotopy Classification of Vector Bundles

Now we come to proving in $\mathbb{A}^1$-homotopy theory the analogues for algebraic vector bundles of the classical result of classification of topological vector bundles in terms of homotopy classes of maps to the Grassmanian varieties. The results and techniques of the next sections is a natural sequel to [49] and is very much inspired from it.

We denote by $\mathbb{G}r_r$ the union of the finite Grassmanian $\mathbb{G}r_{r,m}$ of r-planes in $\mathbb{A}^{m+r}$, which we call the infinite Grassmanian (of r-plans).

Recall from the introduction that if $\mathcal{F}$ is a sheaf of sets, $Sing_\bullet^{\mathbb{A}^1}(\mathcal{F})$ denotes its Suslin Voevodsky construction. We know from [59] that the morphism of spaces

$$\mathcal{F} \to \underline{Sing}_\bullet^{\mathbb{A}^1}(\mathcal{F})$$

is always an $\mathbb{A}^1$-weak equivalence. For any $X \in Sm_k$, the set $\pi_0(\underline{Sing}_\bullet^{\mathbb{A}^1}(\mathcal{F})(X))$ of connected components of the simplicial set $\underline{Sing}_\bullet^{\mathbb{A}^1}(\mathcal{F})(X)$ is by definition the quotient of the set of morphisms $X \to \mathcal{F}$ modulo the equivalence relation generated by naive $\mathbb{A}^1$-homotopy relations; we denote this set also as $\pi_{\mathbb{A}^1}(X; \mathcal{F})$.

Recall also from the introduction that $\Phi_r(X)$ denotes the set of isomorphisms classes of algebraic vector bundles of rank r over X.

The following result implies Theorem 1.29 stated in the introduction:

Theorem 8.1. *Let r be an integer different[1] from 2.*

1) The spaces $\underline{Sing}_\bullet^{\mathbb{A}^1}(\mathbb{G}r_r)$ and $\underline{Sing}_\bullet^{\mathbb{A}^1}(GL_r)$ are $\mathbb{A}^1$-local; more precisely, for any affine smooth k-scheme X the natural maps of simplicial sets

---

[1] Using the results of [60] one may establish the case $r = 2$ as well.

F. Morel, $\mathbb{A}^1$-*Algebraic Topology over a Field*, Lecture Notes in Mathematics 2052, DOI 10.1007/978-3-642-29514-0_8, © Springer-Verlag Berlin Heidelberg 2012

$$\underline{Sing}^{\mathbb{A}^1}_\bullet(\mathbb{G}r_r)(X) \to L_{\mathbb{A}^1}(\mathbb{G}r_r)(X)$$

and

$$\underline{Sing}^{\mathbb{A}^1}_\bullet(GL_r)(X) \to L_{\mathbb{A}^1}(GL_r)(X)$$

are weak-equivalences;

2) For any affine smooth k-scheme X the natural maps

$$\pi_{\mathbb{A}^1}(X, \mathbb{G}r_r) \to Hom_{\mathcal{H}(k)}(X, \mathbb{G}r_r) \cong Hom_{\mathcal{H}_s(k)}(X, \underline{Sing}^{\mathbb{A}^1}_\bullet(\mathbb{G}r_r))$$

and

$$\pi_{\mathbb{A}^1}(X, GL_r) \to Hom_{\mathcal{H}(k)}(X, GL_r) \cong Hom_{\mathcal{H}_s(k)}(X, \underline{Sing}^{\mathbb{A}^1}_\bullet(GL_r))$$

are bijections;

3) For any affine smooth k-scheme X the map "pull-back of the universal rank r vector bundle" $Hom_{Sm_k}(X, \mathbb{G}r_r) \to \Phi_r(X)$ induces a bijection

$$\pi_{\mathbb{A}^1}(X, \mathbb{G}r_r) \cong \Phi_r(X)$$

It follows that for $r \neq 2$ and X affine and k-smooth there are canonical bijections

$$Hom_{\mathcal{H}(k)}(X, BGL_r) \cong Hom_{\mathcal{H}(k)}(X, \mathbb{G}r_r) \cong \Phi_r(X)$$

To prove the previous theorem we will first reduce to the following technical result:

Theorem 8.2. Let r be an integer $\neq 2$ and let A be a smooth k-algebra.

1) Let f and g be two coprime elements in A. The diagram of simplicial sets

$$\begin{array}{ccc}
\underline{Sing}^{\mathbb{A}^1}_\bullet(\mathbb{G}r_r)(Spec(A)) & \to & \underline{Sing}^{\mathbb{A}^1}_\bullet(\mathbb{G}r_r)(Spec(A_f)) \\
\downarrow & & \downarrow \\
\underline{Sing}^{\mathbb{A}^1}_\bullet(\mathbb{G}r_r)(A_g) & \to & \underline{Sing}^{\mathbb{A}^1}_\bullet(\mathbb{G}r_r)(A_{f.g})
\end{array}$$

is homotopy cartesian.

2) Let $A \to B$ an étale A-algebra and $f \in A$ and such that $A/f \to B/f$ is an isomorphism. Then the diagram of simplicial sets

$$\begin{array}{ccc}
\underline{Sing}^{\mathbb{A}^1}_\bullet(GL_r)(Spec(A)) & \to & \underline{Sing}^{\mathbb{A}^1}_\bullet(GL_r)(Spec(A_f)) \\
\downarrow & & \downarrow \\
\underline{Sing}^{\mathbb{A}^1}_\bullet(GL_r)(Spec(B)) & \to & \underline{Sing}^{\mathbb{A}^1}_\bullet(GL_r)(Spec(B_f))
\end{array}$$

is homotopy cartesian.

The proof of this theorem will be given in Sect. 9.3. It relies in an essential way on the solution of the so-called *generalized Serre problem* given by Lindel [41], after Quillen [66] and Suslin [73] for the case of polynomial rings over fields, as well as some works of Suslin [74] and Vorst [82, 83] on the "analogue" of the Serre problem for the general linear group. Observe that for $r = 1$, $GL_1 = \mathbb{G}_m$ so that $\underline{Sing}^{\mathbb{A}^1}_{\bullet}(\mathbb{G}_m)$ is actually simplicially constant and equal to $\mathbb{G}_m$ and this case is trivial, the theorem just follows from the fact that $\mathbb{G}_m$ is a sheaf in the Nisnevich topology.

Proof of Theorem 8.1 assuming Theorem 8.2. By assumption that Theorem 8.2 holds, we see that $\underline{Sing}^{\mathbb{A}^1}_{\bullet}(GL_r)$ has the affine B.G. property in the Nisnevich topology, as well as the affine $\mathbb{A}^1$-invariance property considered in Definition A.7 below. This follows from [59, Corollary 3.5 p. 89].

By Lemma A.8 and Theorem A.19 below, we conclude that for any affine smooth k-scheme X the map $\underline{Sing}^{\mathbb{A}^1}_{\bullet}(GL_r)(X) \to \underline{Sing}^{\mathbb{A}^1}_{\bullet}(GL_r)^{af}(X)$ is a weak equivalence and that $X \mapsto \underline{Sing}^{\mathbb{A}^1}_{\bullet}(GL_r)^{af}(X)$ has the $\mathbb{A}^1$-B.G. property in the Nisnevich topology considered in [59]. From *loc. cit.* Lemma 1.18 p. 101, the claims (1) and (2) for GL_r follow.

The points (1) and (2) for the Grassmannian follows in the same way using the case of the general linear group $\underline{Sing}^{\mathbb{A}^1}_{\bullet}(GL_r)$ and using the Theorem A.9, after observing that $\underline{Sing}^{\mathbb{A}^1}_{\bullet}(GL_r)$ has the homotopy type of the simplicial loop space of $\underline{Sing}^{\mathbb{A}^1}_{\bullet}(\mathbb{G}r_r)$.

This fact as well as the point (3), and the conclusion of the Theorem is proven using exactly the same argument as in the proof of [49, Théorème 4.2.6.], see also the Sect. 9.3 below. □

Remark 8.3. Though we kept our basic assumption that the base field k is perfect, one may extend all the previous results, and that of the next sections, to the case of a general base field. Indeed the simplicial sheaves $\underline{Sing}^{\mathbb{A}^1}_{\bullet}(\mathbb{G}_r)$ and $\underline{Sing}^{\mathbb{A}^1}_{\bullet}(GL_r)$ that we use are defined over the prime field and we are just considering their pull-back to k. The claim follows now from the (unstable version of) "base change" results of [52], and the fact that the previous technical result Theorem 8.2 holds over any field. □

Remark 8.4. At this stage the reason to assume that the base is the spectrum of a field just come from the fact that the works of Quillen, Suslin, Lindel cited above are only known for regular rings containing a field. We observe in contrast that most of Vorst results [83] concerning Hodge algebras over a ring R are available if the corresponding results hold over R. Thus conjecturally our results are also expected over a regular affine base scheme $S = Spec(R)$.

When we let r be infinite, working with stable vector bundles, the $\mathbb{A}^1$-invariant properties are known for any regular ring R by Grothendieck. As a consequence we get that the analogue of Theorems 8.2 and 8.1 hold over a regular affine base scheme if $r = \infty$. □

This remark yields another proof of the following result from [59]:

Corollary 8.5. *Let S be a regular affine scheme. Then for any affine smooth k-scheme X there exist natural bijections*

$$\Phi_\infty(X) \cong Hom_{\mathcal{H}(k)}(X, BGL_\infty) \cong Hom_{\mathcal{H}(k)}(X, \mathbb{G}r_\infty)$$

and

$$K_0(X) \cong Hom_{\mathcal{H}(k)}(X, \mathbb{Z} \times BGL_\infty) \cong Hom_{\mathcal{H}(k)}(X, \mathbb{Z} \times \mathbb{G}r_\infty)$$

Remark 8.6. The proof given here doesn't use Quillen results on algebraic K-theory as opposed to the proof of [59, Theorem 3.13 p. 140] or [49]. As a consequence it is possible to deduce, without using Quillen's results that the K-groups defined by

$$K_n(X) := Hom_{\mathcal{H}_\bullet(k)}(\Sigma^n(X_+), \mathbb{Z} \times BGL_\infty)$$

are the Karoubi–Villamayor groups [36] and that these satisfy Mayer–Vietoris and Nisnevich descent properties for all smooth k-schemes.   □

$\mathbb{G}_m$-Loop spaces for Grassmanians. For a pointed space $\mathcal{X}$, let us denote by $\mathcal{X}^{(\mathbb{G}_m)}$ the pointed space of $\mathbb{G}_m$-loops on $\mathcal{X}$

$$U \mapsto \mathcal{X}^{(\mathbb{G}_m)}(U) = Hom_{\Delta^{op}Shv_\bullet(Sm_k)}((U_+) \wedge (\mathbb{G}_m), \mathcal{X})$$

and by $\Omega^1_{\mathbb{G}_m}(\mathcal{X})$ its $\mathbb{A}^1$-derived version

$$\Omega^1_{\mathbb{G}_m}(\mathcal{X}) := (L_{\mathbb{A}^1}(\mathcal{X}))^{(\mathbb{G}_m)}$$

Theorem 8.7. *For any $n \neq 2$ and any smooth affine k-scheme X, the canonical map*

$$\underline{Sing}^{\mathbb{A}^1}_\bullet(\mathbb{G}r_r)^{(\mathbb{G}_m)}(X) \to \Omega^1_{\mathbb{G}_m}(GL_r)(X)$$

is a simplicial weak equivalence.

Proof. Apply Theorem 8.1 to $\mathbb{G}_m \times X$ and just care about what happens to the base point! We leave the details to the reader.   □

Remark 8.8. We observe that following Lemma 6.47, one can construct an $\mathbb{A}^1$-localization functor which commutes to finite products, thus the morphism in the theorem is indeed a morphism of simplicial groups.   □

An other interesting consequence of Theorem 8.1 is the following Theorem. Observe that $\underline{Sing}^{\mathbb{A}^1}_\bullet(GL_r)$ naturally splits (as a space) as $\mathbb{G}_m \times \underline{Sing}^{\mathbb{A}^1}_\bullet(SL_r)$ and that the morphism $SL_r/SL_{r-1} \to GL_r/GL_{r-1}$ is an isomorphism. Thus in the next statement we could use GL_r instead of SL_r.

Theorem 8.9. *Let $r \neq 2$ be an integer.*

1) The morphism of spaces

$$\underline{Sing}^{\mathbb{A}^1}_\bullet(SL_r)/\underline{Sing}^{\mathbb{A}^1}_\bullet(SL_{r-1}) \to \underline{Sing}^{\mathbb{A}^1}_\bullet(SL_r/SL_{r-1})$$

is an isomorphism;

2) For any smooth affine k-scheme X the map of simplicial sets

$$\underline{Sing}^{\mathbb{A}^1}_\bullet(SL_r/SL_{r-1})(X) \to \underline{Sing}^{\mathbb{A}^1}_\bullet(\mathbb{A}^r - \{0\})(X)$$

is a weak equivalence (in fact a trivial cofibration).

3) The space $\underline{Sing}^{\mathbb{A}^1}_\bullet(\mathbb{A}^r - \{0\})$ is $\mathbb{A}^1$-local and for any smooth affine k-scheme X the map of simplicial sets

$$\underline{Sing}^{\mathbb{A}^1}_\bullet(\mathbb{A}^r - \{0\})(X) \to L_{\mathbb{A}^1}(\mathbb{A}^r - \{0\})(X)$$

is a weak equivalence.

Proof. The fact that the maps in (1) are monomorphism is immediate. To check it is an epimorphism of scheves let A be the henselization of a point in a smooth k-scheme. Let $\sigma : \Delta^n_A \to SL_r/SL_{r-1}$ be a morphism corresponding to an n-simplex of $\underline{Sing}^{\mathbb{A}^1}_\bullet(\mathbb{A}^r - \{0\})(A)$. The obstruction to lift this morphism to a morphism $\Delta^n_A \to SL_r$ is an SL_{r-1} torsor. By Lindel's Theorem [41] it is trivial (induce from one over A). This proves (1).

For (2) one proceeds as follows. Let A be a smooth k-algebra and $\sigma : \Delta^n_A \to \mathbb{A}^r - \{0\}$ be a morphism together with a lifting to SL_r/SL_{r-1} of the restriction to $\partial\Delta^n_A$. We have to show that we may lift σ itself in a compatible way.

Everything amounts in fact to understanding the fiber product Π through σ of SL_r/SL_{r-1} and Δ^n_A over $\mathbb{A}^r - \{0\}$. Now consider the fiber product T' of SL_r and Δ^n_A over $\mathbb{A}^r - \{0\}$; it is a torsor over the group $SL'_{r-1} \subset SL_r$ of matrices of the form

$$\begin{pmatrix} SL_{r-1} & ? \\ 0 & 1 \end{pmatrix}$$

As everything is affine and any torsor over a product of additive group over an affine basis is trivial, it follows that T is induced from an SL_{r-1} torsor T over Δ^n_A. Thus Π is isomorphic to $\mathbb{A}^{r-1} \times \Delta^n_A$; the rest is easy.

The point (3) follows from (1) and (2) and from Theorem 8.1. □

Remark 8.10. It follows from this that for X affine smooth over k, the map

$$\pi_{\mathbb{A}^1}(X, \mathbb{A}^r - \{0\}) \to Hom_{\mathcal{H}(k)}(X, \mathbb{A}^r - \{0\})$$

is a bijection. In particular any morphism in the right handside comes from an actual morphism $X \to \mathbb{A}^r - \{0\}$. □

8.2   The Theory of the Euler Class for Affine Smooth Schemes

For a given a smooth affine k-scheme X and an integer $r \geq 4$ we will use the previous results to study the map "adding the trivial line bundle":

$$\Phi_{r-1}(X) \to \Phi_r(X)$$

following the classical obstruction method in homotopy theory: the previous map is induced by the map

$$BGL_{r-1} \to BGL_r$$

corresponding to the inclusion $GL_{r-1} \subset GL_r$ and everything amounts to understanding the $\mathbb{A}^1$-connectivity of its $\mathbb{A}^1$-homotopy fiber $\mathcal{F}_r$.

We start by observing:

Proposition 8.11. *Let $H \subset G$ be a monomorphism of sheaves of groups and assume that $\pi_0^{\mathbb{A}^1}(G)$ is a strongly $\mathbb{A}^1$-invariant sheaf. Then the simplicial fibration sequence*

$$G/H \to BH \to BG$$

is an $\mathbb{A}^1$-fibration sequence.

Proof. This is an immediate application of Theorem 6.53.   □

We thus get for any r an $\mathbb{A}^1$-fibration sequence

$$GL_r/GL_{r-1} \to BGL_{r-1} \to BGL_r$$

as $\pi_0^{\mathbb{A}^1}(GL_r) = \mathbb{G}_m$. We observe that $SL_r/SL_{r-1} \cong GL_r/GL_{r-1}$ is an isomorphism. Thus the $\mathbb{A}^1$-homotopy fiber $\mathcal{F}_r$ is canonically isomorphic to $L_{\mathbb{A}^1}(SL_r/SL_{r-1}) \cong L_{\mathbb{A}^1}(\mathbb{A}^r - \{0\})$.

Proposition 8.12. *For any $r \geq 1$ the diagram of spaces*

$$SL_{r-1} \to SL_r \to SL_r/SL_{r-1} \cong \mathbb{A}^r - \{0\}$$

is an $\mathbb{A}^1$-fibration sequence between $\mathbb{A}^1$-local spaces.

Proof. This follows from the previous results and Sect. 6.3. Indeed the $\mathbb{A}^1$-fibration sequence above produces on the left the required $\mathbb{A}^1$-fibration sequence

$$SL_{r-1} \to SL_r \to SL_r/SL_{r-1} \cong \mathbb{A}^r - \{0\}$$

by using Theorem 6.46 and Lemma 6.45.   □

For any integer $n \geq 1$, the space $\mathbb{A}^n - \{0\}$ is $(n-2)$-connected by Theorem 6.40. An interesting consequence of this connectivity statement, of Theorem 8.1 and the obstruction theory from the Appendix B.3 is the following:

Corollary 8.13. *Let X be a smooth affine k-scheme and assume that X is isomorphic in $\mathcal{H}(k)$ to a smooth k-scheme Y of Krull dimension d. Then any algebraic vector bundle of rank $r > d$ on $X = Spec(A)$ splits off a free vector bundle of rank $r - d$, at least if $r \geq 4$.*

We observe that the fact that Y is or is not affine doesn't play any role in the statement. This occurs for instance in case X is an affine smooth k-scheme which is a torsor over a vector bundle on a smooth projective Y. We may also take d to be the Krull dimension of A and this case is a particular case of a result by Serre.

Now we know from Theorem 6.40 that, moreover, there exists a canonical isomorphism of sheaves of abelian groups:

$$\pi_{n-1}(\underline{Sing}^{\mathbb{A}^1}_{\bullet}(\mathbb{A}^n - \{0\})) \cong \mathbf{K}^{MW}_n$$

Theorem 8.1 and the obstruction theory from the Appendix B.3, then produce the theory of the Euler class mentioned in the introduction:

Theorem 8.14 (Theory of Euler class). *Assume $r \geq 4$. Let X be a smooth affine k-scheme of dimension $\leq r$, and let ξ be an oriented algebraic vector bundle of rank r (recall that an orientation of ξ means a trivialization of $\Lambda^r(\xi)$). Define its Euler class*

$$e(\xi) \in H^r_{Nis}(X; \mathbf{K}^{MW}_r)$$

to be the obstruction class in

$$H^r_{Nis}(X; \pi^{\mathbb{A}^1}_{r-1}(\underline{Sing}^{\mathbb{A}^1}_{\bullet}(\mathbb{A}^r - \{0\}))) \cong H^r_{Nis}(X; \mathbf{K}^{MW}_r)$$

obtained from the $\mathbb{A}^1$-fibration sequence

$$\mathbb{A}^r - \{0\} \to BGL_{r-1} \to BGL_r$$

or even, because ξ is oriented, from:

$$\mathbb{A}^r - \{0\} \to BSL_{r-1} \to BSL_r$$

Then:

$$\xi \text{ splits off a trivial line bundle} \Leftrightarrow e(\xi) = 0 \in H^r_{Nis}(X; \mathbf{K}^{MW}_r)$$

Remark 8.15. 1) We showed in Theorem 5.47 of Sect. 5.3 that the group $H^r(X; \underline{\mathbf{K}}_r^{MW})$ coincides with the oriented Chow group $\tilde{C}H^r(X)$ defined in [7]. It can be shown that our Euler class coincides with the one defined in *loc. cit.*. This proves the main conjecture of *loc. cit.*, the case $r = 2$ was already established there, at least if $char(k) \neq 2$. See also [23].

2) If ξ is an algebraic vector bundle of rank r over X, let $\lambda = \Lambda^r(\xi) \in Pic(X)$ be its maximal external power. The obstruction class $e(\xi)$ obtained by the $\mathbb{A}^1$-fibration sequence

$$\mathbb{A}^r - \{0\} \to BGL_{r-1} \to BGL_r$$

lives now in the corresponding "twisted" group $H^r(X; \underline{\mathbf{K}}_r^{MW}(\lambda))$ obtained by conveniently twisting the sheaf $\underline{\mathbf{K}}_r^{MW}$ by λ (see Sect. 5.3). In the same way we have that, if the dimension of X is $\leq r$:

ξ splits off a trivial line bundle $\Leftrightarrow e(\xi) = 0 \in H^r(X; \underline{\mathbf{K}}_r^{MW}(\lambda))$. $\square$

8.3   A Result Concerning Stably Free Vector Bundles

We also obtain as a direct consequence of our main results, the following theorem.[2]

Theorem 8.16. *(Stably free vector bundles) Assume $r \geq 3$. Let $X = Spec(A)$ be a smooth affine k-scheme. The canonical composition*

$$Hom_{\mathcal{H}(k)}(X, \mathbb{A}^{r+1} - \{0\})/(SL_{r+1}(A)) \to Hom_{\mathcal{H}(k)}(X, BSL_r)$$

$$\to Hom_{\mathcal{H}(k)}(X, BGL_r)$$

obtained from Theorems 8.1 and the homotopy exact sequence for the fibration sequence

$$\mathbb{A}^{r+1} - \{0\} \to BSL_r \to BSL_{r+1}$$

is injective and its image is exactly the set $\Phi_r^{st.free}(X)$ of isomorphism classes of algebraic vector bundles ξ of rank r over X such that $\xi \oplus \theta^1$ is trivial.

If the Krull dimension[3] of X is $\leq r$ the natural map

$$Hom_{\mathcal{H}(k)}(X, \mathbb{A}^{r+1} - \{0\}) \to H^r_{Nis}(X; \underline{\mathbf{K}}_{r+1}^{MW})$$

---

[2]I want to thank J. Barge and J. Lannes and independently J. Fasel who observed some mistakes in the previous version of the theorem.

[3]As observed by Jean Fasel, one may here assume only that X is isomorphic in the $\mathbb{A}^1$-homotopy category to a smooth k-scheme of Krull dimension $\leq r$.

is a bijection. Thus we obtain a bijection

$$H^r(X; \underline{\mathbf{K}}_{r+1}^{MW})/(SL_{r+1}(A)) \cong \Phi_r^{st.free}(X)$$

Proof. It follows from our previous results applied to the $\mathbb{A}^1$-fibration sequence:

$$\mathbb{A}^{r+1} - \{0\} \to BSL_r \to BSL_{r+1}$$

from Theorem 8.9 and the observation that for any X, $Hom_{\mathcal{H}(k)}(X, BSL_r) \to Hom_{\mathcal{H}(k)}(X, B\mathbb{G}L_r)$ is always injective as $BSL_r \to BGL_r$ is a $\mathbb{G}_m$-torsor, thus an $\mathbb{A}^1$-covering. The second statement is proven by an argument of Postnikov tower, and then by using our computations of $\pi_r^{\mathbb{A}^1}(\mathbb{A}^{r+1} - \{0\})$. $\square$

Remark 8.17. 1) In general the action of $SL_{r+1}(A)$ factors through the group of $\mathbb{A}^1$-homotopy classes $Hom_{\mathcal{H}(k)}(X, SL_{r+1})$, and one can show that in case the Krull dimension of X is $\leq r$, the above quotient has a canonical abelian group structure and is isomorphic to the cokernel of the group homomorphism

$$Hom_{\mathcal{H}(k)}(X, SL_{r+1}) = Hom_{\mathcal{H}(k)}(\Sigma(X_+), BSL_{r+1})$$

$$\to Hom_{\mathcal{H}(k)}(\Sigma(X_+), K(\underline{\mathbf{K}}_{r+1}^{MW}, r+1))$$

$$= H^r(X; \underline{\mathbf{K}}_{r+1}^{MW})$$

induced by the relative Postnikov k-invariant $BSL_{r+1} \to K(\underline{\mathbf{K}}_{r+1}^{MW}, r+1)$ of the morphism $BSL_r \to BSL_{r+1}$.

2) W. Van der Kallen made quite analogous considerations in [78]. In particular he obtained, under certain conditions on the stable rank of the commutative ring A, an abelian group structure on the set $Um_{r+1}(A)/E_{r+1}(A)$ of unimodular vectors of rank $(r+1)$ modulo the action of the group of elementary matrices $E_{r+1}(A)$. Jean Fasel pointed out to us that for a smooth affine k-scheme $X = Spec(A)$, the obvious map

$$Um_{r+1}(A)/E_{r+1}(A) \to Hom_{\mathcal{H}(k)}(X, \mathbb{A}^{r+1} - \{0\})$$

is a bijection, and in case $dim(X) \leq r$, that the group structure defined by Van der Kallen coincides with ours through the bijection

$$Hom_{\mathcal{H}(k)}(X, \mathbb{A}^{r+1} - \{0\}) \cong H_{Nis}^r(X; \underline{\mathbf{K}}_{r+1}^{MW})$$

of the Theorem. $\square$

Remark 8.18. One should observe that the group $H^n(X; \underline{\mathbf{K}}_{n+1}^{MW})$, for X of dimension $\leq n$ is a natural quotient of

$$\oplus_{x \in X^{(n)}} K_{n+1}^{MW}(\kappa(x); \lambda_x)$$ $\square$

Chapter 9
The Affine B.G. Property
for the Linear Groups and the
Grassmanian

In this section, for a commutative ring A and an integer $r \geq 0$ we will simply let $\Phi_r(A)$ denote $H^1_{Zar}(Spec(A); GL_r) = H^1_{Nis}(Spec(A); GL_r)$. We extend this notation to $r = \infty$ by setting $\Phi_\infty(A) := colim_r \Phi_r(A)$ where the colimit is taken through the maps of adding the free A-module of rank one A.

If $B \to A$ is a ring homomorphism and M is a B-module we denote by M_A the A-module $A \otimes_B M$. An A-module N is said to be *extended* from B if there exists a B-module M such that M_A is isomorphic to N as an A-module. If M is an A-module, we let $Aut_A(M)$ denote the group of A-automorphisms of M.

9.1   Preliminaries and Recollections on Regularity

Definition 9.1. Let $r \in \mathbb{N} \cup \{\infty\}$. A commutative ring A is said to be Φ_r-regular if for any integer $n \geq 1$ the map

$$\Phi_r(A) \to \Phi_r(A[T_1, \ldots, T_n])$$

is bijective.

Remark 9.2. 1) Quillen [66] and Suslin [73] have proven that a commutative field is Φ_r-regular for any $r \geq 1$. Then Lindel [41] has proven that any regular ring which is an algebra of finite type over a field is also Φ_r-regular for any $r \geq 1$.

2) Grothendieck proved that regular rings are Φ_∞-regular. Observe indeed that for A an integral domain, $\Phi_\infty(A)$ is isomorphic to the functor $A \mapsto \tilde{K}_0(A)$ of reduced K_0. The notion of being Φ_∞-regular is in fact equivalent to being K_0-regular.   □

One of the fundamental lemma of Quillen's paper [66] is the following fact as reformulated in [40]:

F. Morel, $\mathbb{A}^1$-*Algebraic Topology over a Field*, Lecture Notes in Mathematics 2052,   209
DOI 10.1007/978-3-642-29514-0_9, © Springer-Verlag Berlin Heidelberg 2012

Lemma 9.3. *Let A be a commutative ring, $n \geq 1$ be an integer and P be a projective $A[T_1, \ldots, T_n]$-module of rank r.*
Then the set $\{a \in A | P$ is extended from $A_a \}$ is an ideal of A.
Consequently:

1) *If for each maximal ideal $\mathcal{M} \subset A$ of A the local ring $A_{\mathcal{M}}$ is Φ_r-regular then so is A itself.*
2) *If there exists a family $(f_1, \ldots, f_t)$ of elements of A which are coprime and such that A_{f_i} is Φ_r-regular for any i then A is Φ_r-regular.*

In fact Roitman [67, Proposition 2 p. 54] has proven the converse:

Proposition 9.4. *Let A be a noetherian commutative ring. The following conditions are equivalent:*

(i) A is Φ_r-regular;
(ii) For each maximal ideal $\mathcal{M} \subset A$ of A the local ring $A_{\mathcal{M}}$ is Φ_r-regular.

We will use the following facts which follow easily from the results of Milnor [45, Sect. 2]:

Lemma 9.5. *Let $r \in \mathbb{N} \cup \{\infty\}$. Let*

$$\begin{array}{ccc} \Lambda & \to & \Lambda_1 \\ \downarrow & & \downarrow \\ \Lambda_2 & \to & \Lambda' \end{array}$$

be a cartesian square of commutative rings in which the morphism $\Lambda_1 \to \Lambda'$ (and thus $\Lambda \to \Lambda_2$) is an epimorphism. Assume further that for any projective Λ_1-module P (of rank r when r is finite) the group homomorphism

$$Aut_{\Lambda_1}(P) \to Aut_{\Lambda'}(P_{\Lambda'}) \tag{9.1}$$

is onto. Then the map

$$\Phi_r(\Lambda) \to \Phi_r(\Lambda_1) \times_{\Phi_r(\Lambda')} \Phi_r(\Lambda_2)$$

is a bijection.

We will also need the following which also holds for $r = \infty$:

Lemma 9.6. *Let*
$$\begin{array}{ccc} \Lambda & \to & \Lambda_1 \\ \downarrow & & \downarrow \\ \Lambda_2 & \to & \Lambda' \end{array}$$

be a cartesian square of commutative rings in which the morphism $\Lambda_1 \to \Lambda'$ (and thus $\Lambda \to \Lambda_2$) is a split epimorphism.

1) If $\Phi_r(\Lambda_1) \to \Phi_r(\Lambda')$ is a bijection, then $\Phi_r(\Lambda) \to \Phi_r(\Lambda_2)$ is also a bijection.

2) If moreover the rings Λ_1, Λ_2 and Λ' are Φ_r-regular then so is Λ.

Proof. For the first statement we observe that if $\Lambda_1 \to \Lambda'$ is a split epimorphism which induces a bijection $\Phi_r(\Lambda_1) \to \Phi_r(\Lambda')$ then the homomorphisms (9.1) in Lemma 9.5 are indeed epimorphisms. Thus from that Lemma 9.5 we conclude that $\Phi_r(\Lambda) \to \Phi_r(\Lambda_2)$ is also a bijection.

We next observe that for any integer $n \geq 0$ the following diagram of commutative rings is still cartesian

$$
\begin{array}{ccc}
\Lambda[X_1,\ldots,X_n] & \to & \Lambda_1[X_1,\ldots,X_n] \\
\downarrow & & \downarrow \\
\Lambda_2[X_1,\ldots,X_n] & \to & \Lambda'[X_1,\ldots,X_n]
\end{array}
$$

The homomorphism $\Lambda_1[X_1,\ldots,X_n] \to \Lambda'[X_1,\ldots,X_n]$ are also split epimorphisms. If moreover Λ_1 and Λ' are Φ_r-regular then $\Phi_r(\Lambda_1[X_1,\ldots,X_n]) \to \Phi_r(\Lambda'[X_1,\ldots,X_n])$ are also bijection so that from the previous point it follows that $\Phi_r(\Lambda[X_1,\ldots,X_n]) \to \Phi_r(\Lambda_2[X_1,\ldots,X_n])$ is also a bijection. If moreover Λ_2 is also Φ_r-regular, then it follows that Λ is Φ_r-regular. $\square$

Let f and g be two coprime elements of a ring A (i.e. the ideal (f,g) is A itself). We let $A_{<f,g>}$ denote the ring defined by the following cartesian diagram of commutative rings:

$$
\begin{array}{ccc}
A_{<f,g>} & \to & A_f[Y] \\
\downarrow & & \downarrow \\
A_g[X] & \to & A_{fg}
\end{array}
\tag{9.2}
$$

in which the morphism $A_g[X] \to A_{fg}$ is the obvious one mapping the variable X to $\frac{1}{f}$ and the morphism $A_f[Y] \to A_{fg}$ maps Y to $\frac{1}{g}$. Thus all arrows in the previous diagram are epimorphisms.

Lemma 9.7. *Assume that A is Φ_r-regular:*

1) *Then the ring $A_{<f,g>}$ is Φ_r-regular.*

2) *If f or g is a unit then $\Phi_r(A) \to \Phi_r(A_{<f,g>})$ is a bijection.*

Proof. 1) We observe that by Proposition 9.4, A_f and A_g are also Φ_r-regular. We also observe that for any $n \geq 0$, $A_{<f,g>}[X_1,\ldots,X_n]$ fits in the following cartesian diagram:

$$
\begin{array}{ccc}
A_{<f,g>}[X_1,\ldots,X_n] & \to & A_f[Y][X_1,\ldots,X_n] \\
\downarrow & & \downarrow \\
A_g[X][X_1,\ldots,X_n] & \to & A_{fg}[X_1,\ldots,X_n]
\end{array}
\tag{9.3}
$$

By proposition 9.4 again, as f and g are coprime, we may assume that f or g is a unit. For instance f. Then the bottom horizontal morphism is a split epimorphism which induces a bijection $\Phi_r(A_g[X][X_1,\ldots,X_n]) \cong \Phi_r(A_g[X_1,\ldots,X_n])$ by the fact that A_g is Φ_r-regular. We conclude that (1) holds using 9.6. The same Lemma also implies that (2) holds as well.
$\square$

Let A be a commutative ring and let $n \geq 0$ be an integer. We set

$$A[\Delta^n] := A[T_0,\ldots,T_n]/(\Sigma_i T_i - 1)$$

and

$$A[\partial\Delta^n] := A[\Delta^n]/(\Pi_{i=0}^n T_i)$$

with the convention that $A[\partial\Delta^0] = 0$, and finally

$$A[\Lambda^n] := A[\Delta^n]/(\Pi_{i=1}^n T_i)$$

Observe that $A[\Delta^n] \cong A[T_1,\ldots,T_n]$ and that through this identification, the epimorphism $A[\Delta^n] \twoheadrightarrow A[\partial\Delta^n]$ is the quotient morphism $A[T_1,\ldots,T_n] \twoheadrightarrow A[T_1,\ldots,T_n]/(\Pi_{i=0}^n T_i)$ with the convention that $T_0 = 1 - \Sigma_{i=1}^n T_i$, and that the epimorphism $A[\Delta^n] \twoheadrightarrow A[\Lambda^n]$ is the quotient morphism $A[T_1,\ldots,T_n] \twoheadrightarrow A[T_1,\ldots,T_n]/(\Pi_{i=1}^n T_i)$.

For a pair (f,g) of coprime elements in A and any integer $n \geq 0$ we let B_n be the commutative ring defined by the following cartesian square of R-algebras:

$$\begin{array}{ccc} B_n & \to A[X,Y][\partial\Delta^n] \\ \downarrow & \downarrow \\ A_{<f,g>}[\Delta^n] \to & A_{<f,g>}[\partial\Delta^n] \end{array} \qquad (9.4)$$

so that the canonical morphism of rings

$$A[X,Y][T_1,\ldots,T_n] \to B_n$$

is an epimorphism.

Our main technical result in this section is the following one. This is where we need to assume $r \neq 2$.

Theorem 9.8. *Assume r is finite and $\neq 2$ and assume that A is an essentially smooth k-algebra or $r = \infty$ and A is a regular ring. Then B_n is Φ_r-regular for any $n \geq 1$.*

We first prove the following variant of Lemma 9.6:

Lemma 9.9. *Let*

$$\begin{array}{ccc} \Lambda & \to \Lambda_1 \\ \downarrow & \downarrow \\ \Lambda_2 & \to \Lambda' \end{array}$$

be a cartesian square of commutative rings in which the morphism $\Lambda_1 \to \Lambda'$ (and thus $\Lambda \to \Lambda_2$) is an epimorphism. Assume further that:

1) *Λ_1, Λ_2 and Λ' are Φ_r-regular*
2) *Any projective Λ_2-module of rank r is free.*
3) *The map $\Phi_r(\Lambda_1) \to \Phi_r(\Lambda')$ is a bijection.*
4) *For any integer $n \geq 0$ the group homomorphism*

$$GL_r(\Lambda_1[T_1, \ldots, T_n]) \to GL_r(\Lambda'[T_1, \ldots, T_n])$$

is onto.

Then for any integer $n \geq 0$, any projective $\Lambda[T_1, \ldots, T_n]$-module is free and thus Λ is Φ_r-regular.

Proof. Let $n \geq 1$ be an integer and let P be a projective module of rank r over $\Lambda[T_1, \ldots, T_n]$. We want to prove that P is actually free. By the assumption (2) and the fact that Λ_2 is Φ_r-regular, the restriction P_2 of P to $\Lambda_2[T_1, \ldots, T_n]$ is free. Let P_1 be the restriction of P to Λ_1. As its restriction to $\Lambda'[T_1, \ldots, T_n]$ is the same as the restriction of P_2, it is free. By assumption (3), we see that P_1 is also free.

By Milnor's results [45, Sect. 2], P is thus obtained up to isomorphism by gluing a free $\Lambda_1[T_1, \ldots, T_n]$-module of rank r and a free $\Lambda_2[T_1, \ldots, T_n]$-module of rank r along an element α of $GL_r(\Lambda'[T_1, \ldots, T_n])$.

But by assumption (4), $GL_r(\Lambda_1[T_1, \ldots, T_n]) \to GL_r(\Lambda'[T_1, \ldots, T_n])$ is onto and we can lift α to $\alpha_1 \in GL_r(\Lambda_1[T_1, \ldots, T_n])$. This implies again using Milnor's results that P itself is trivial. □

Proof of the Theorem 9.8. By Quillen's Lemma 9.3, we may assume A is a local ring. Thus either f or g must thus be a unit. We assume thus throughout this proof that for instance f is a unit (so that $A_f = A$).

We wish to apply Lemma 9.9 to the cartesian square (9.4) which defines B_n. We thus have to prove that assumptions (1)–(4) of that Lemma hold in that case.

We first prove (1). By Lindel's theorem [41] A and thus also $A[X, Y]$ are Φ_r-regular for any r, and by Lemma 9.7, $A_{<f,g>}$ is Φ_r-regular. We conclude using the following result which is a particular case of Vorst's [82, Corollary 4.1 (i)]:

Theorem 9.10. *(Vorst) Assume C is a noetherian ring and is Φ_r-regular for any r. Then $C[\partial \Delta]$ and $C[\Lambda^n]$ are Φ_r-regular for any r.*

To prove assumption (2) we observe that we already know that $A_{<f,g>}$ is Φ_r-regular. Thus $\Phi_r(A_{<f,g>}[\Delta^n]) = \Phi_r(A_{<f,g>})$. By Lemma 9.7 we know that $\Phi_r(A_{<f,g>}) = \Phi_r(A)$, thus this latter set is a point as A is local.

We know prove assumption (3). In the following diagram:

$$A[X,Y][\partial\Delta^n]$$
$$\downarrow$$
$$A_{<f,g>}[\partial\Delta^n] \to A_f[Y][\partial\Delta^n] = A[Y][\partial\Delta^n] \qquad (9.5)$$
$$\downarrow \qquad\qquad \downarrow \qquad\qquad \downarrow$$
$$A_g[X][\partial\Delta^n] \to A_{fg}[\partial\Delta^n] = A_g[\partial\Delta^n]$$

the composition

$$A[X,Y][\partial\Delta^n] \to A_{<f,g>}[\partial\Delta^n] \to A[Y][\partial\Delta^n] \qquad (9.6)$$

is the morphism obtained by mapping X to $\frac{1}{f}$. Thus setting $X' := X - \frac{1}{f}$, this composition (9.6) is the morphism

$$A[X,Y][\partial\Delta^n] \cong A[X',Y][\partial\Delta^n] \to A[Y][\partial\Delta^n], X' \mapsto 0$$

By Theorem 9.10 we know that $A[Y][\partial\Delta^n]$ is Φ_r-regular. Thus the composition (9.6) induces a bijection on Φ_r. Thus to prove assumption (3) it suffices by the commutativity of diagram (9.5) to prove that the morphism

$$A_{<f,g>}[\partial\Delta^n] \to A_f[Y][\partial\Delta^n] = A[Y][\partial\Delta^n]$$

induces a bijection on Φ_r. But this follows from Lemma 9.6 statement (1) because the bottom horizontal morphism of diagram (9.5) can be identified to the morphism

$$A_g[\partial\Delta^n][X'] \to A_g[\partial\Delta^n], X' \mapsto 0$$

where again $X' := X - \frac{1}{f}$.

It remains to prove assumption (4) that for any integer $m \geq 0$ the group homomorphism

$$GL_r(A[X,Y][\partial\Delta^n][T_1,\ldots,T_m]) \to GL_r(A_{<f,g>}[\partial\Delta^n][T_1,\ldots,T_m]) \quad (9.7)$$

is onto. We observe that $GL_r(A_{<f,g>}[\partial\Delta^n][T_1,\ldots,T_m])$ fits into the following cartesian square of groups:

$$GL_r(A_{<f,g>}[\partial\Delta^n][T_1,\ldots,T_m]) \to GL_r(A_f[Y][\partial\Delta^n][T_1,\ldots,T_m])$$
$$\downarrow \qquad\qquad\qquad\qquad \downarrow \qquad (9.8)$$
$$GL_r(A_g[X][\partial\Delta^n][T_1,\ldots,T_m]) \to GL_r(A_{fg}[\partial\Delta^n][T_1,\ldots,T_m])$$

Moreover the composition

$$GL_r(A[X,Y][\partial\Delta^n][T_1,\ldots,T_m]) \to GL_r(A_f[Y][\partial\Delta^n][T_1,\ldots,T_m])$$

is a split epimorphism (as $A_f = A$). Thus it suffices to prove the surjectivity of (9.7) to show that the kernel of

$$\Psi : GL_r(A_g[X][\partial\Delta^n][T_1,\ldots,T_m]) \to GL_r(A_g[\partial\Delta^n][T_1,\ldots,T_m]) \qquad (9.9)$$

$$X \mapsto \frac{1}{f}$$

is contained in the image of $GL_r(A[X,Y][\partial\Delta^n][T_1,\ldots,T_m])$.

For this we will use two results. The first one is a result of Vorst which follows directly from [83, Corollary 4.1], see *loc. cit.* Example 4.2 (ii):

Theorem 9.11. *(Vorst) Let A be a smooth k-algebra, $r \neq 2$ and $m \geq 0$ be integers. Then any $M \in GL_r(A[\partial\Delta^n][T_1,\ldots,T_m])$ is a product $E.C$ where $E \in E_r(A[\partial\Delta^n][T_1,\ldots,T_m])$ and C is constant, i.e. in the image of $GL_r(A[\partial\Delta^n]) \to GL_r(A[\partial\Delta^n][T_1,\ldots,T_m])$.*

The case $r = 1$ of the result is clear. Observe the previous result is the first place where we have to assume that $r \neq 2$.

We now recall some well-known facts (for instance from [31]). For R a commutative ring and $\mathcal{I} \subset R$ an ideal, let's denote by $E_r(R;\mathcal{I}) \subset E_r(R)$ the *normal subgroup* generated by the elementary matrices of the form $E_{i,j}(x)$ with coefficients x in the ideal $\mathcal{I}$. The second result we will use is:

Lemma 9.12. *1) For any epimorphism $\pi : R \twoheadrightarrow S$ the group homomorphism $E_r(R;\mathcal{I}) \to E_r(S;\pi(\mathcal{I}))$ is also onto.*

2) For any ring R and any ideal $\mathcal{I} \subset R$ such that the ring homomorphism $R \to R/\mathcal{I}$ admits a (ring) section, the following is a short exact sequence of groups

$$1 \to E_r(R;\mathcal{I}) \to E_r(R) \to E_r(R/\mathcal{I}) \to 1$$

Proof. In fact (1) is clear from the definition as $E_r(R;\mathcal{I})$ is the group generated by elements of the form $e^{-1}.E_{i,j}(x).e$, with $x \in \mathcal{I}$ and $e \in E_r(R)$.

To prove (2), let us denote by G the quotient group $E_r(R)/E_r(R;\mathcal{I})$. We wish to prove that the epimorphism $G \twoheadrightarrow E_r(R/\mathcal{I})$ is an isomorphism. Let $\sigma : R/\mathcal{I} \to R$ be a ring section and let also $\sigma : E_r(R/\mathcal{I}) \to G$ denote the obvious section induced by $\sigma : E_r(R/\mathcal{I}) \to E_r(R)$). It suffices to prove that this section is an epimorphism. But G is generated by (classes of) elementary matrices $E_{i,j}(x)$, $x \in R$. For any such x write $x = \sigma(\overline{x}) + y$, where $\overline{x}$ is the class of x in $R/\mathcal{I}$ and $y \in \mathcal{I}$. Then in G,

$$E_{i,j}(x) = E_{i,j}(\sigma(\overline{x})).E_{i,j}(y) = E_{i,j}(\sigma(\overline{x}))$$

thus indeed $\sigma : E_r(R) \to G$ is onto. $\square$

Now set $R := A_g[\partial\Delta^n][T_1,\ldots,T_m]$. The homomorphism Ψ (9.9) is just the evaluation at $\frac{1}{f}$

$$GL_r(R[X]) \to GL_r(R)$$

We claim that the kernel of this homomorphism is $E_r((X - \frac{1}{f}).R[X]) \subset E_r(R[X])$. This will prove the claim on $Ker\Psi$ by Lemma 9.12 point (1) as $A[X,Y][\partial\Delta^n][T_1, \ldots, T_m] \to R[X]$ is onto, and this will also finish the proof of Theorem 9.8.

Now let $M \in GL_r(R[X])$ be in the kernel of the homomorphism Ψ (9.9). By Vorst result 9.11 there is $C \in GL_r(R)$ (in $GL_r(A_g[\partial\Delta^n]) \subset GL_r(R)$ indeed) and $E \in E_r(R[X])$ with $M = E.C$. As Ψ is the evaluation at $\frac{1}{f}$, we get that $C = (ev_{\frac{1}{f}}(E))^{-1}$ is actually an elementary matrix as well. Thus M lies indeed in $E_r(R[X])$ and by the point (2) of Lemma 9.12 we obtain that $M \in E_r(R[X]; (X - \frac{1}{f}))$ as required.   □

Remark 9.13. 1) As a particular case of [74, Lemma 3.3] one can show that indeed for $r \geq 3$ the group $E_r(X.R[X])$ is the group generated by the elementary matrices with coefficients in the ideal $X.R[X]$.

2) It is shown in general in [31] that $E_r(R; \mathcal{I})$ is always a normal subgroup of $GL_r(R)$.

3) The case $r = \infty$ follows analogously from [83, Corollary 4.4], which was also proved in [22].   □

9.2   Preliminaries and Recollections on Patching Projective Modules

Lemma 9.14. *Let $r \in \mathbb{N} \cup \{\infty\}$ and $\pi : B \twoheadrightarrow A$ be an epimorphism of commutative rings. Let f_0 and f_1 be elements in B such that $(f_0, f_1) = B$. Assume furthermore:*

1. For $i \in \{0, 1\}$ the epimorphism $B_{f_i} \twoheadrightarrow A_{f_i}$ induces a surjection

$$\Phi_r(B_{f_i}) \twoheadrightarrow \Phi_r(A_{f_i})$$

2. The epimorphism $B_{f_0.f_1} \twoheadrightarrow A_{f_0.f_1}$ induces an injection

$$\Phi_r(B_{f_0.f_1}) \hookrightarrow \Phi_r(A_{f_0.f_1})$$

3. Moreover for any projective $B_{f_0.f_1}$-module Q (of rank r if r is finite) the group homomorphism

$$Aut_{B_{f_0.f_1}}(Q) \to Aut_{A_{f_0.f_1}}(Q_A)$$

is onto.

Then the map $\Phi_r(B) \to \Phi_r(A)$ induced by $\pi : B \twoheadrightarrow A$ is surjective.

Remark 9.15. We observe that assumption (2) and (3) together are in fact equivalent to requiring that if Q_1 and Q_2 are projective $B_{f_0 \cdot f_1}$-modules of rank r and if $\alpha : (Q_1)_A \cong (Q_2)_A$ is an isomorphism of $A_{f_0 \cdot f_1}$-modules then there exists an isomorphism $\beta : Q_1 \cong Q_2$ of $B_{f_0 \cdot f_1}$-modules which lifts α, in the obvious sense. Indeed, from (2) it is clear that Q_1 and Q_2 are isomorphic $B_{f_0 \cdot f_1}$-modules. Choose an isomorphism $\gamma : Q_1 \cong Q_2$. If (3) holds, then the automorphism

$$(Q_1)_A \overset{(\gamma)_A}{\cong} (Q_2)_A \overset{\alpha^{-1}}{\cong} (Q_1)_A$$

can be lifted to an automorphism $\beta' : Q_1 \cong Q_1$ of $B_{f_0 \cdot f_1}$-modules. Then clearly the isomorphism of $B_{f_0 \cdot f_1}$-modules $\beta : Q_1 \overset{(\beta')^{-1}}{\cong} Q_1 \overset{\gamma}{\cong} Q_2$ lifts α. □

Proof of Lemma 9.14. Let P be a projective A-module of rank r. From assumption (1) we know that for $i \in \{0,1\}$ there exists a projective B_{f_i}-module Q_i of rank r and an isomorphism of A_{f_i}-modules

$$\alpha_i : (Q_i)_A \cong P_{f_i}$$

Let us denote by

$$\alpha : ((Q_0)_{f_1})_A \cong ((Q_1)_{f_0})_A$$

the isomorphism of $A_{f_0 \cdot f_1}$-modules

$$((Q_0)_{f_1})_A \overset{(\alpha_0)_{f_1}}{\cong} P_{f_0 \cdot f_1} \overset{(\alpha_1)_{f_0}^{-1}}{\cong} ((Q_1)_{f_0})_A$$

By assumption (2) and (3) together and the previous remark, α can be lifted to an isomorphism of $B_{f_0 \cdot f_1}$-modules

$$\beta : (Q_0)_{f_1} \cong (Q_1)_{f_0}$$

Now let Q be the projective B-module of rank r obtained by gluing Q_1 and Q_2 along β (using again Milnor's results from [45, Sect. 2]). Then Q_A is the projective A-module of rank r obtained by gluing $(Q_0)_A$ and $(Q_1)_A$ along α, but the latter is clearly isomorphic to P. □

The following is the main technical result of this section:

Theorem 9.16. *We keep the notations of the previous section. Let $r \in \mathbb{N} \cup \{\infty\}$ and $r \neq 2$. Assume A is an essentially smooth k-algebra if r is finite or that A is regular if r is infinite. Let f and g be two coprime elements of A. Then for any $n \geq 0$ the natural epimorphism of A-algebras $A[X,Y][\Delta^n] \twoheadrightarrow B_n$ (cf diagram (9.4)), induces a surjection:*

$$\Phi_r(A) = \Phi_r(A[X,Y][\Delta^n]) \twoheadrightarrow \Phi_r(B_n)$$

Proof. In the following proof we simply denote by A_n the ring $A[X,Y][\Delta^n]$. We wish to apply Lemma 9.14 to the epimorphism $A_n \twoheadrightarrow B_n$ and to f and g which are also coprime in A_n.

We thus have to prove that conditions (1)–(3) from that Lemma hold.

Let us prove first (1) and (2). To do this it is clearly sufficient to assume that one of f or g is invertible, f for instance, and to prove that the map

$$\Phi_r(A) = \Phi_r(A_n) \to \Phi_r(B_n)$$

is in fact bijective.   □

We thus assume that f is a unit. We first observe that by Lemma 9.7 the following maps are bijections: $\Phi_r(A) \cong \Phi_r(A_{<f,g>}) \cong \Phi_r(A_{<f,g>}[\Delta^n])$. To prove that $\Phi_r(A) \cong \Phi_r(B_n)$ it is thus sufficient to prove that the morphism $B_n \twoheadrightarrow A_{<f,g>}[\Delta^n]$ induces a bijection $\Phi_r(B_n) \cong \Phi_r(A_{<f,g>}[\Delta^n])$.

To do this we make first the following definition:

Definition 9.17. 1) Two morphisms of rings $\phi_0, \phi_1 : R \twoheadrightarrow S$ are said to be homotopic if there exists a morphism

$$\eta : R \to S[T]$$

with $\eta(i) = \phi_i$, for $i \in \{0,1\}$.

2) A morphism of (commutative) rings $\phi : R \twoheadrightarrow S$ is called a homotopy equivalence if there exists a morphism of rings $\psi : S \to R$ such that both $\psi \circ \phi$ and $\phi \circ \psi$ are homotopic to the identity.

Then we observe:

Lemma 9.18. *A homotopy equivalence $R \to S$ between rings which are both Φ_r-regular induces a bijection*

$$\Phi_r(R) \cong \Phi_r(S)$$

By Theorem 9.8 B_n is Φ_r-regular and by Lemma 9.7 $A_{<f,g>}[\Delta^n]$ is Φ_r-regular. Our next observation is that the epimorphism $B_n \twoheadrightarrow A_{<f,g>}[\Delta^n]$ is a homotopy equivalence (assuming that f is a unit !), so we will conclude by the previous Lemma that $\Phi_r(B_n) \cong \Phi_r(A_{<f,g>}[\Delta^n])$ is a bijection.

Let us now prove that $B_n \twoheadrightarrow A_{<f,g>}[\Delta^n]$ is a homotopy equivalence.

We first give a name to some morphisms. We denote by $\pi : A[X,Y] \twoheadrightarrow A_{<f,g>}$ the obvious epimorphism. We denote by $s : A[Y] \to A_{<f,g>}$ the homomorphism of rings induced through the cartesian square (9.2) by the canonical morphism $A_g \subset A_g[X]$. Clearly, s is a section of the canonical epimorphism $p : A_{<f,g>} \twoheadrightarrow A_f[Y] = A[Y]$ (recall that f is invertible) and is also the composition $A[Y] \hookrightarrow A[X,Y] \overset{\pi}{\twoheadrightarrow} A_{<f,g>}$.

We next denote by $\gamma : A_{<f,g>} \to A[X,Y]$ the composition:

$$A_{<f,g>} \twoheadrightarrow A[Y] \hookrightarrow A[X,Y]$$

The composition $A_{<f,g>} \xrightarrow{g} A[X,Y] \xrightarrow{\pi} A_{<f,g>}$ is equal to $s \circ p$ and the composition $\gamma \circ \pi : A[X,Y] \to A[X,Y]$ is the morphism of A-algebras $X \mapsto \frac{1}{f}$, $Y \mapsto Y$.

In what follows, if $f : R \to S$ is a morphism of rings, we simply denote by $f[\Delta^n] : R[\Delta^n] \to S[\Delta^n]$ and $f[\partial\Delta^n] : R[\partial\Delta^n] \to S[\partial\Delta^n]$ the obvious extensions.

We then define a morphism of rings $\sigma : A_{<f,g>}[\Delta^n] \to B_n$. To do this it is sufficient by definition of B_n to define a morphism $A_{<f,g>}[\Delta^n] \to A[X,Y][\partial\Delta^n]$ and a morphism $A_{<f,g>}[\Delta^n] \to A_{<f,g>}[\Delta^n]$ which agree when composed to $A_{<f,g>}[\partial\Delta^n]$.

For the morphism $A_{<f,g>}[\Delta^n] \to A[X,Y][\partial\Delta^n]$ we take the composition

$$\alpha : A_{<f,g>}[\Delta^n] \twoheadrightarrow A_{<f,g>}[\partial\Delta^n] \xrightarrow{\gamma[\partial\Delta^n]} A[X,Y][\partial\Delta^n]$$

We observe then that $\pi[\partial\Delta^n] \circ \alpha$ is the composition

$$A_{<f,g>}[\Delta^n] \twoheadrightarrow A_{<f,g>}[\partial\Delta^n] \to A_{<f,g>}[\partial\Delta^n]$$

where the morphism $A_{<f,g>}[\partial\Delta^n] \to A_{<f,g>}[\partial\Delta^n]$ is $(\pi \circ \gamma)[\partial\Delta^n]$.

We thus take for the morphism $A_{<f,g>}[\Delta^n] \to A_{<f,g>}[\Delta^n]$ the morphism $\beta := (\pi \circ \gamma)[\Delta^n]$.

By construction $\pi[\partial\Delta^n] \circ \alpha$ is equal to the composition $A_{<f,g>}[\Delta^n] \xrightarrow{\beta} A_{<f,g>}[\Delta^n] \xrightarrow{p} A_{<f,g>}[\partial\Delta^n]$ which shows that α and β induce the morphism

$$\sigma : A_{<f,g>}[\Delta^n] \to B_n$$

we were looking for.

Denote by $q : B_n \twoheadrightarrow A_{<f,g>}[\Delta^n]$ the canonical epimorphism. We claim that $\sigma \circ q$ and $q \circ \sigma$ are homotopic to the (corresponding) identity morphisms.

Let $H : A[X,Y] \to A[X,Y][T]$ be the morphism of A-algebras defined by: $X \mapsto T.X + (1-T)\frac{1}{f}$, $Y \mapsto Y$. This is clearly a homotopy between $Id_{A[X,Y]}$ and $g \circ \pi$. We claim that the composition

$$A[X,Y] \xrightarrow{H} A[X,Y][T] \xrightarrow{\pi[T]} A_{<f,g>}[T]$$

induces a (unique) morphism

$$\eta : A_{<f,g>} \to A_{<f,g>}[T]$$

To prove this one uses the cartesian square of rings (9.3) with $n = 1$ and $X_1 = T$, and check that the composition $A[X,Y] \overset{H}{\to} A[X,Y][T] \overset{\pi[T]}{\twoheadrightarrow} A_{<f,g>}[T] \to A[Y][T]$ is the morphism of A-algebras $X \mapsto \frac{1}{f}$ and $Y \mapsto Y$, which thus factors as $A[X,Y] \overset{X \mapsto \frac{1}{f}}{\to} A[Y] \subset A[Y][T]$, and also that the composition $A[X,Y] \overset{H}{\to} A[X,Y][T] \overset{\pi[T]}{\twoheadrightarrow} A_{<f,g>}[T] \to A_g[X][T]$ is the morphism of A-algebras $X \mapsto T.X + (1-T)\frac{1}{f}$, $Y \mapsto \frac{1}{g}$, which also factors as $A[X,Y] \overset{Y \mapsto \frac{1}{g}}{\to} A_g[X] \overset{H'}{\to} A_g[X][T]$, where H' maps X to $T.X + (1-T)\frac{1}{f}$. This proves that H induces a unique morphism η) from the cartesian square

$$
\begin{array}{ccc}
A_{<f,g>} & \to & A[Y] \\
\downarrow & & \downarrow \\
A_g[X] & \to & A_g
\end{array}
$$

to

$$
\begin{array}{ccc}
A_{<f,g>}[T] & \to & A[Y][T] \\
\downarrow & & \downarrow \\
A_g[X][T] & \to & A_g[T]
\end{array}
$$

which induces the previous homotopy on each corner. This morphism thus defines a homotopy $\eta : A_{<f,g>} \to A_{<f,g>}[T]$ with the required property.

Now the morphism of the diagram

$$
\begin{array}{c}
A[X,Y][\partial\Delta^n] \\
\downarrow \\
A_{<f,g>}[\Delta^n] \to A_{<f,g>}[\partial\Delta^n]
\end{array}
$$

to the diagram

$$
\begin{array}{c}
A[X,Y][\partial\Delta^n][T] \\
\downarrow \\
A_{<f,g>}[\Delta^n][T] \to A_{<f,g>}[\partial\Delta^n][T]
\end{array}
$$

which is induced by $H[\partial\Delta^n]$ on the top right corner, by $\eta[\partial\Delta^n]$ on the bottom right corner and by $\eta[\Delta^n]$ on the bottom left corner induces a homotopy

$$
h : B_n \to B_n[T]
$$

from Id_{B_n} to $\sigma \circ q$. Moreover, $\eta[\Delta^n]$ gives a homotopy between $Id_{A_{<f,g>}[\Delta^n]}$ and $q \circ \sigma$.

This finishes the proof of (1) and (2).

We now prove axiom (3) to finish the proof of our theorem. Let us thus assume that both f and g are invertible and let P be a projective $A[X,Y][\partial\Delta]$-module of rank r. By Theorem 9.10 we know that P is extended from a projective $A[\partial\Delta]$-module Q of rank r. Clearly the following commutative square

$$Aut_{A_{<f,g>}[\partial\Delta^n]}(Q_{A_{<f,g>}[\partial\Delta^n]}) \to Aut_{A[Y][\partial\Delta^n]}(Q_{A[Y][\partial\Delta^n]})$$
$$\downarrow \qquad\qquad\qquad\qquad \downarrow$$
$$Aut_{A[X][\partial\Delta^n]}(Q_{A[X][\partial\Delta^n]}) \quad \to \quad Aut_{A[\partial\Delta^n]}(Q_{A[\partial\Delta^n]})$$

is a cartesian square of groups. But now we consider the following commutative square

$$Aut_{A[X,Y][\partial\Delta^n]}(Q_{A[X,Y][\partial\Delta^n]}) \to Aut_{A[Y][\partial\Delta^n]}(Q_{A[Y][\partial\Delta^n]})$$
$$\downarrow \qquad\qquad\qquad\qquad \downarrow \qquad\qquad\qquad\qquad (9.10)$$
$$Aut_{A[X][\partial\Delta^n]}(Q_{A[X][\partial\Delta^n]}) \quad \to \quad Aut_{A[\partial\Delta^n]}(Q_{A[\partial\Delta^n]})$$

The vertical morphisms are induced by evaluation of Y at $\frac{1}{g}$. As a consequence, the obvious morphisms of rings $A[X][\partial\Delta^n] \to A[X,Y][\partial\Delta^n]$ and $A[\partial\Delta^n] \to A[Y][\partial\Delta^n]$ induce compatible vertical sections of the vertical morphisms of (9.10). As a consequence the induced vertical morphism on horizontal Kernels in the previous diagram is also split surjective. It follows at once that the induced morphism

$$Aut_{A[X,Y][\partial\Delta^n]}(P) \to Aut_{A_{<f,g>}[\partial\Delta^n]}(P_{A_{<f,g>}[\partial\Delta^n]})$$

is onto. $\square$

9.3   The Affine B-G Properties for the Grassmanian and the General Linear Groups

Our aim in this section is to prove our main technical result Theorem 8.2. We start with the following one.

Theorem 9.19. *Let $r \in \mathbb{N}\cup\{\infty\}$ and $r \neq 2$. Let $B \twoheadrightarrow A$ be an epimorphism between essentially smooth k-algebras if r is finite and of regular rings if r is infinite.*

1) Then for any integer $n \geq 0$ the morphism

$$B[\Delta^n] \twoheadrightarrow A[\Delta^n] \times_{A[\Lambda^n]} B[\Lambda^n]$$

induces a bijection

$$\Phi_r(B[\Delta^n]) \cong \Phi_r(A[\Delta^n] \times_{A[\Lambda^n]} B[\Lambda^n])$$

2) The map of simplicial sets

$$\underline{Sing}_{\bullet}^{\mathbb{A}^1}(\mathbb{G}r_r)(B) \to \underline{Sing}_{\bullet}^{\mathbb{A}^1}(\mathbb{G}r_r)(A)$$

is a Kan fibration.

3) For pair (f, g) of coprime elements in A the diagram of simplicial sets

$$\begin{array}{ccc} \underline{Sing}_{\bullet}^{\mathbb{A}^1}(\mathbb{G}r_r)(A_{<f,g>}) & \to & \underline{Sing}_{\bullet}^{\mathbb{A}^1}(\mathbb{G}r_r)(A_f[Y]) \\ \downarrow & & \downarrow \\ \underline{Sing}_{\bullet}^{\mathbb{A}^1}(\mathbb{G}r_r)(A_g[X]) & \to & \underline{Sing}_{\bullet}^{\mathbb{A}^1}(\mathbb{G}r_r)(A_{f.g}) \end{array}$$

is homotopy cartesian.

Proof. (1) To prove that the map

$$\Phi_r(B[\Delta^n]) \to \Phi_r(A[\Delta^n] \times_{A[\Lambda^n]} B[\Lambda^n])$$

is bijective we will apply Lemma 9.5 to compute the right hand side. As A is Φ_r-regular by Lindel's Theorem in the finite case, or by Grothendieck result in the infinite case, it follows from Vorst [83, Theorem 1.1 and Corollary 4.4] that any projective $A[\Lambda^n]$-module of finite type is extended (stably in case r is infinite) from A. It clearly follows that the surjection $\Phi_r(A[\Delta^n]) \to \Phi_r(A[\Lambda^n])$ is bijective as the ring homomorphism $A[\Lambda^n] \to A$ which maps T_i to 0 for $i > 0$ and T_0 to 1 is a retraction of $A \to A[\Delta^n] \twoheadrightarrow A[\Lambda^n]$.

Thus if we can check the assumption of the Lemma we will get that $\Phi_r(A[\Delta^n] \times_{A[\Lambda^n]} B[\Lambda^n]) \to \Phi_r(B[\Lambda^n])$ is a bijection; again using Vorst *loc. cit.* and the same argument we know that $\Phi_r(B[\Delta^n]) \to \Phi_r(B[\Lambda^n])$ is bijective so that altogether we get the statement.

Let's check the assumptions of the Lemma 9.5 with respect to the epimorphism

$$A[\Delta^n] \twoheadrightarrow A[\Lambda^n]$$

By Lindel's or Grothendieck's Theorem, accordingly, any projective $A[\Delta^n]$-module of finite type is (stably) extended from a projective A-module of finite type P. We have to show that the group homomorphism

$$Aut_{A[\Delta^n]}(P[\Delta^n]) \to Aut_{A[\Lambda^n]}(P[\Lambda^n])$$

is an epimorphism (where we simply denote $P \otimes_A A[\Delta^n]$ by $P[\Delta^n]$ and $P \otimes_A A[\Lambda^n]$ by $P[\Lambda^n]$). Clearly the morphism of Endomorphism rings

$$End_{A[\Delta^n]}(P[\Delta^n]) \to End_{A[\Lambda^n]}(P[\Lambda^n])$$

is onto; this follows from the fact that P is a projective A-module and that $A[\Delta^n] \to A[\Lambda^n]$ is onto. Now we claim that for any projective A-module of finite type P, an element $\alpha \in End_{A[\Delta^n]}(P[\Delta^n])$ is invertible if and only if its image $\overline{\alpha}$ in $End_{A[\Lambda^n]}(P[\Lambda^n])$ is invertible. This clearly implies the surjectivity on the automorphisms groups level.

To prove the claim, we observe that if P is a summand of Q and if the property is known for Q, it holds for P. Indeed write $Q = P \oplus P'$. For $\alpha \in End_{A[\Delta^n]}(P[\Delta^n])$ let $\beta \in End_{A[\Delta^n]}(Q[\Delta^n])$ be the sum of α and of the identity of $P'[\Delta^n]$. Clearly the image of β in $End_{A[\Lambda^n]}(Q[\Lambda^n])$ is $\overline{\alpha} \oplus Id_{P'[\Lambda^n]}$, a sum of two automorphisms. Thus it is an automorphism and $\beta = \alpha \oplus Id_{P'[\Delta^n]}$ must be an automorphism of $A[\Delta^n]$-modules. Thus α is an automorphism as well.

Thus we may reduce to proving the claim when Q is free, isomorphic to some A^m. In that case the claim is a trivial consequence of the fact that the morphism of rings $A[\Delta^n] \to A[\Lambda^n]$ induces an isomorphism on the group of units. This is proven in [49].

(2) and (3) are easy consequences of (1): for (2) just do the same thing as in [49], proof of Theorem 4.2.6 p. 63. For (3), just observe that by definition of $A_{<f,g>}$ the diagram is cartesian. By (2) it is a cartesian square in which each morphism is a fibration, it is thus homotopy cartesian. □

Corollary 9.20. *Keeping the same assumptions as in the theorem:*

1) The epimorphism of rings

$$A[X,Y] \twoheadrightarrow A_{<f,g>}$$

induces a trivial fibration

$$\underline{Sing}_{\bullet}^{\mathbb{A}^1}(\mathbb{G}r_r)(A[X,Y]) \to \underline{Sing}_{\bullet}^{\mathbb{A}^1}(\mathbb{G}r_r)(A_{<f,g>})$$

2) The morphism $A \to A_{<f,g>}$ induces a weak-equivalence

$$\underline{Sing}_{\bullet}^{\mathbb{A}^1}(\mathbb{G}r_r)(A) \to \underline{Sing}_{\bullet}^{\mathbb{A}^1}(\mathbb{G}r_r)(A_{<f,g>})$$

3) The following diagram

$$\underline{Sing}_{\bullet}^{\mathbb{A}^1}(\mathbb{G}r_r)(Spec(A)) \to \underline{Sing}_{\bullet}^{\mathbb{A}^1}(\mathbb{G}r_r)(Spec(A_f))$$
$$\downarrow \qquad\qquad\qquad \downarrow$$
$$\underline{Sing}_{\bullet}^{\mathbb{A}^1}(\mathbb{G}r_r)(A_g) \quad \to \quad \underline{Sing}_{\bullet}^{\mathbb{A}^1}(\mathbb{G}r_r)(A_{f.g})$$

is homotopy cartesian.

Proof. (1) Follows from Theorem 9.19 and (2) follows from (1). (3) follows from Theorem 9.19 point (3). □

The last property is called the affine Zariski B.G. property for $\underline{Sing}_\bullet^{\mathbb{A}^1}(\mathbb{G}r_n)$. It is one of our main technical result. Let's now prove the affine Nisnevich B.G. property (see Definition A.7) for $\underline{Sing}_\bullet^{\mathbb{A}^1}(GL_r)$:

Theorem 9.21. *Assume $r \geq 3$. Let A be a smooth k-algebra $A \to B$ an étale A-algebra and $f \in A$ and such that $A/f \to B/f$ is an isomorphism. Then the diagram of simplicial groups*

$$\begin{array}{ccc}
\underline{Sing}_\bullet^{\mathbb{A}^1}(GL_r)(Spec(A)) & \to & \underline{Sing}_\bullet^{\mathbb{A}^1}(GL_r)(Spec(A_f)) \\
\downarrow & & \downarrow \\
\underline{Sing}_\bullet^{\mathbb{A}^1}(GL_r)(Spec(B)) & \to & \underline{Sing}_\bullet^{\mathbb{A}^1}(GL_r)(Spec(B_f))
\end{array} \qquad (9.11)$$

is homotopy cartesian.

We first make some simple recollections and observations which from standard homotopy theory and which we will use later. Given a morphism a simplicial groups $H_\bullet \to G_\bullet$ one may define its homotopy quotient $G_\bullet/^h H_\bullet$ to be the homotopy fiber at the base point of the induced morphism of pointed simplicial sets $B(H_\bullet) \to B(G_\bullet)$. This is a pointed simplicial set. When $H_\bullet \to G_\bullet$ is injective there exists a natural pointed weak equivalence between $G_\bullet/H_\bullet$ and $G_\bullet/^h H_\bullet$.

Let us define a *very weak equivalence* $X \to Y$ to be a map of simplicial sets which induces a weak equivalence from X to a sum of some of the connected components of Y. For a morphism of simplicial groups $H_\bullet \to G_\bullet$ this is equivalent to inducing an injection $\pi_0(H) \to \pi_0(G)$ and an isomorphism $\pi_i(H) \to \pi_i(G)$ for $i > 0$.

Lemma 9.22. *1) A commutative square of simplicial groups*

$$\begin{array}{ccc}
G_{3,\bullet} & \to & G_{2,\bullet} \\
\downarrow & & \downarrow \\
G_{1,\bullet} & \to & G_{0,\bullet}
\end{array}$$

is homotopy cartesian if and only if the induced morphism

$$G_{1,\bullet}/^h G_{3,\bullet} \to G_{0,\bullet}/^h G_{2,\bullet}$$

is a very weak equivalence.
2) Assume the cartesian square of simplicial groups

$$\begin{array}{ccc}
G_{3,\bullet} & \subset & G_{2,\bullet} \\
\cap & & \cap \\
G_{1,\bullet} & \subset & G_{0,\bullet}
\end{array} \qquad (9.12)$$

has the following property: each of its morphism is a monomorphism and moreover there exists a sub-simplicial group $H_{0,\bullet} \subset G_{0,\bullet}$ such that the quotient simplicial set is (simplicially) constant and such that setting $H_i := H \cap G_i$ gives a (automatically cartesian) square

$$\begin{array}{ccc} H_{3,\bullet} & \subset & H_{2,\bullet} \\ \cap & & \cap \\ H_{1,\bullet} & \subset & H_{0,\bullet} \end{array} \qquad\qquad (9.13)$$

which is homotopy cartesian.

Then the square (9.12) is homotopy cartesian.

Proof. The point (1) is easy. The proof of the second statement will use diagram chase and the observation:

For any i, $G_{i,\bullet}/H_{i,\bullet}$ being a sub-simplicial set of $G_{0,\bullet}/H_{0,\bullet}$ is also simplicially constant and thus for any integer $n \in \mathbb{N}$, $\pi_n(H_{i,\bullet}) \cong \pi_n(G_{i,\bullet})$ for $n > 0$ and for $n = 0$ one has a short exact sequence (of groups and sets):

$$1 \to \pi_0(H_{i,\bullet}) \subset \pi_0(G_{i,\bullet}) \twoheadrightarrow \pi_0(G_{i,\bullet}/H_{i,\bullet}) = G_{i,\bullet}/H_{i,\bullet} \to *$$

The last equality is a reformulation of the assumption that the quotients $G_{i,\bullet}/H_{i,\bullet}$ are assumed to be simplicially constant. Now by construction the diagram of (constant simplicial) sets

$$\begin{array}{ccc} G_{3,\bullet}/H_{3,\bullet} & \subset & G_{2,\bullet}/H_{2,\bullet} \\ \cap & & \cap \\ G_{1,\bullet}/H_{1,\bullet} & \subset & G_{0,\bullet}/H_{0,\bullet} \end{array}$$

is cartesian. A diagram chase using this and the above exact sequences, then produces an exact Mayer-Vietoris type diagram of the form:

$$\pi_1(H_{1,\bullet}) \oplus \pi_1(H_{2,\bullet}) \to \pi_1(H_{0,\bullet}) \to \pi_0(G_{3,\bullet}) \to \pi_0(G_{1,\bullet}) \times_{\pi_0(G_{0,\bullet})} \pi_0(G_{2,\bullet}) \to 1$$

Taking into account that $\pi_n(H_{i,\bullet}) \to \pi_n(G_{i,\bullet})$ is an isomorphism for $n \geq 1$ and the assumption on the square (9.13) is homotopy cartesian, this exact sequence can be completed in a long Mayer-Vietoris type exact sequence of the form

$$\cdots \to \pi_{n+1}(G_{1,\bullet}) \oplus \pi_{n+1}(G_{2,\bullet}) \to \pi_{n+1}(G_{0,\bullet}) \to \pi_n(G_{3,\bullet})$$

$$\to \pi_n(G_{1,\bullet}) \oplus \pi_n(G_{2,\bullet})$$

$$\to \pi_n(G_{n,\bullet}) \to \cdots$$

This easily implies that if $\Pi_\bullet$ denotes the homotopy fiber product of $G_{1,\bullet}$ and $G_{2,\bullet}$ over $G_{0,\bullet}$ then the morphism (of simplicial groups)

$$G_{3,\bullet} \to \Pi_\bullet$$

induces an isomorphism on each π_n, thus proving that (9.12) is homotopy cartesian. $\square$

Proof of the Theorem. We may clearly assume that A and B are integral domains, and thus that $A \to B$ is injective. As a consequence, the diagram (9.11) consists of injections.

Let us denote as usual by $E_r(A) \subset GL_r(A)$ the subgroup generated by elementary matrices. We introduce inside $\underline{Sing}_\bullet^{\mathbb{A}^1}(\mathbb{G}L_r)(Spec(A))$ the sub-simplicial group $\underline{Sing}_\bullet^{\mathbb{A}^1}(E_r)(Spec(A)) \subset \underline{Sing}_\bullet^{\mathbb{A}^1}(GL_r)(Spec(A))$, which in degree n equals $\overline{E_r(A[\Delta^n])}$.

We claim that $\underline{Sing}_\bullet^{\mathbb{A}^1}(E_r)(Spec(B_f)) \subset \underline{Sing}_\bullet^{\mathbb{A}^1}(\mathbb{G}L_r)(Spec(B_f))$ satisfies the above assumption of (2) for (the bottom right corner of) Diagram (9.11), proving our claim.

We claim in fact that

$$H_{1,\bullet}/H_{3,\bullet} \to H_{0,\bullet}/H_{2,\bullet}$$

is an isomorphism. It is clearly a monomorphism, by construction. From [82, Lemma 2.4 (i)] (here we use $r \geq 3$) for any $n \geq 0$, any element $\alpha \in E_r(B_f[\Delta^n]) = H_{0,n}$ can be written $\gamma\beta$ with $\beta \in E_r(A_f[\Delta^n])$ and $\gamma \in E_r(B[\Delta^n])$. But clearly $E_r(A_f[\Delta^n]) \subset H_{2,\bullet}$ and $E_r(B[\Delta^n]) \subset H_{1,\bullet}$. Thus α can be written $\gamma\beta$ with $\beta \in H_{2,\bullet}$ and $\gamma \in H_{1,\bullet}$. This proves the surjectivity.

To apply the (2) of the previous Lemma, it remains to show that each simplicial set $G_{i,\bullet}/H_{i,\bullet}$ is in fact simplicially constant. As each is a sub-simplicial set of $G_{0,\bullet}/H_{0,\bullet}$, it is sufficient to prove this for the latter one. This follows from [82] in which it is proven that for any smooth k-algebra A, any integer $n \geq 0$ and again for any $r \geq 3$, any element $\alpha \in \mathbb{G}L_r(A[\Delta^n])$ can be written $\gamma\beta$ with $\gamma \in \mathbb{G}L_r(A)$ a "constant" element and $\beta \in E_r(A[\Delta^n])$. This statement proves that the map of simplicial sets

$$GL_r(A) \to \underline{Sing}_\bullet^{\mathbb{A}^1}(\mathbb{G}L_r)(A)/\underline{Sing}_\bullet^{\mathbb{A}^1}(E_r)(A)$$

is a surjective map for each smooth k-algebra A, thus proving the target is always simplicially constant. $\square$

Altogether we finally proved the Theorem 8.2.

Appendix A
The (Affine) B.G. Property for Simplicial Sheaves

A.1 Some Recollections on the B.G. Property

We make free use of some notions and results from [15, 59].

Definition A.1. Let $\mathcal{B}$ be a presheaf of simplicial sets on Sm_k.

1) [15] We say that it satisfies the *B.G.-property in the Zariski topology* if for each $X \in Sm_k$ and each open covering of X by two open subschemes U and V the following diagram of simplicial sets is homotopy cartesian:

$$
\begin{array}{ccc}
\mathcal{B}(X) & \to & \mathcal{B}(V) \\
\downarrow & & \downarrow \\
\mathcal{B}(U) & \to & \mathcal{B}(U \cap V)
\end{array}
$$

2) We say that it satisfies the $\mathbb{A}^1$-*B.G. property in the Zariski topology* if $\mathcal{B}$ satisfies the B.G. property in the Zariski topology and if moreover, for any $X \in Sm_k$ the map

$$\mathcal{B}(X) \to \mathcal{B}(X \times \mathbb{A}^1)$$

induced by the projection $X \times \mathbb{A}^1 \to X$ is a weak equivalence.

The notion in (1) was introduced by Brown and Gersten in [15]. One of their main Theorem is:

Theorem A.2. [15] *A morphism $\mathcal{X} \to \mathcal{Y}$ of presheaves of simplicial sets on a Zariski site (for instance on Sm_k) which satisfy both the B.G. property in the Zariski topology and which is a local simplicial weak equivalence in the Zariski topology[1] induces, for any $U \in Sm_k$ a weak equivalence $\mathcal{X}(U) \to \mathcal{Y}(U)$ of simplicial sets.*

---

[1]That is, induces a weak equivalence on each Zariski stalk.

F. Morel, $\mathbb{A}^1$-*Algebraic Topology over a Field*, Lecture Notes in Mathematics 2052,   227
DOI 10.1007/978-3-642-29514-0, © Springer-Verlag Berlin Heidelberg 2012

As an application they endowed the category of sheaves of simplicial sheaves on some space with a model category structure (which is called the B.G. structure). One can do the same for the category $\Delta^{op}Shv(Sm_k)_{Zar}$ of simplicial sheaves of sets on Sm_k in the Zariski topology. We will denote by R_{Zar} a chosen fibrant resolution functor. As a consequence of their result, one obtains:

Lemma A.3. *Let $\mathcal{B}$ be a simplicial presheaf of sets on Sm_k which satisfies the B.G. property in the Zariski topology. Then the canonical morphism of simplicial presheaves of sets $\mathcal{B} \to R_{Zar}(a_{Zar}(\mathcal{B}))$ induces for any $X \in Sm_k$ a weak equivalence of simplicial sets*

$$\mathcal{B}(X) \to R_{Zar}(a_{Zar}(\mathcal{B}))(X)$$

Here a_{Zar} denotes the sheafification functor in the Zariski topology.

Indeed both terms satisfy the B.G. property in the Zariski topology and the morphism clearly induces a weak equivalence on Zariski stalks.

Remark A.4. This result allows one to compute maps in the associated simplicial homotopy category $\mathcal{H}_s(Sm_k)_{Zar}$ from any $X \in Sm_k$ to $a_{Zar}(\mathcal{B})$: indeed $\pi_0(R_{Zar}(a_{Zar}(\mathcal{B}))(X)$ is by the very definition of the model category structure equal to the set $Hom_{\mathcal{H}_s((Sm_k)_{Zar})}(X, a_{Zar}(\mathcal{B}))$.

Definition A.5. Let $\mathcal{B}$ be a presheaf of simplicial sets on Sm_k.

1) [59] We say that $\mathcal{B}$ satisfies the *B.G.-property in the Nisnevich topology* if and only if for any distinguished square[2] in Sm_k of the form

$$
\begin{array}{ccc}
W & \subset & V \\
\downarrow & & \downarrow \\
U & \subset & X
\end{array}
$$

the diagram of simplicial sets

$$
\begin{array}{ccc}
\mathcal{B}(X) & \to & \mathcal{B}(V) \\
\downarrow & & \downarrow \\
\mathcal{B}(U) & \to & \mathcal{B}(W)
\end{array}
$$

is homotopy cartesian.

2) We say it satisfies the $\mathbb{A}^1$-*B.G. property in the Nisnevich topology* if it satisfies the B.G. property in the Nisnevich topology and if moreover, for any $X \in Sm_k$ the map

---

[2]In the sense of [59, Definition 1.3 p.96].

$$\mathcal{B}(X) \to \mathcal{B}(X \times \mathbb{A}^1)$$

induced by the projection $X \times \mathbb{A}^1 \to X$ is a weak equivalence.

One of the technical result in [59] involving the B.G. property in the Nisnevich topology is quite analogous to the original result of Brown and Gersten: a morphism $\mathcal{X} \to \mathcal{Y}$ of presheaves of simplicial sets on Sm_k which satisfy both the B.G. property in the Nisnevich topology and which is a local simplicial weak equivalence in the Nisnevich topology[3] induces, for any $U \in Sm_k$ a weak equivalence $\mathcal{X}(U) \to \mathcal{Y}(U)$ of simplicial sets.

Remark A.6. Let $\mathcal{X}$ be a simplicial presheaf which satisfies the $\mathbb{A}^1$-B.G. property in the Nisnevich topology. Denote by $a_{Nis}(\mathcal{X})$ its sheafification in the Nisnevich topology. Then it is $\mathbb{A}^1$-local in the sense of [59, Definition 2.1 p. 106]. Indeed take a simplicially fibrant resolution $a_{Nis}(\mathcal{X}) \to \mathcal{Y}$; by the previous result just quoted (see also [59, Proposition 1.16 p. 100]) then for any $U \in Sm_k$ the morphism $\mathcal{X}(U) \to \mathcal{Y}(U)$ is a simplicial weak equivalence. As a consequence for any $U \in Sm_k$ the morphism

$$\mathcal{Y}(U) \to \mathcal{Y}(U \times \mathbb{A}^1) = \mathcal{Y}^{\mathbb{A}^1}(U)$$

is a weak equivalence. This implies at once that the morphism $\mathcal{Y} \to \mathcal{Y}^{\mathbb{A}^1}$ is a simplicial weak equivalence. Thus $\mathcal{Y}$ is $\mathbb{A}^1$-local and so is $\mathcal{X}$.

From a technical point of view we will be interested in slightly weaker conditions, only involving affine smooth k-schemes.

Definition A.7. Let $\mathcal{B}$ be a presheaf of simplicial sets on Sm_k.

1) We say that $\mathcal{B}$ satisfies the *affine B.G. property in the Zariski topology* if for any smooth k-algebra A and any coprime elements f and g of A the diagram

$$\begin{array}{ccc} \mathcal{B}(Spec(A)) & \to & \mathcal{B}(Spec(A_f)) \\ \downarrow & & \downarrow \\ \mathcal{B}(Spec(A_g)) & \to & \mathcal{B}(Spec(A_{f.g})) \end{array}$$

is homotopy cartesian.

2) We say that $\mathcal{B}$ satisfies the *affine B.G. property in the Nisnevich topology* if for any smooth k-algebra A, any étale A-algebra $A \to B$ and $f \in A$ such that $A/f \to B/f$ is an isomorphism, the diagram

$$\begin{array}{ccc} \mathcal{B}(Spec(A)) & \to & \mathcal{B}(Spec(A_f)) \\ \downarrow & & \downarrow \\ \mathcal{B}(Spec(B)) & \to & \mathcal{B}(Spec(B_f)) \end{array}$$

is homotopy cartesian.

---

[3]That is, induces a weak equivalence on each Nisnevich stalk.

3) We say that $\mathcal{B}$ satisfies the *affine $\mathbb{A}^1$-invariance property* if for any smooth k-algebra A the morphism

$$\mathcal{B}(Spec(A)) \rightarrow \mathcal{B}(Spec(A) \times \mathbb{A}^1)$$

induced by the projection $Spec(A) \times \mathbb{A}^1 \rightarrow Spec(A)$, is a weak equivalence.

Observe that the affine B.G. property in the Nisnevich topology implies the affine B.G. property in the Zariski topology.

A.2   The Affine Replacement of a Simplicial Presheaf

For X a smooth k-scheme, we denote by Sm_k^{af}/X the category of smooth affine k-schemes over X that is to say the category whose objects are morphism of smooth k-schemes $Y \rightarrow X$ with Y affine, with the obvious notion of morphisms.

Let $\mathcal{B}$ be a presheaf of simplicial sets on Sm_k. For any $X \in Sm_k$ we denote by $\mathcal{B}^{af}$ the presheaf of simplicial sets on Sm_k defined for $X \in Sm_k$ by the formula:

$$\mathcal{B}^{af}(X) := holim_{(Y \rightarrow X) \in Sm_k^{af}/X} Ex(\mathcal{B}(Y))$$

where Ex denotes a fixed choice of a functorial fibrant resolution in the category of simplicial sets (see [59] for instance). We call it the affinisation of $\mathcal{B}$.

We observe that by definition [13], this is indeed a presheaf of simplicial sets on Sm_k and moreover that there is a morphism of presheaf of simplicial sets on Sm_k:

$$\mathcal{B} \rightarrow \mathcal{B}^{af}$$

Lemma A.8. *The previous morphism induces for each affine smooth k-scheme X a weak equivalence*

$$\mathcal{B}(X) \rightarrow \mathcal{B}^{af}(X)$$

Proof. Because X is affine, the category Sm_k^{af}/X admits a final object and is thus contractible. By [13], the morphism is thus a weak equivalence. $\square$

In particular this morphism of simplicial presheaves is a local weak equivalence (i.e induces a weak equivalence on each stalk in the Zariski topology—or any topology with affine local point).

Theorem A.9. *Let $\mathcal{B}$ be a presheaf of simplicial sets on Sm_k. Assume that $\mathcal{B}$ satisfies the affine B.G. property and the affine $\mathbb{A}^1$-invariance property.*

Then the affine replacement $\mathcal{B}^{af}$ satisfies the $\mathbb{A}^1$-B.G. property in the Zariski topology.

Proof. The proof follows an idea of Weibel [84] in the same way as the proof of [49, Théorème 3.1.6 p. 37]. The details are left to the reader. □

An immediate consequence is the following affine version of the result of Brown–Gersten (Lemma A.3) which thus says that a presheaf of simplicial sets on Sm_k which satisfies the affine B.G. property in the Zariski topology and the affine $\mathbb{A}^1$-invariance property computes the "right thing" for affine smooth k-schemes. Observe however that we use the $\mathbb{A}^1$-invariance property which is not used in the classical case.

Lemma A.10. *Let $\mathcal{B}$ be a simplicial presheaf of sets on Sm_k which satisfies the affine B.G. property in the Zariski topology and the affine $\mathbb{A}^1$-invariance. Then the canonical morphism of simplicial presheaves of sets $\mathcal{B} \to R_{Zar}(a_{Zar}(\mathcal{B}))$ induces for any affine $X \in Sm_k$ a weak equivalence of simplicial sets*

$$\mathcal{B}(X) \to R_{Zar}(a_{Zar}(\mathcal{B}))(X)$$

Proof. We use the following commutative square of simplicial presheaves of sets

$$\begin{array}{ccc} \mathcal{B} & \to & R_{Zar}(a_{Zar}(\mathcal{B})) \\ \downarrow & & \downarrow \\ \mathcal{B}^{af} & \to & R_{Zar}(a_{Zar}(\mathcal{B}^{af})) \end{array}$$

The left vertical morphism induces a weak equivalence on sections on affine smooth k-schemes by Lemma A.8. The right vertical morphism induces a weak equivalence on sections on any smooth k-schemes because $a_{Zar}(\mathcal{B}) \to a_{Zar}(\mathcal{B}^{af})$ is a local weak equivalence in the Zariski topology. The bottom horizontal morphism induces a weak equivalence on sections on any smooth k-schemes by Theorem A.2 because it is a local weak equivalence in the Zariski topology between presheaves with the B.C. property in the Zariski topology. For the latter one it is clear and for the first one it is Theorem A.9. This gives the result. □

A.3   The Affine B.G. Property in the Nisnevich Topology

Slightly more difficult will be the analogue in the Nisnevich topology of Theorem A.9. In fact we do not know how to prove the analogue. We will have to assume that $\mathcal{G}$ is a presheaf of simplicial groups.

Theorem A.11. *Let $\mathcal{G}$ be a simplicial presheaf of groups on Sm_k. Assume that it satisfies the affine B.G. property in the Nisnevich topology as well as the affine $\mathbb{A}^1$-invariance property.*

Then the affine replacement $\mathcal{G}^{af}$ satisfies the $\mathbb{A}^1$-B.G. property in the Nisnevich topology.

We observe the following immediate consequence which is proven in the same way as Lemma A.10:

Corollary A.12. *Let $\mathcal{G}$ be a simplicial presheaf of groups on Sm_k satisfying the assumption of the previous theorem. Then its associated sheaf in the Nisnevich topology $a_{Nis}(\mathcal{G})$ is $\mathbb{A}^1$-local and moreover for any smooth affine k-scheme U the map*

$$\mathcal{G}(U) \to R_{Nis}(a_{Nis}(\mathcal{G}))(U)$$

is a weak equivalence.

Remark A.13. Using [59, Theorem 1.66 p. 70] gives a resolution functor R_{Nis} (or R_{Zar}) which commutes to finite products. In particular it takes group object to group objects and in the statement of the corollary we may assume the morphism is a morphism of simplicial presheaves of groups.

Proof of the Theorem A.11. From Theorem A.9 we already know that the affine replacement $\mathcal{G}^{af}$ satisfies the $\mathbb{A}^1$-B.G. property in the Zariski topology. Moreover by Lemma A.8 it also still satisfies the affine Nisnevich property. Finally, $\mathcal{G}^{af}$ is a simplicial presheaf of groups. Thus it suffices to prove Theorem A.14 below.   □

Theorem A.14. *Let $\mathcal{G}$ be a simplicial presheaf of groups on Sm_k. Assume that it satisfies the $\mathbb{A}^1$-B.G. property in the Zariski topology as well as the affine B.G. property in the Nisnevich topology. Then it satisfies the B.G. property in the Nisnevich topology.*

Remark A.15. The assumption that $\mathcal{G}$ satisfies the $\mathbb{A}^1$-invariance is crucial in our argument. We do not know whether the statement of the previous Theorem holds if we only assume $\mathcal{G}$ satisfies the B.G. property in the Zariski topology as well as the affine B.G. property in the Nisnevich topology.

To prove the theorem we need some preliminaries. We will prove the following crucial lemma:

Lemma A.16. *For any distinguished square of the form*

$$\begin{array}{ccc} W & \subset & V \\ \downarrow & & \downarrow \\ U & \subset & X \end{array}$$

for which the closed complement $Z := X - U$ with its reduced induced structure is k-smooth, the diagram

$$\begin{array}{ccc} \mathcal{G}(X) & \to & G(V) \\ \downarrow & & \downarrow \\ \mathcal{G}(U) & \to & \mathcal{G}(W) \end{array}$$

is homotopy cartesian.

We postpone the proof to the end of the section. We now prove how to deduce Theorem A.14 from this statement. We recall the following facts from [13]:

Lemma A.17. *1) For any commutative diagram of simplicial sets of the form*

$$\begin{array}{ccc} \mathcal{B}_1 \rightarrow \mathcal{B}_2 \rightarrow \mathcal{B}_3 \\ \downarrow \ (1) \ \downarrow \ (2) \ \downarrow \\ \mathcal{C}_1 \rightarrow \mathcal{C}_2 \rightarrow \mathcal{C}_3 \end{array} \qquad (A.1)$$

if the diagram (1) and (2) are homotopy cartesian the diagram

$$\begin{array}{ccc} \mathcal{B}_1 \rightarrow \mathcal{B}_3 \\ \downarrow \ (3) \ \downarrow \\ \mathcal{C}_1 \rightarrow \mathcal{C}_3 \end{array}$$

is homotopy cartesian.
2) Consider a functor $\mathcal{F} : \mathcal{I} \rightarrow Squares$ from a small category $\mathcal{I}$ to the category of commutative squares of simplicial sets. Suppose that for any $i \in \mathcal{I}$, the square $F(i)$ is homotopy cartesian. Then the square of simplicial sets $holim_{\mathcal{I}}\mathcal{F}$ is still homotopy cartesian.

Proof of the Theorem A.14. We wish to prove that for any distinguished square of the form

$$\begin{array}{ccc} W \subset V \\ \downarrow \quad \downarrow \\ U \subset X \end{array}$$

the diagram of simplicial groups

$$\begin{array}{ccc} \mathcal{G}(X) \rightarrow \mathcal{G}(V) \\ \downarrow \qquad \downarrow \\ \mathcal{G}(U) \rightarrow \mathcal{G}(W) \end{array} \qquad (A.2)$$

is homotopy cartesian.

Denote by Z the complement closed subset of U in X endowed with its reduced induced structure. As k is perfect there exists a flag of increasing closed subschemes

$$\emptyset = Z_0 \subset Z_1 \subset \cdots \subset Z_d = Z$$

with each $Z_{i+1} - Z_i$ smooth over k. Define the corresponding decreasing flag of open subsets

$$X = U_0 \supset U_1 \supset \cdots \supset U_d = U$$

by setting $U_i := X - Z_i$. Observe that $U_{i+1} = U_i - (Z_{i+1} - Z_i)$ with $Z_{i+1} - Z_i$ closed in U_i and k-smooth. For each i we thus have an elementary distinguished square (with obvious notations)

$$V_{i+1} \subset V_i$$
$$\downarrow \qquad \downarrow \qquad\qquad\qquad \text{(A.3)}$$
$$U_{i+1} \subset U_i$$

and we know from Lemma A.16 that the associated commutative square

$$\mathcal{G}(U_i) \to \mathcal{G}(U_{i+1})$$
$$\downarrow \qquad\qquad \downarrow$$
$$\mathcal{G}(V_i) \to \mathcal{G}(V_{i+1})$$

is homotopy cartesian. By Lemma A.17, and an easy induction, this implies that the square

$$\mathcal{G}(X) \to \mathcal{G}(U)$$
$$\downarrow \qquad\quad \downarrow$$
$$\mathcal{G}(V) \to \mathcal{G}(W)$$

is homotopy cartesian. The theorem is proven. $\qquad\qquad\qquad\qquad$ $\square$

Proof of the Lemma A.16. We will use the notations from [59, p. 115] concerning deformation to the normal cone. The morphism denoted there by $\tilde{g}_{X,Z}$ induces a morphism of simplicial sets denoted by $\mathcal{G}(\tilde{g}_{X,Z})$:

$$\mathcal{G}(B(X,Z) - f(Z \times \mathbb{A}^1))/^h\mathcal{G}(B(X,Z)) \to \mathcal{G}(X - Z)/^h\mathcal{G}(X) \qquad \text{(A.4)}$$

and the morphism denoted by $\alpha_{X,Z}$ in *loc. cit.* induces a morphism of simplicial sets denoted by $\mathcal{G}(\alpha_{X,Z})$

$$\mathcal{G}(B(X,Z) - f(Z \times \mathbb{A}^1))/^h\mathcal{G}(B(X,Z)) \to \mathcal{G}(N_{X,Z}^*)/\mathcal{G}(N_{X,Z}) \qquad \text{(A.5)}$$

Here $N_{X,Z}$ denotes the normal bundle of Z in Y and $N_{X,Z}^*$ the complement of the zero section.

We introduce two assumptions depending on an integer $d \geq 1$:

(A1)(d) The statement of the Lemma A.16 is true if the codimension of Z in X is $\leq d$;

(A2)(d) For any closed immersion $Z \hookrightarrow X$, of codimension $\leq d$, between smooth k-schemes the two maps of simplicial sets

$$\mathcal{G}(\tilde{g}_{X,Z}) : \mathcal{G}(B(X,Z) - f(Z \times \mathbb{A}^1))/^h\mathcal{G}(B(X,Z)) \to \mathcal{G}(X - Z)/^h\mathcal{G}(X)$$

and

$$\mathcal{G}(\alpha_{X,Z}) : \mathcal{G}(B(X,Z) - f(Z \times \mathbb{A}^1))/^h\mathcal{G}(B(X,Z)) \to \mathcal{G}(N_{X,Z}^*)/\mathcal{G}(N_{X,Z})$$

are very weak equivalences.

We recall that a very weak equivalence is a map of simplicial sets that induces a weak equivalence to a sum of some of the connected components of the target. Observe that a composition of very weak equivalences is still a very weak equivalence.

We now make the following observation. Given an open subset $\Omega \subset X$, we may form an other commutative square

$$
\begin{array}{ccc}
W_\Omega & \subset & V_\Omega \\
\downarrow & & \downarrow \\
U_\Omega & \subset & \Omega
\end{array}
\tag{A.6}
$$

with $U_\Omega = U \cap \Omega$, and V_Ω (resp. W_Ω) being the inverse image of Ω (resp. of U_Ω through $V \to X$. This diagram is obviously still distinguished. As a consequence, because $\mathcal{G}$ satisfies the B.G. property in the Zariski topology and by Lemma A.17, to check the fact that the Lemma for a given diagram, it suffices to check it for the diagrams of the form (A.6) where the Ω run over an open covering of X as well as the intersections between the members of the covering. Thus to prove any of our two assumptions, me may choose X as small as we want around a given point in Z.

This implies using the techniques from [59, p. 115] that for any d:

$$
(\mathbf{A1})(\mathbf{d}) \Leftrightarrow (\mathbf{A2})(\mathbf{d})
$$

We left the details to the reader.

We also recall from Lemma 9.22 that a commutative square .

We will prove these equivalent properties for any $\mathcal{B}$ and any $d \geq 1$ by induction on d. We observe that $(\mathbf{A1})(\mathbf{1})$ holds: by the affine B.G. property in the Nisnevich topology it holds for $X = Spec(A)$ affine k-smooth and $Z = Spec(A/f)$ a closed subscheme of X smooth over k. Let us assume $d \geq 2$ and that both $(\mathbf{A1})(\mathbf{d-1})$ and $(\mathbf{A2})(\mathbf{d-1})$ hold for any presheaf of groups satisfying the assumptions of the Lemma. We now want to prove $(\mathbf{A1})(\mathbf{d})$.

Assume

$$
\begin{array}{ccc}
W & \subset & V \\
\downarrow & & \downarrow \\
U & \subset & X
\end{array}
$$

is a distinguished square in Sm_k, with $Z := (X - U)_{red}$ k-smooth and of codimension d. From our above localization principe we may assume that X is affine and that $Z \hookrightarrow X$ is a regular closed immersion defined by a regular sequence $(x_1, \ldots, x_d)$ of regular functions on X such that moreover each closed k-subscheme of X of the form $\{x_1 = 0, \ldots, x_i = 0\}$, for $i \in \{1, \ldots, d\}$, is still smooth over k, in particular Z itself is smooth. Let us set $Y = X/(x_1 = 0, \ldots, x_{d-1} = 0)$; Y is also smooth over k and we have a diagram of closed immersion of the form: $Z \hookrightarrow Y \hookrightarrow X$, with Y of codimension $d - 1$ in X and Z of codimension 1 in Y.

Denote by $Y' \subset V$ the pull back of Y through $V \to X$. To prove that the diagram of simplicial groups of the Lemma is homotopy cartesian we know from Lemma 9.22 that it suffices to prove that

$$\mathcal{G}(U)/{}^h\mathcal{G}(X) \to \mathcal{G}(W)/{}^h\mathcal{G}(V) \tag{A.7}$$

is a very weak equivalence.

We observe that for any diagram of simplicial groups of the form

$$\mathcal{G}_1 \to \mathcal{G}_2 \to \mathcal{G}_3$$

the obvious diagram of homotopy quotients

$$\mathcal{G}_2/{}^h\mathcal{G}_1 \to \mathcal{G}_3/{}^h\mathcal{G}_1 \to \mathcal{G}_3/{}^h\mathcal{G}_2$$

is a homotopy fibration sequence between pointed simplicial sets.

Taking this into account, we get a commutative diagram in which the lines are homotopy fiber sequences of pointed simplicial sets

$$\mathcal{G}(X-Z)/{}^h\mathcal{G}(X) \to \mathcal{G}(X-Y)/{}^h\mathcal{G}(X) \to \mathcal{G}(X-Y)/{}^h\mathcal{G}(X-Z)$$
$$\downarrow \qquad\qquad\qquad \downarrow \qquad\qquad\qquad \downarrow$$
$$\mathcal{G}(V-Z)/{}^h\mathcal{G}(V) \to \mathcal{G}(V-Y')/{}^h\mathcal{G}(V) \to \mathcal{G}(X-Y)/{}^h\mathcal{G}(X-Z)$$

To prove that (A.7) is a very weak equivalence, it is thus sufficient to prove that the square of simplicial sets on the right

$$\mathcal{G}(X-Y)/{}^h\mathcal{G}(X) \to \mathcal{G}(X-Y)/{}^h\mathcal{G}(X-Z)$$
$$\downarrow \qquad\qquad\qquad\qquad \downarrow \tag{A.8}$$
$$\mathcal{G}(V-Y')/{}^h\mathcal{G}(V) \to \mathcal{G}(X-Y)/{}^h\mathcal{G}(X-Z)$$

is homotopy cartesian. In fact this implies slightly more, because we prove that (A.7) is a weak equivalence.

Now to prove that statement, observe that each closed immersion $Y \hookrightarrow X$, $Y' \hookrightarrow V$ and thus $Y - Z \hookrightarrow X - Z$ and $Y' - Z \hookrightarrow V - Z$ are of codimension $\leq d - 1$. Moreover there is a "closed immersion" of the distinguished square

$$Y' - Z \to Y'$$
$$\downarrow \qquad\quad \downarrow$$
$$Y - Z \to Y$$

into the distinguished square

$$V - Z \to V$$
$$\downarrow \qquad\quad \downarrow$$
$$X - Z \to X$$

As each of the morphisms in these diagrams are étale the functoriality of the deformation to the normal bundle discussed in [59, p.117] together with our inductive assumption **(A2)(d-1)** implies that the morphisms of the type $\tilde{g}$ and α induce in this case an explicit weak equivalence[4] between the diagram (A.8) and the diagram:

$$\mathcal{G}(E(N_{X,Y})^*)/{}^h\mathcal{G}(E(N_{X,Y})) \to \mathcal{G}(E(N^*_{X-Z,Y-Z}))/{}^h\mathcal{G}(E(N_{X-Z,Y-Z}))$$
$$\downarrow \qquad\qquad\qquad\qquad\qquad\qquad \downarrow$$
$$\mathcal{G}(E(N_{V,Y'})^*)/{}^h\mathcal{G}(E(N_{V,Y'})) \to \mathcal{G}(E(N^*_{V-Z,Y'-Z}))/{}^h\mathcal{G}(E(N_{V-Z,Y'-Z}))$$

But of course each of the normal bundles are trivialized in a compatible way (using the fixed regular sequence) and it is not hard, using finally the $\mathbb{A}^1$-invariance of $\mathcal{G}$ and the computations from [59, p. 112], to show that the last diagram is weakly equivalent to (with "obvious notations")

$$\Omega_s^{d-1}(\mathcal{G}((\mathbb{G}_m)^{\wedge d} \wedge (Y_+))) \to \Omega_s^{d-1}(\mathcal{G}((\mathbb{G}_m)^{\wedge d} \wedge ((Y-Z)_+)))$$
$$\downarrow \qquad\qquad\qquad\qquad\qquad \downarrow$$
$$\Omega_s^{d-1}(\mathcal{G}((\mathbb{G}_m)^{\wedge d} \wedge (Y'_+))) \to \Omega_s^{d-1}(\mathcal{G}((\mathbb{G}_m)^{\wedge d} \wedge ((Y'-Z)_+)))$$

But the presheaf $Y \mapsto \Omega_s^{d-1}(\mathcal{G}((\mathbb{G}_m)^{\wedge d} \wedge (Y_+)))$ also satisfies the assumptions of the Lemma. And moreover the commutative square

$$Y'-Z \to Y'$$
$$\downarrow \qquad\quad \downarrow$$
$$Y-Z \to Y$$

is distinguished and $Z \hookrightarrow Y$ of codimension 1. Thus by the property **(A1)(1)** this is homotopy cartesian.   □

A.4   A Technical Result

We start with the following lemma:

Lemma A.18. *Let $\mathcal{X}$ be a simplicial presheaf of pointed sets on Sm_k. Assume the associated simplicial sheaf $a_{Nis}(\mathcal{X})$ in the Nisnevich topology is $\mathbb{A}^1$-local. Assume also that the associated sheaf in the Zariski topology to the presheaf $U \mapsto \pi_0(\mathcal{X}(U))$ is trivial (Thus $\mathcal{X}$ is also 0-connected in the Nisnevich topology). Assume further that $\mathcal{X}$ satisfies the B.G. property in the Zariski topology. Then it satisfies the B.G. property in the Nisnevich topology.*

---

[4]This means a zig-zag—in fact only two morphisms in two different directions—of morphisms of diagrams whose morphisms are weak equivalences at each corner.

Proof. Let $\mathcal{X} \to R_{Nis}(a_{Nis}(\mathcal{X}))$ denotes a simplicially fibrant resolution of $a_{Nis}(\mathcal{X})$ in the sense of [34, 59]. Then $R_{Nis}(a_{Nis}(\mathcal{X}))$ is of course still $\mathbb{A}^1$-local and satisfies the Brown–Gersten property in the Nisnevich topology. It is thus sufficient to prove that

$$\mathcal{X} \to R_{Nis}(a_{Nis}(\mathcal{X}))$$

induces a local weak equivalence of presheaves in the Zariski topology, because as both sides satisfy the Brown–Gersten property in the Zariski topology, it will induce a termwise weak-equivalence of simplicial presheaves by the main result of [15], thus proving the result.

To prove the above morphism is a local weak equivalence in the Zariski topology it is sufficient to prove that for any localization X_x of some smooth k-variety X at some point $x \in X$ the map of simplicial sets

$$\mathcal{X}(X_x) \to R_{Nis}(a_{Nis}(\mathcal{X}))(X_x)$$

is a weak equivalence.

To do this we use Corollaries 6.2 and 6.9 which assert that under, our assumptions, for $n \geq 1$ the n-homotopy sheaf associated to the presheaf $U \mapsto \pi_n(\mathcal{X}(U))$ in the Nisnevich topology is strongly $\mathbb{A}^1$-invariant, and is also the Zariski sheaf associated to this presheaf (from the proof. As this is also true by assumption for $n = 0$ as everything is assumed to be trivial, this implies[5] that $\mathcal{X} \to R_{Nis}(a_{Nis}(\mathcal{X}))(X_x)$ induces a weak equivalence on sections over any smooth local k-scheme.   □

Here is our main application in this section:

Theorem A.19. *Let $\mathcal{B}$ be a pointed presheaf of Kan simplicial sets on Sm_k. Assume $\mathcal{B}$ satisfies the affine B.G. property in the Zariski topology, the affine $\mathbb{A}^1$-invariance property and that the presheaf of (simplicial) loop spaces $\Omega_s^1(\mathcal{B})$ satisfies the affine B.G. property in the Nisnevich topology. Assume further that the sheaf associated to the presheaf $\pi_0(\mathcal{B})$ is trivial in the Zariski topology (and thus also in the Nisnevich topology) and that the associated sheaf to $\pi_1(\mathcal{B}) = \pi_0(\Omega_s^1(\mathcal{B}))$ in the Zariski topology is the same as the one associated in the Nisnevich topology and is a strongly $\mathbb{A}^1$-invariant sheaf of groups.*

Then $a_{Nis}(\mathcal{B})$ is $\mathbb{A}^1$-local and for any smooth affine k-scheme U the morphism

$$\mathcal{B}(U) \to R(a_{Nis}(\mathcal{B}))(U)$$

is a weak equivalence of simplicial sets.

---

[5]Using comparison of the Postnikov towers in both the Zariski and the Nisnevich topology.

Remark A.20. The main example of application we have in mind is of course $\mathcal{B} = \underline{Sing}^{\mathbb{A}^1}_{\bullet}(\mathbb{G}r_n)$, $n \neq 2$, which is actually a simplicial sheaf in the Nisnevich topology; see the proof of Theorem 8.1. $\square$

Proof. By our assumptions the presheaf of simplicial sets $\mathcal{B}$ satisfy the hypotheses of Theorem A.9. Thus its affine replacement $\mathcal{B}^{af}$ satisfies the B.G. property in the Zariski topology and is also $\mathbb{A}^1$-invariant.

Now consider the presheaf of (simplicial) loop spaces $\Omega^1_s(\mathcal{B}^{af})$; we claim it is nothing but the affine replacement of the presheaf $\Omega^1_s(\mathcal{B})$. This comes from the commutation between loops space and homotopy limits [13]. Moreover using Kan's construction, we can find an equivalent[6] presheaf of simplicial groups. By our hypothesis it satisfies the assumptions of Theorem A.11 and thus satisfies the B.G. property in the Nisnevich topology and is $\mathbb{A}^1$-invariant.

We claim that the presheaf $\mathcal{B}^{af}$ satisfies the assumptions of Lemma A.18. Indeed, from what we have just said the sheaf $a_{Nis}(\mathcal{B}^{af})$ is $\emptyset$-connected and its loop space $\Omega^1_s(a_{Nis}(\mathcal{B}^{af}))$ satisfies the $\mathbb{A}^1$-B.G. property in the Nisnevich topology: it is thus $\mathbb{A}^1$-local by Remark A.6. Now by assumption, $a_{Nis}(\mathcal{B}^{af})$ is 0-connected, and its π_1 is a strongly $\mathbb{A}^1$-invariant sheaf of groups. By our results from Chap. 6 we conclude that $a_{Nis}(\mathcal{B}^{af})$ is $\mathbb{A}^1$-local.

Now by Lemma A.18, we conclude that $\mathcal{B}^{af}$ satisfies the B.G property in the Nisnevich topology and moreover that $a_{Nis}(\mathcal{B}^{af})$ is $\mathbb{A}^1$-local. As

$$a_{Nis}(\mathcal{B}) \to a_{Nis}(\mathcal{B}^{af})$$

is a weak equivalence, the Theorem is proven. $\square$

---

[6]Meaning together with a zig-zag of morphisms of presheaves of pointed simplicial sets, each morphism in the zig-zag being a global weak equivalence of presheaves.

Appendix B
Recollection on Obstruction Theory

What usually refers to obstruction theory is the following situation. Given a diagram

$$E$$
$$\downarrow$$
$$X \to B$$

in some category, can we find a sequence of "obstructions" whose triviality guarantees the existence of a morphism $X \to E$ which makes the obvious triangle commutative?

The main examples come from homotopy theory, in a reasonable closed model category [66] which has an appropriate notion of "truncated t-structure", in other words, in which objects admit a "reasonable" Postnikov tower. We will not try to formalize this further, in this appendix we will only recall[1] how the theory works in the homotopy category of simplicial sheaves on a site T (with enough points) which is of finite type in the sense of [59, Definition 1.31 p. 58]. This is the case for the sites Sm_k either in the Zariski or Nisnevich topology as it follows from Theorem 1.37 p. 60 from *loc. cit.*. Let us fix such a site T of finite type.

B.1 The Postnikov Tower of a Morphism

Given a morphism of simplicial sheaves of sets of the form $f : \mathcal{E} \to \mathcal{B}$ which is also a simplicial fibration,[2] we introduce the tower of simplicial sheaves

$$\mathcal{E} \to \cdots \to P^{(n)}(f) \to \cdots \to P^{(1)}(f) \to \mathcal{B}$$

---

[1]Actually we didn't find any reference for this.

[2]Any morphism in the simplicial homotopy category $\mathcal{H}_s(T)$ of simplicial sheaves of sets on T, see [59, Sect. 2.1 p. 46] for a recollection, can be represented up to an isomorphism in $\mathcal{H}_s(T)$ by a simplicial fibration [34, 59].

F. Morel, A¹-*Algebraic Topology over a Field*, Lecture Notes in Mathematics 2052,   241
DOI 10.1007/978-3-642-29514-0, © Springer-Verlag Berlin Heidelberg 2012

such that $P^{(n)}(f)$ is the associated simplicial sheaf to the presheaf $U \mapsto P^{(n)}(\mathcal{E}(U) \to \mathcal{B}(U))$; the latter is the usual Postnikov section construction associated to the Kan fibration, as f is assumed to be a simplicial fibration, $\mathcal{E}(U) \to \mathcal{B}(U)$ described for instance in [44]. When $\mathcal{B} = *$ is a point this is the construction described in [59, p. 57].

This tower is always a tower of local fibrations, and evaluating at each point x of the site T gives exactly the Postnikov tower of the stalk $x^*(f)$ of the morphism f at x. As a consequence, each morphism $P^{(n+1)}(f) \to P^{(n)}(f)$ is n-connected in the following sense:

Definition B.1. Let $n \geq -1$ be an integer. A morphism of simplicial sheaves of sets $\mathcal{E} \to \mathcal{B}$ is said to be n-*connected* if and only if for any points x of the site, the map of simplicial sets $\mathcal{E}_x \to \mathcal{B}_x$ is n-connected in the classical sense.

We simply recall that a map simplicial sets $f : E \to B$ is n-connected in the classical sense means that $\pi_0(E) \to \pi_0(B)$ is onto and that given any base point $y \in E$ the morphism

$$\pi_i(E; y) \to \pi_i(B; f(y))$$

is an epimorphism for $i = n + 1$ and isomorphism for $i \leq n$. When B is 0-connected with a base point $x \in B$, this is equivalent to the homotopy fiber Γ_x of $E \to B$ at x being an n-connected space: $\pi_i(\Gamma_x)$ trivial for $i \leq n$.

We have the following easy observation:

Lemma B.2. *Let $f : \mathcal{E} \to \mathcal{B}$ be a morphism of simplicial sheaves of sets. If $\mathcal{B}$ is 0-connected and $\mathcal{E}$ pointed, then f is n-connected if and only if the (homotopy) fiber $\mathcal{F} = f^{-1}(*)$ is an n-connected simplicial sheaf of sets.*

We will use the following Lemma which is a generalization of [59, Corollary 1.41 p.61].

Lemma B.3. *Assume that $\mathcal{E} \to \mathcal{B}$ is simplicial fibration between simplicially fibrant simplicial sheaves. Assume it is n-connected. Given an object $X \in T$ of cohomological dimension $\leq d$ for some integer $d \geq 0$ then the map of simplicial sets $\mathcal{E}(X) \to \mathcal{B}(X)$ is $(n - d)$-connected.*

Proof. cf [59, p. 60 and 61].   □

Remark B.4. 1) For X of cohomological dimension $\leq (n + 1)$, the lemma implies in that $\mathcal{E}(X) \to \mathcal{B}(X)$ is (-1)-connected. This means that map is surjective on the π_0 or in other words that any $\mathcal{H}_s(T)$-morphism $X \to \mathcal{B}$ can be lifted to a morphism $X \to \mathcal{E}$. If moreover X is of cohomological dimension $\leq n$, the map

$$Hom_{\mathcal{H}_s(T)}(X, \mathcal{E}) \to Hom_{\mathcal{H}_s(T)}(X, \mathcal{B})$$

is a bijection.

2) In fact for $f : \mathcal{E} \to \mathcal{B}$, the morphisms $P^{(n+1)}(f) \to P^{(n)}(f)$ in the Postnikov tower of f are more than n-connected: the fiber has "its homotopy sheaves concentrated in dimension $n + 1$".

3) The morphisms $\mathcal{E} \to P^{(n)}(f)$ are also n-connected. A consequence of the previous Lemma is the property that the obvious morphism

$$\mathcal{E} \to holim_n P^{(n)}(f)$$

is a simplicial weak equivalence. $\square$

Corollary B.5. *Given an object $X \in T$ of cohomological dimension $\leq d$ for some integer $d \geq 0$ then the map*

$$Hom_{\mathcal{H}_s(T)}(X, \mathcal{E}) \to Hom_{\mathcal{H}_s(T)}(X, P^{(n)}(f))$$

is surjective for $d \leq n + 1$ and bijective for $d \leq n$.

B.2   Twisted Eilenberg–MacLane Objects

Let G be a sheaf of groups on T. Given a sheaf of G-modules M and an integer $n \geq 2$ we define a simplicial sheaf of sets $K^G(M; n)$ in the following way. Take the model of Eilenberg–MacLane simplicial sheaf $K(M, n)$ of type (M, n) constructed in [59, p. 56] for instance. This construction being functorial, the simplicial sheaf $K(M, n)$ is endowed with a canonical action of the sheaf of groups G. Let $E(G)$ the weakly contractible simplicial sheaf of sets on which G acts "freely" so that the quotient $E(G)/G$ is the classifying object $B(G)$ (see [59, p. 128]). Then we set

$$K^G(M, n) := E(G) \times_G K(M, n)$$

This simplicial sheaf is 0-connected and pointed, its π_1 is canonically isomorphic to G and there is an obvious morphism $K^G(M, n) \to B(G)$, which is a local fibration and whose fiber is $K(M, n)$. Observe that the base point of $K(M, n)$ provides a canonical section to $K^G(M, n) \to B(G)$.

We consider a simplicial fibration $f : \mathcal{E} \to \mathcal{B}$. We assume $\mathcal{E}$ is pointed, $\mathcal{E}$ and $\mathcal{B}$ are 0-connected and that the induced morphism on the π_1 is an isomorphism (we point $\mathcal{B}$ by the image through f of the base point of $\mathcal{E}$). We simply denote by G the sheaf of groups $\pi_1(\mathcal{E}) = \pi_1(\mathcal{B})$. Observe that for $n \geq 2$, the sheaf $\pi_n(f) := \pi_n(P^{(n)}(f))$ is thus in a canonical way a G-module, because $G = \pi_1(P^{(n)}(f))$. The following is the basic technical lemma needed to the obstruction theory we will use.

Lemma B.6. *We keep the previous assumptions and notations. Then $P^{(0)}(f)$ is weakly equivalent to the point, $P^{(1)}(f)$ is weakly equivalent to $B(G)$, and for each $n \geq 2$ there exists a canonical morphism in $\mathcal{H}_{s,\bullet}(T)/B(G)$*

$$P^{(n-1)}(f) \to K^G(\pi_n(f), n+1)$$

such that the square

$$
\begin{array}{ccc}
P^{(n)}(f) & \to & B(G) \\
\downarrow & & \downarrow \\
P^{(n-1)}(f) & \to & K^G(\pi_n(f), n+1)
\end{array}
\tag{B.1}
$$

is a homotopy cartesian square.

Proof. The first statement is clear. The second statement follows at once from part (1) of Lemma B.7 below. We now observe that from the axioms of closed model categories, we can always obtain a commutative square of the form

$$
\begin{array}{ccc}
P^{(n)}(f) & \to & E' \\
\downarrow & & \downarrow \\
P^{(n-1)}(f) & \to & B(G)
\end{array}
$$

where $E' \to B(G)$ is a weak equivalence and $P^{(n)}(f) \to E'$ an inclusion (a cofibration). Form the amalgamate sum of the diagram

$$
\begin{array}{ccc}
P^{(n)}(f) & \to & E' \\
\downarrow & & \\
P^{(n-1)}(f) & &
\end{array}
$$

and call it E''. Of course we obtain a commutative diagram of simplicial sheaves of sets over $B(G)$:

$$
\begin{array}{ccc}
P^{(n)}(f) & \to & E' \\
\downarrow & & \downarrow \\
P^{(n-1)}(f) & \to & E''
\end{array}
$$

It is clear now that $\pi_1(E'') = G$, that $\pi_i(E'') = 0$ for $i \in \{2, n\}$ and that $\pi_{n+1}(E'') = \pi_n(\mathcal{B})$: these facts follow from the observation that the cone of $(P^{(n)}(f) \to P^{(n-1)}(f))$ is by definition equal to the cone of $E' \to E''$, and the fact that $E' \cong B(G)$. From part (2) of Lemma B.7 below we get a canonical pointed morphism (in $\mathcal{H}_{s,\bullet}(T)$): $E'' \to K^G(\pi_n(\mathcal{B}), n+1)$ over $B(G)$. The composition

$$P^{(n-1)}(f) \to E'' \to K^G(\pi_n(\mathcal{B}), n+1)$$

has the required property: this follows using points of the site because it is known in classical algebraic topology. $\square$

Lemma B.7. *Let $\mathcal{X}$ be a pointed 0-connected simplicial sheaf of sets, and denote simply by G its π_1-sheaf.*

1) There exists a canonical morphism in $\mathcal{H}_{s,\bullet}(T)$

$$\mathcal{X} \to B(G)$$

which induces the identity on π_1 (and thus a weak equivalence $P^{(1)}(\mathcal{X}) \cong B(G)$). Moreover this morphism induces for any sheaf of groups H a map:

$$Hom_{\mathcal{H}_{s,\bullet}(T)}(\mathcal{X}, B(H)) \to Hom(G, H)$$

which is a bijection. Here the right hand side means the set of morphisms of sheaves of groups and the map is evaluation at the π_1.

2) Assume that the $\mathcal{H}_{s,\bullet}(T)$-morphism $\mathcal{X} \to B(G)$ has a section $s : B(G) \to \mathcal{X}$ in $\mathcal{H}_{s,\bullet}(T)$ which we fix, and that $n \geq 2$ is an integer such that $\pi_i(\mathcal{X}) = 0$ for $1 < i < n$. Then there exists a canonical morphism in $\mathcal{H}_{s,\bullet}(T)$

$$\mathcal{X} \to K^G(\pi_n(\mathcal{X}), n)$$

which induces the identity morphism on π_i for $i \leq n$ and which is compatible in the obvious sense to both the projection to $B(G)$ and the section from $B(G)$. In particular, it induces a weak equivalence $P^{(n)}(\mathcal{X}) \cong K^G(\pi_n(\mathcal{X}), n)$).

Proof. First recall that for $n \geq 1$ and for an $(n-1)$-reduced simplicial set[3] L there exists a natural map $L \to K(\pi_n(L), n)$, the base point being the canonical one. This comes from the definition of $K(M, n)$: a morphism $X \to K(M, n)$ (for M a group for $n = 1$, an abelian group for $n \geq 2$) is the same thing as a an n-cocycle of the normalized cochain complex $C_N^*(X; M)$; see [44] for instance.

We also remind that for a simplicial set L with base point ℓ_0, one denotes by $L^{(n)} \subset L$ the sub-simplicial set consisting in dimension q of the simplexes of L whose n-skeleton is constant equal to the base point ℓ_0. When L is an n-connected Kan simplicial set, the inclusion $L^{(n)} \subset L$ is a weak-equivalence (See also [44] for instance). These facts at once generalize to locally fibrant pointed simplicial sheaves in an obvious way.

1) Having that in mind, let us denote by $\mathcal{X}^{(0)} \subset \mathcal{X}$ the sub-simplicial sheaf associated to the presheaf $U \mapsto (\mathcal{X}(U))^{(0)}$. The inclusion

$$\mathcal{X}^{(0)} \subset \mathcal{X}$$

---

[3] $(n-1)$-reduced means with only one i-simplex for $i \leq (n-1)$.

is thus a simplicial weak equivalence. Moreover, we have for each U a canonical map of pointed simplicial sets $(\mathcal{X}(U))^{(0)} \to B(\pi_1((\mathcal{X}(U))^{(0)}))$. Sheafification of this morphism of simplicial presheaves defines a morphism of simplicial sheaves $\mathcal{X}^{(0)} \to B(\pi_1(\mathcal{X})) = B(G)$. The diagram

$$\mathcal{X}^{(0)} \to B(G)$$
$$\downarrow$$
$$\mathcal{X}$$

in which the vertical morphism is a simplicial weak equivalence defines a $\mathcal{H}_{s,\bullet}(T)$-morphism $\mathcal{X} \to B(G)$. The assertion on the π_1 is clear. To check it has the second property we have to check surjectivity and injectivity. Given a morphism of sheaves of groups $G \to H$, it defines by the functoriality of the construction $G \mapsto BG$, a morphism of pointed simplicial sheaves of sets $BG \to BH$. Composition with the $\mathcal{H}_{s,\bullet}(T)$-morphism $\mathcal{X} \to BG$ just constructed proves surjectivity. Let's prove injectivity. Take two morphisms α_1 and $\alpha_2 \colon \mathcal{X} \to B(H)$ in $\mathcal{H}_{s,\bullet}(T)$. By the pointed version of [59, Proposition 1.13 p. 52] we may represent each of these morphisms by a diagram of pointed simplicial sheaves of sets

$$\mathcal{Y}_{\alpha_i} \to B(H)$$
$$\downarrow$$
$$\mathcal{X}$$

where $\mathcal{Y}_i \to \mathcal{X}$ is a (pointed) hypercovering. Taking the fiber product, we may further assume that $\mathcal{Y}_{\alpha_1} = \mathcal{Y}_{\alpha_2} = \mathcal{Y}$. We may further assume that $\mathcal{X}$ is locally fibrant (or in fact fibrant if we wish). As a consequence $\mathcal{Y}$ can also be assumed locally fibrant (and pointed) because $\mathcal{Y} \to \mathcal{X}$ is a trivial local fibration. From what we already saw, $\mathcal{Y}^{(0)} \to \mathcal{Y}$ is a simplicial weak equivalence. Now the diagram

$$\mathcal{Y}^{(0)} \rightrightarrows B(H)$$
$$\downarrow$$
$$\mathcal{X}$$

factors through

$$\mathcal{Y}^{(0)} \to BG \rightrightarrows BH$$
$$\downarrow$$
$$\mathcal{X}$$

by the functoriality of the Postnikov tower and the fact that for $\mathcal{Y}$ 0-reduced $P^{(1)}(\mathcal{Y}) = B\pi_1(\mathcal{Y})$. As a morphism of simplicial sheaves of the form $BG \to BH$ is always of the form $B(\rho : G \to H)$ for $\rho = \pi_1(BG \to BH)$, we get injectivity.

For the point (2) we proceed as follows. Let us denote by G the sheaf $\pi_1(\mathcal{X})$. We may assume that the morphism $f : \mathcal{X} \to B(G)$ (given by 1) is a simplicial fibration. We factor it as $\mathcal{X} \to P^{(n)}(f) \to B(G)$, so that $P^{(n)}(f) \to B(G)$ is a local fibration. We may clearly reduce to the case $\mathcal{X} \cong P^{(n)}(f)$ and as $B(G)$ is locally fibrant, we may assume that $\mathcal{X}$ is also locally fibrant. Moreover, again by the pointed version of [59, Proposition 1.13 p. 52], we may represent the section s by an actual diagram of pointed simplicial sheaves of sets

$$\mathcal{Y} \to \mathcal{X}$$
$$\downarrow$$
$$BG$$

in which $\mathcal{Y} \to BG$ is a pointed trivial local fibration. Let us denote by $\tilde{\mathcal{X}}$ the fiber product $EG \times_{BG} \mathcal{X}$ (see [59, p. 128] for an explicit definition of the simplicially weakly contractible simplicial sheaf EG as well as the morphism $EG \to BG$), by $\tilde{\mathcal{Y}}$ the fiber product $EG \times_{BG} \mathcal{Y}$ and by $\tilde{\mathcal{Y}} \to \tilde{\mathcal{X}}$ the induced morphism of G-simplicial sheaf of sets. As $\mathcal{Y} \to BG$ is a simplicial weak equivalence, the induced map $\tilde{\mathcal{Y}} \to EG$ is a G-equivariant simplicial weak equivalence. Thus $\tilde{\mathcal{Y}}$ is weakly contractible. Denote by $\mathcal{C}$ the cone of the morphism of simplicial sheaves of sets

$$\tilde{\mathcal{Y}} \to \tilde{\mathcal{X}}$$

by which we means the amalgamate sum $\tilde{\mathcal{X}} \coprod_{\tilde{\mathcal{Y}}} C(\tilde{\mathcal{Y}})$, where $C(\tilde{\mathcal{Y}}) := (\tilde{\mathcal{Y}} \times \delta^1)/(\tilde{\mathcal{Y}} \times \{0\})$. By the very construction, the morphism of simplicial sheaves of sets

$$\tilde{\mathcal{X}} \to \mathcal{C}$$

is a G-equivariant morphism, a simplicial weak equivalence, and moreover, $\mathcal{C}$ is pointed as a G-object. Consider the resolution functor $Ex^{\mathcal{G}}$ constructed in [59, Theorem 1.66 p. 69]. As it commutes to finite products by loc. cit., then $Ex^{\mathcal{G}}(\mathcal{C})$ is clearly endowed with a G-action and the simplicial weak equivalence $\mathcal{C} \to Ex^{\mathcal{G}}(\mathcal{C})$ is G-equivariant. As $\mathcal{C}$ is pointed as a G-object, so is $Ex^{\mathcal{G}}(\mathcal{C})$. As the latter is simplicially fibrant, the morphism $Ex^{\mathcal{G}}(\mathcal{C})^{(n-1)} \to Ex^{\mathcal{G}}(\mathcal{C})$ is a G-equivariant pointed weak equivalence. Observe that $\pi_n(Ex^{\mathcal{G}}(\mathcal{C})^{(n-1)}) = \pi_n(\mathcal{X})$. As $Ex^{\mathcal{G}}(\mathcal{C})^{(n-1)}$ is an $(n-1)$-reduced simplicial sheaf of sets, from what we have recall above, there exists a natural G-equivariant map

$$Ex^{\mathcal{G}}(\mathcal{C})^{(n-1)} \to K(\pi_n(\mathcal{X}), n)$$

obtained by sheafification of the classical one. Now perform the Borel construction $EG \times_G (-)$ and remember the definition $K^G(\pi_n(\mathcal{X}), n) := EG \times_G (K(\pi_n(\mathcal{X}), n))$ of twisted Eilenberg–MacLane objects, to produce a diagram of simplicial sheaves of sets

$$EG \times_G (Ex^{\mathcal{G}}(\mathcal{C})^{(n-1)} \to K^G(\pi_n(\mathcal{X}), n)$$
$$\downarrow$$
$$EG \times_G \tilde{\mathcal{X}} \to \quad EG \times_G Ex^{\mathcal{G}}(\mathcal{C})$$
$$\downarrow$$
$$\mathcal{X}$$

in which all the vertical morphisms are simplicial weak equivalences. This clearly defines the $\mathcal{H}_{s,\bullet}(T)$-morphism we are seeking, because by construction using points of the site it is compatible to the classical construction. $\square$

Remark B.8. In fact with a little bit more work, one can prove that the canonical morphism in $\mathcal{H}_{s,\bullet}(T)$

$$\mathcal{X} \to K^G(\pi_n(\mathcal{X}), n)$$

given by point (2) of the previous lemma is the unique one with these properties. $\square$

B.3   The Obstruction Theory We Need

We can now explain the obstruction theory we will use. Let us fix a diagram in the category of simplicial sheaves of sets of the form

$$\mathcal{E}$$
$$\downarrow \qquad\qquad\qquad (\text{B.2})$$
$$X \to \mathcal{B}$$

with $X \in T$. Our aim is to give, under some assumptions, both a criterium for the existence and/or uniqueness of a morphism in $\mathcal{H}_s(T)$

$$X \to \mathcal{E}$$

which makes the triangle

$$\mathcal{E}$$
$$\nearrow \downarrow$$
$$X \to \mathcal{B}$$

commutative in $\mathcal{H}_s(T)$. We may clearly assume that $f : \mathcal{E} \to \mathcal{B}$ is a simplicial fibration.

We make the following assumptions:

1. $\mathcal{B}$ is 0-connected and pointed.
2. The morphism $\mathcal{E} \to \mathcal{B}$ is $(n-2)$-connected, or in other words (by Lemma B.2) the homotopy fiber of $\mathcal{E} \to \mathcal{B}$ is $(n-2)$-connected.

We observe that $P^{(i)}(f) \to \mathcal{B}$ is a weak equivalence for $i \leq n - 2$ and that from Lemma B.6 there exists a canonical homotopy cartesian square in $\mathcal{H}_{s,\bullet}(T)$:

$$
\begin{array}{ccc}
P^{(n-1)}(f) \to & & B(G) \\
\downarrow & & \downarrow \\
\mathcal{B} = P^{(n-2)}(f) \to & K^G(\pi_{n-1}(f), n)
\end{array}
\tag{B.3}
$$

Given $X \in T$ the previous homotopy cartesian square gives a surjection

$$
Hom_{\mathcal{H}_s(T)}(X, P^{(n-1)}(f)) \twoheadrightarrow Hom_{\mathcal{H}_s(T)}(X, \mathcal{B}) \times_{Hom_{\mathcal{H}_s(T)}(X, K^G(\pi_{n-1}(f), n))}
$$

$$
Hom_{\mathcal{H}_s(T)}(X, BG)
$$

Obstruction to lifting. By Corollary B.5 the map

$$
Hom_{\mathcal{H}_s(T)}(X, \mathcal{E}) \to Hom_{\mathcal{H}_s(T)}(X, P^{(n-1)}(f))
$$

is a surjection for all X of cohomological dimension $\leq n$.

We thus obtained the following:

Theorem B.9. *Keeping the previous assumptions on $f : \mathcal{E} \to \mathcal{B}$, for any X of cohomological dimension $\leq n$ the map*

$$
Hom_{\mathcal{H}_s(T)}(X, \mathcal{E}) \to Hom_{\mathcal{H}_s(T)}(X, \mathcal{B}) \times_{Hom_{\mathcal{H}_s(T)}(X, K^G(\pi_{n-1}(f), n))}
$$

$$
Hom_{\mathcal{H}_s(T)}(X, BG)
\tag{B.4}
$$

is surjective.

We deduce the following obstruction theory:

Corollary B.10. *Keeping the previous assumptions on $f : \mathcal{E} \to \mathcal{B}$, for any X of cohomological dimension $\leq n$ and any morphism $g : X \to \mathcal{B}$ in $\mathcal{H}_s(T)$, there exists a morphism $h : X \to \mathcal{E}$ which lifts g in $\mathcal{H}_s(T)$ if and only if the composition $X \to \mathcal{B} \to K^G(\pi_{n-1}(f), n))$ lifts through $BG \to K^G(\pi_{n-1}(f), n))$.*

Now for any sheaf of G-modules M, for any $\lambda \in Hom_{\mathcal{H}_s(T)}(X, BG) \cong H^1(X; G)$ let us consider the subset

$$
E^n_\lambda(X; M) \subset Hom_{\mathcal{H}_s(T)}(X, K^G(M, n))
$$

of elements $X \to K^G(M, n)$ whose composition to BG gives λ. This set is pointed by $X \to BG$ composed with canonical section $BG \to K^G(M, n)$.

Given $g : X \to \mathcal{B}$ in $\mathcal{H}_s(T)$ one gets by composition with $\mathcal{B} \to BG$ a morphism

$$
\lambda_g : X \to BG
$$

Clearly a reformulation of the corollary is to say that the element $e(g) \in E^n_{\lambda_g}(X; \pi_{n-1}(f))$ is the obstruction to the lifting of g:

$$(g \text{ lifts through } \mathcal{E} \to \mathcal{B}) \Leftrightarrow (e(g) \text{ is the base point in } E^n_{\lambda_g}(X; \pi_{n-1}(f)))$$

On the "Kernel" of the Map $Hom_{\mathcal{H}_s(T)}(X, \mathcal{E}) \to Hom_{\mathcal{H}_s(T)}(X, \mathcal{B})$

By "Kernel" of that pointed map, we mean the subset K^n of $Hom_{\mathcal{H}_s(T)}(X, \mathcal{E})$ consisting of morphisms whose composition with $f : \mathcal{E} \to \mathcal{B}$ is trivial (the base point of $Hom_{\mathcal{H}_s(T)}(X, \mathcal{B})$).

We want to study this kernel in the critical case. By Corollary B.5 and our assumptions, the map

$$Hom_{\mathcal{H}_s(T)}(X, \mathcal{E}) \to Hom_{\mathcal{H}_s(T)}(X, \mathcal{B})$$

is a surjection for all X of cohomological dimension $\leq n - 1$ and a bijection for all X of cohomological dimension $\leq n - 2$. By the critical case we mean when the cohomological dimension is $\leq n - 1$ so that the map is surjective. We just want to use the pointed simplicial fibration sequence

$$\Gamma \to \mathcal{E} \to \mathcal{B}$$

where Γ is the (homotopy) fiber at the base point, and which is $(n - 2)$-connected by assumption. We will use the natural action up to homotopy of the h-group $\Omega^1_s(\mathcal{B})$ on the fiber Γ. For any X we obtain a exact sequence of pointed sets and groups

$$Hom_{\mathcal{H}_s(T)}(X, \Omega^1_s(\mathcal{B})) \to Hom_{\mathcal{H}_s(T)}(X, \Gamma) \to K^n \to *$$

The left hand side is indeed a group which acts on the middle set, and exactness means that K^n is the orbit set.

For X of cohomological dimension $\leq n - 1$ we can express in a simpler way the middle term; by Corollary B.5 the maps

$$Hom_{\mathcal{H}_s(T)}(X, \Gamma) \to Hom_{\mathcal{H}_s(T)}(X, P^{(n-1)}(\Gamma))$$

$$\to Hom_{\mathcal{H}_s(T)}(X, K(\pi_{n-1}(f), n - 1))$$

are all bijective (the latter one uses point 2) of Lemma B.7). But has it is well known, the right hand side is an abelian group isomorphic to $H^{n-1}(X; \pi_{n-1}(f))$. At the end we get an exact sequence as above of the form

$$Hom_{\mathcal{H}_s(T)}(X, \Omega^1_s(\mathcal{B})) \to H^{n-1}(X; \pi_{n-1}(f)) \to K^n \to *$$

Remark B.11. Beware that in general the action on the left is not given by a homomorphism: take the universal situation, in the category of simplicial sets, that is to say the fibration

$$K(M, n) \to BG \to K^G(M, n)$$

In that case the loop space in question is $\Omega_s^1(K^G(M, n))$ which is easily seen to be equivalent to the semi-direct product $G \ltimes K(M, n-1)$ and the action of $\Omega_s^1(K^G(M, n)) = G \ltimes K(M, n-1)$ on $K(M, n-1)$ is the standard action of a semi-direct product $G \ltimes N$ onto a G-module N. This action can't induce in general a group homomorphism. □

Cohomological Interpretation of the Obstruction Sets $E_\lambda^n(X; M)$

The pointed sets of the form $E_\lambda^n(X; M)$ have a natural cohomological interpretation, and in particular are abelian groups, for each fixed λ. We end this appendix by explaining this fact.

First by [59, Proposition 1.15 p. 130], there is a canonical bijection

$$Hom_{\mathcal{H}_s(T)}(X, BG) \simeq H^1(X; G)$$

identifying morphism $X \to BG$ and isomorphism classes of G-torsors over X. Thus our λ corresponds to an isomorphism class of G-torsors over X. Pick up one G-torsor $Y \to X$ in the class of λ. Consider the sheaf of sets M_λ on T obtained as

$$M_\lambda := {}_G\backslash(Y \times M)$$

The quotient being computed in the category of sheaves of sets on T. The obvious morphism $M_\lambda \to {}_G\backslash Y = X$ defines it as a sheaf of sets on X. It is called the sheaf obtained by twisting M by λ. Our aim is to prove that this sheaf is in a canonical way an abelian sheaf on X and that for each $n \geq 2$ the pointed set $E_\lambda^n(X; M)$ is canonically in bijection with the n-th cohomology group

$$H_{T/X}^n(X; M_\lambda)$$

Remark B.12. 1) Given λ and M, all these constructions only depend on the choice of a representative $Y \to X$ of λ.

2) If λ is the trivial G-torsor θ, the result is quite clear and in fact $E_\theta^n(X; M)$ is canonically in bijection with $H_{T/X}^n(X; M_\theta) = H_T^n(X; M)$.

The following two lemmas are easy to prove.

Lemma B.13. *Given a sheaf of groups G on T, a G-torsor Y over the final sheaf, and a G-sheaf of sets M the canonical morphism $Y \times M \to Y \times (Y \times_G M)$ is an isomorphism of sheaves on T.*

Lemma B.14. *Given a sheaf of groups G on T, a G-torsor Y over the point and two G-sheaves of sets M and N, the canonical morphism of sheaves on T: $Y \times M \times N \to (Y \times_G M) \times (Y \times_G N)$ induces an isomorphism*

$$Y \times_G (M \times N) \to (Y \times_G M) \times (Y \times_G N)$$

As a consequence, if M is a sheaf of G-module, the sheaf $Y \times_G M$ has a canonical structure of sheaf of abelian groups.

Lemma B.15. *Keeping the obvious notations, the X-sheaf M_λ admits a canonical structure of abelian X-sheaf. Let us denote by $K(M_\lambda, n)$ the usual simplicial Eilenberg–MacLane object in the category of sheaves over X. Then there exists a canonical isomorphism of simplicial sheaves over X of the form*

$$Y \times_G K(M, n) \cong K(M_\lambda, n)$$

Proof. We only use the explicit definition of $K(-, n)$ [44] which show that in each degree $K(M_\lambda, n)$ is a product (over X) of copies of M_λ. The conclusion follows from the two previous lemmas. □

For any $X \in T$, for any integer ≥ 0, for any sheaf of G-modules M and for any $\lambda \in Hom_{\mathcal{H}_s(T)}(X, BG) = H^1(X; G)$ we now describe a natural map of pointed sets

$$H^n(X; M_\lambda) \to E^n_\lambda(X; M) \tag{B.5}$$

We will use Verdier's formula to compute the left hand side; see [14] or also [59, Proposition 1.13 p. 52 and Proposition 1.26 p. 57], from which we freely use the notation and results.

Let $\mathcal{U} \to X$ be a hypercovering (a local trivial fibration) and $\mathcal{U} \to K(M_\lambda, n)$ a morphism of simplicial sheaves over X which represent a class $\alpha \in H^n(X; M_\lambda)$. From [59, Lemma 1.12 p.128], we may assume (up to taking a refinement of $\mathcal{U}$) that there exists a morphism of simplicial sheaves $\mathcal{U} \to BG$ such that the Pull-back of the G-torsor $EG \to BG$ is isomorphic to the G-torsor $Y \times_X \mathcal{U} \to \mathcal{U}$.

Now from Lemmas B.13 and B.14, there exists a canonical isomorphism of simplicial sheaves (over X)

$$Y \times_X K(M_\lambda, n) = Y \times K(M, n)$$

Beware there is no X in index on the right: this comes from the fact that we apply the Lemmas to the X-sheaf of abelian groups $M|_X X \to X$. Our isomorphism then follows from the tautological one $Y \times K(M, n) = Y \times_X K(M \times X \to X, n)$. The previous isomorphism is moreover G-equivariant.

We thus obtain a G-equivariant morphism

$$Y \times_X \mathcal{U} \to Y \times_X K(M_\lambda, n) = Y \times K(M, n)$$

By assumption on $\mathcal{U}$ there exists a cartesian square

$$
\begin{array}{ccc}
Y \times \mathcal{U} & \to & EG \\
\downarrow & & \downarrow \\
\mathcal{U} & \to & BG
\end{array}
$$

the top horizontal one being G-equivariant. We now consider the G equivariant morphism

$$Y \times \mathcal{U} \to EG \times K(M, n)$$

and pass to the quotient by G to get a morphism os simplicial sheaves:

$$\mathcal{U} \to K^G(M, n)$$

As the composition $\mathcal{U} \to K^G(M, n) \to BG$ represents λ its associated class (by Verdier's formula [59, Proposition 1.13 p. 52]) in $Hom_{\mathcal{H}_s(T)}$ $(X, K^G(M, n))$ actually lies in $E^n_\lambda(X; M)$. It is not hard to see that this class only depends on the class of α so that we have indeed constructed the expected map (B.5).

We are now ready to prove the following result:

Theorem B.16. *The map just constructed:*

$$H^n(X; M_\lambda) \to E^n_\lambda(X; M)$$

is a bijection.

Proof. We will indicate a way to construct a map the other way and will let it to the reader to check both maps are inverse to each other.

Take an element $\beta \in E^n_\lambda(X; M)$. That is to say a morphism in $\mathcal{H}_s(T)$: $X \to K^G(M, n)$ inducing λ.

Because $K^G(M, n)$ is locally fibrant, by the Verdier formula already used, we may represent β by an actual morphism

$$\mathcal{U} \to K^G(M, n)$$

where $\mathcal{U} \to X$ is an hypercovering. Denote by $\tilde{\mathcal{U}}$ the fiber product $EG \times_{BG} \mathcal{U}$ through $\mathcal{U} \to K^G(M, n) \to BG$. because the square

$$
\begin{array}{ccc}
EG \times K(M, n) & \to & EG \\
\downarrow & & \downarrow \\
K^G(M, n) & \to & BG
\end{array}
$$

is cartesian, there is a canonical G-equivariant morphism

$$\tilde{\mathcal{U}} \to EG \times K(M,n) \to K(M,n)$$

By the assumption the G-torsor $\tilde{\mathcal{U}} \to \mathcal{U}$ is also the pull-back of $Y \to X$ because β induces λ (use also [59, Proposition 1.15 p. 130]). This means that $\tilde{\mathcal{U}}$ is isomorphic to $Y \times_X \mathcal{U}$. We thus get a G equivariant morphism $\tilde{\mathcal{U}} \to Y$. We now claim that the G equivariant morphism product of the two previous morphisms

$$\tilde{\mathcal{U}} \to Y \times K(M,n)$$

induces after passing to quotient by G a morphism over X

$$\mathcal{U} \to K(M_\lambda, n)$$

We claim that the class of this map in $H^n(X; M_\lambda)$ only depends on β and that the map

$$E_\lambda^n(X; M) \to H^n(X; M_\lambda)$$

is the inverse to the map of the Theorem. $\square$

References

1. J. Arason, R. Elman, Powers of the fundamental ideal in the Witt ring, J. Algebra **239**(1), 150–160 (2001)
2. A. Asok, B. Doran, Vector bundles on contractible smooth schemes. Duke Math. J. **143**(3), 513–530 (2008)
3. A. Asok, B. Doran, A1-homotopy groups, excision, and solvable quotients. Adv. Math. **221**(4), 1144–1190 (2009)
4. A. Asok, F. Morel, Smooth varieties up to A^1-homotopy and algebraic h-cobordisms. Adv. Math. **227**(5), 1990–2058 (2011)
5. J. Ayoub, Un contre-exemple à la conjecture de A^1-connexité de F. Morel. C. R. Acad. Sci. Paris Sér I **342**, 943–948 (2006)
6. J. Barge, J. Lannes, in *Suites de Sturm, indice de Maslov et priodicit de Bott (French). (Sturm Sequences, Maslov Index and Bott Periodicity)*. Progress in Mathematics, vol. 267 (Birkhäuser, Basel, 2008)
7. J. Barge, F. Morel, Groupe de Chow des cycles orientés et classe d'Euler des fibrés vectoriels. C. R. Acad. Sci. Paris Sér I t. **330**, 287–290 (2000)
8. J. Barge et F. Morel, Cohomologie des groupes linéaires, K-théorie de Milnor et groupes de Witt. C. R. Acad. Sci. Paris Sér I t. **328**, 191–196 (1999)
9. H. Bass, J. Tate, *The Milnor Ring of a Global Field. In Algebraic K-Theory. II: "Classical" Algebraic K-Theory and Connections with Arithmetic*. Proceeding Conference, Battelle Memorial Institute, Seattle, Washington, 1972. Lecture Notes in Mathematics, vol. 342 (Springer, Berlin, 1973), pp. 349–446
10. A.A. Beilinson, J. Bernstein, P. Deligne, Faisceaux pervers (French). (Perverse sheaves). Analysis and topology on singular spaces, I (Luminy, 1981). Astrisque (Soc. Math., Paris, France) **100**, 5–171 (1982)
11. S.M. Bhatwadekar, R. Sridharan, The Euler class group of a noetherian ring. Compos. Math. **122**(2), 183–222 (2000)
12. S. Bloch, A. Ogus, Gersten's conjecture and the homology of schemes. Ann. Sci. École Norm. Sup. (4) **7**, 181–201 (1974/1975)
13. A.K. Bousfield, D.M. Kan, Homotopy limits, in *Completions and Localizations*. Lecture Notes in Mathematics, vol. 304 (Springer, Berlin, 1972), pp. v + 348
14. K.S. Brown, Abstract homotopy theory and generalized sheaf cohomology. Trans. Amer. Math. Soc. **186**, 419–458 (1974)
15. K.S. Brown, S.M. Gersten, *Algebraic K-Theory as Generalized Sheaf Cohomology. In Algebraic K-Theory. I: Higher K-Theories*. Proceeding Conference, Battelle Memorial Institute, Seattle, Washington, 1972. Lecture Notes in Mathematics, vol. 341 (Springer, Berlin, 1973), pp. 266–292

16. C. Cazanave, Théorie homotopique des schémas d'Atiyah et Hitchin. Thèse de Doctorat, Université Paris XIII, September 2009
17. C. Cazanave, Classes d'homotopie de fractions rationnelles (French). C. R. Math. Acad. Sci. Paris **346**(3–4), 129–133 (2008)
18. C. Cazanave, Algebraic homotopy classes of rational functions. Preprint (2009)
19. J.-L. Colliot-Thélène, R.T. Hoobler, B. Kahn, *The Bloch-Ogus-Gabber Theorem (English. English summary)*. In Algebraic K-Theory (Toronto, ON, 1996). Fields Institute Communications, vol. 16 (American Mathematical Society, Providence, 1997), pp. 31–94
20. E. Curtis, Simplicial homotopy theory. Adv. Math. **6**, 107–209 (1971)
21. F.Déglise, Modules homotopiques avec transferts et motifs génériques. Ph.D. Thesis, University of Paris VII, 2002
22. B. Dayton, C. Weibel, KK-theory of hyperplanes. Trans. Amer. Math. Soc. **257**(1), 119–141 (1980)
23. J. Fasel, Groupes de Chow orientés. Thèse EPFL, 3256 (2005)
24. J. Fasel, The Chow-Witt ring. Doc. Math. **12**, 275–312 (2007)
25. O. Gabber, Gersten's conjecture for some complexes of vanishing cycles. Manuscripta Math. **85**(3–4), 323–343 (1994)
26. R. Godement, *Topologie algébrique et théorie des faisceaux* (Hermann, Paris, 1958)
27. A. Grothendieck, Sur quelques points d'algèbre homologique. (French) Tôhoku Math. J. (2)9, 119–221 (1957)
28. A. Grothendieck, La théorie des classes de Chern. Bull. Soc. Math. France **86**, 137–154 (1958)
29. A. Grothendieck, J. Dieudonné, Éléments de Géométrie Algébrique IV. Étude locale des schémas et des morphismes de schémas (Troisième Partie) (French). Inst. Hautes Études Sci. Publ. Math. **28**, 361 (1967)
30. A. Grothendieck, J. Dieudonné, Éléments de Géométrie Algébrique IV. Étude locale des schémas et des morphismes de schémas (Quatrième Partie) (French). Inst. Hautes Études Sci. Publ. Math. **32**, 361 (1967)
31. A.J. Hahn, O.T. O'Meara, The classical groups and K-theory. With a foreword by J. Dieudonné. in *Grundlehren der Mathematischen Wissenschaften (Fundamental Principles of Mathematical Sciences)*, vol. 291 (Springer, Berlin, 1989), pp. xvi + 576
32. M. Hovey, Model category structures on chain complexes of sheaves. Trans. Am. Math. Soc. **353**, 2441–2457 (2001)
33. P. Hu, I. Kriz, The Steinberg relation in stable $\mathbb{A}^1$-homotopy. Int. Math. Res. Not. (17), 907–912 (2001)
34. J.F. Jardine, Simplicial presheaves. J. Pure Appl. Algebra **47**(1), 35–87 (1987)
35. A. Joyal, A letter to Grothendieck, April 1983
36. M. Karoubi, O. Villamayor, Foncteurs K_n en algèbre et en topologie. C. R. Acad. Sci. Paris A–B **269**, 416–419 (1969)
37. K. Kato, Symmetric bilinear forms, quadratic forms and Milnor K-theory in characteristic two. Invent. Math. **66**(3), 493–510 (1982)
38. K. Kato, A generalization of local class field theory by using KK-groups, II. J. Fac. Sci. Univ. Tokyo Sect. IA Math. **27**(3), 603–683 (1980)
39. M. Knus, Max-Albert. *Quadratic and Hermitian Forms Over Rings*. Grundlehren der Mathematischen Wissenschaften [Fundamental Principles of Mathematical Sciences], 294. Springer-Verlag, Berlin, 1991. xii+524 pp. ISBN: 3-540-52117-8
40. T.Y. Lam, in *Serre's Conjecture*. Lecture Notes in Mathematics, vol. 635 (Springer, Berlin, 1978), pp. xv + 227
41. H. Lindel, On the Bass-Quillen conjecture concerning projective modules over polynomial rings. Invent. Math. **65**(2), 319–323 (1981/1982)
42. H. Matsumara, *Commutative Ring Theory* (Cambridge University Press, Cambridge, 1989)

43. H. Matsumoto, Sur les sous-groupes arithmétiques des groupes semi-simples déployés. (French) Ann. Sci. École Norm. Sup. (4) **2**, 1–62 (1969)

44. J.P. May, *Simplicial Objects in Algebraic Topology*. Reprint of the 1967 original. Chicago Lectures in Mathematics. University of Chicago Press, Chicago, IL, 1992. viii+161 pp.

45. J. Milnor, *Introduction to Algebraic K-Theory* (Princeton University Press, Princeton)

46. J. Milnor, Construction of universal bundles I. Ann. Math. **63**, 272–284 (1956)

47. J. Milnor, Algebraic K-theory and quadratic forms. Invent. Math. **9**, 318–344 (1970)

48. J. Milnor, D. Husemoller, in *Symmetric Bilinear Forms*. Ergebnisse der Mathematik, Band 73 (Springer, Berlin, 1973)

49. F. Morel, *Théorie homotopique des schémas*. (French) [Homotopy theory of schemes] Astérisque No. 256 (1999), vi+119 pp.

50. F. Morel, An introduction to $\mathbb{A}^1$-homotopy theory. *Contemporary developments in algebraic K-theory*, 357–441 (electronic), ICTP Lect. Notes, XV, Abdus Salam Int. Cent. Theoret. Phys., Trieste, 2004.

51. F. Morel, in *On the Motivic π_0 of the Sphere Spectrum*, ed. by J.P.C. Greenlees. Axiomatic, Enriched and Motivic Homotopy Theory (Kluwer, Boston, 2004) pp. 219–260

52. F. Morel, in *The Stable $\mathbb{A}^1$-Connectivity Theorems*. K-Theory, vol. 35 (2005), pp. 1–68

53. F. Morel, Milnor's conjecture on quadratic forms and mod 2 motivic complexes. Rend. Sem. Mat. Univ. Padova **114**, 63–101 (2005)

54. F. Morel, Sur les puissances de l'idéal fondamental de l'anneau de Witt. Comment. Math. Helv. **79**(4), 689 (2004)

55. F. Morel, in $\mathbb{A}^1$-*Algebraic Topology*. Proceedings of the International Congress of Mathematicians (Madrid, 2006)

56. F. Morel, On the Friedlander-Milnor conjecture for groups of small rank. in *Current Developments in Mathematics*, ed. by D. Jerison, B. Mazur, T. Mrowka, W. Schmid, R. Stanley, S.T. Yau (International Press, Boston, 2010)

57. F. Morel, L'immeuble simplicial d'un groupe semi-simple simplement connexe déployé. Sur la structure et l'homologie des groupes de Chevalley sur les anneaux de polynômes (in preparation)

58. F. Morel, Generalized transfers and the rigidity property for the $C_*^{\mathbb{A}^1}((\mathbb{G}_m)^{\wedge n})$'s (in preparation)

59. F. Morel, V. Voevodsky, $\mathbb{A}^1$-homotopy theory of schemes, Inst. Hautes Études Sci. Publ. Math. No. 90 (1999), 45–143 (2001)

60. L.-F. Moser, $\mathbb{A}^1$-locality results for linear algebraic groups (in preparation)

61. M.P. Murthy, Zero cycles and projective modules. Ann. Math. **140**(2), 405–434 (1994)

62. Y. Nisnevich, The completely decomposed topology on schemes and associated descent spectral sequences in algebraic K-theory. in *Algebraic K-Theory: Connections with Geometry and Topology (Lake Louise, AB, 1987)*. NATO Adv. Sci. Inst. Ser. C Math. Phys. Sci., vol. 279 (Kluwer, Dordrecht, 1989), pp. 241–342

63. M. Ojanguren, I. Panin, A purity theorem for the Witt group. Ann. Sci. École Norm. Sup. (4) **32**(1), 71–86 (1999)

64. I. Panin, Homotopy invariance of the sheaf WNis and of its cohomology, In Quadratic forms, linear algebraic groups, and cohomology, 325–335, Dev. Math., 18, Springer, New York, 2010

65. D. Quillen, in *Homotopical Algebra*. Lecture Notes in Mathematics, vol. 43 (Springer, Berlin, 1967), pp. iv + 156

66. D. Quillen, Projective modules over polynomial rings. Invent. Math. **36**, 167–171 (1976)

67. M. Roitman, On projective modules over polynomial rings. J. Algebra **58**(1), 51–63 (1979)

68. M. Rost, Chow groups with coefficients. Doc. Math. **1**(16), 319–393 (1996) (electronic)

69. W. Scharlau, Quadratic and Hermitian forms. *Grundlehren der Mathematischen Wissenschaften (Fundamental Principles of Mathematical Sciences)*, vol. 270 (Springer, Berlin, 1985)

70. Schmid, Wittring Homology. Ph.D. Dissertation, Universitaet Regensburg (1998)

71. J.-P. Serre, *Corps locaux (French). Deuxième édition.* Publications de l'Université de Nancago, no. VIII (Hermann, Paris, 1968)

72. J.-P. Serre, *Algèbre locale multiplicités.* (French) Cours au Collège de France, 1957–1958. Seconde édition, 1965. Lecture Notes in Mathematics, 11 Springer-Verlag, Berlin-New York 1965 vii+188 pp.

73. A. Suslin, Projective modules over polynomial rings are free. (Russian) Dokl. Akad. Nauk SSSR **229**(5), 1063–1066 (1976)

74. A. Suslin, The structure of the special linear group over rings of polynomials. Izv. Akad. Nauk SSSR Ser. Mat. **41**(2), 235–252, 477 (1977) (Russian)

75. A. Suslin, V. Voevodsky, Singular homology of abstract algebraic varieties. Invent. Math. **123**(1), 61–94 (1996)

76. A. Suslin, V. Voevodsky, Bloch-Kato conjecture and motivic cohomology with finite coefficients. in *The Arithmetic and Geometry of Algebraic Cycles (Banff, AB, 1998).* NATO Sci. Ser. C Math. Phys. Sci., vol. 548 (Kluwer, Dordrecht, 2000), pp. 117–189

77. B. Totaro, The Chow ring of a classifying space. in *Algebraic K-Theory*, ed. by W. Raskind, C. Weibel. Proceedings of Symposia in Pure Mathematics, vol. 67 (American Mathematical Society, Providence, 1999), pp. 249–281 (33 pp.)

78. W. Van der Kallen, A module structure on certain orbit sets of unimodular rows. J. Pure Appl. Algebra **57**(3), 281–316 (1989)

79. V. Voevodsky, Cohomological theory of presheaves with transfers. in *Cycles, Transfers, and Motivic Homology Theories.* Ann. of Math. Stud., vol. 143 (Princeton University of Press, Princeton, 2000), pp. 87–137

80. V. Voevodsky, Triangulated categories of motives over a field. in *Cycles, Transfers, and Motivic Homology Theories.* Ann. of Math. Stud., vol. 143 (Princeton University of Press, Princeton, 2000), pp. 188–238

81. V. Voevodsky, The $\mathbb{A}^1$-homotopy theory. in *Proceedings of the International Congress of Mathematicians*, Berlin, 1998

82. T. Vorst, The general linear group of polynomial rings over regular rings. Com. Algebra **9**(5), 499–509 (1981)

83. T. Vorst, The Serre problem for discrete Hodge algebras. Math. Z. **184**(3), 425–433 (1983)

84. C. Weibel, Homotopy algebraic K-theory. in *Algebraic K-Theory and Algebraic Number Theory (Honolulu, HI, 1987).* Contemporary Mathematics, vol. 83 (American Mathematical Society, Providence, 1989), pp. 461–488

85. M. Wendt, On Fibre Sequences in Motivic Homotopy Theory, Ph.D. Universität Leipzig, 2007

86. M. Wendt, On the $\mathbb{A}^1$-fundamental groups of smooth toric varieties. Adv. Math. **223**(1), 352–378 (2010)

87. M. Wendt, $\mathbb{A}^1$-Homotopy of Chevalley Groups. J. K-Theory **5**(2), 245–287 (2010)

Index

F. Morel, $\mathbb{A}^1$-*Algebraic Topology over a Field*, Lecture Notes in Mathematics 2052,   259
DOI 10.1007/978-3-642-29514-0, © Springer-Verlag Berlin Heidelberg 2012

LECTURE NOTES IN MATHEMATICS

Edited by J.-M. Morel, B. Teissier; P.K. Maini

Editorial Policy (for the publication of monographs)

1. Lecture Notes aim to report new developments in all areas of mathematics and their applications - quickly, informally and at a high level. Mathematical texts analysing new developments in modelling and numerical simulation are welcome.

   Monograph manuscripts should be reasonably self-contained and rounded off. Thus they may, and often will, present not only results of the author but also related work by other people. They may be based on specialised lecture courses. Furthermore, the manuscripts should provide sufficient motivation, examples and applications. This clearly distinguishes Lecture Notes from journal articles or technical reports which normally are very concise. Articles intended for a journal but too long to be accepted by most journals, usually do not have this "lecture notes" character. For similar reasons it is unusual for doctoral theses to be accepted for the Lecture Notes series, though habilitation theses may be appropriate.

2. Manuscripts should be submitted either online at www.editorialmanager.com/lnm to Springer's mathematics editorial in Heidelberg, or to one of the series editors. In general, manuscripts will be sent out to 2 external referees for evaluation. If a decision cannot yet be reached on the basis of the first 2 reports, further referees may be contacted: The author will be informed of this. A final decision to publish can be made only on the basis of the complete manuscript, however a refereeing process leading to a preliminary decision can be based on a pre-final or incomplete manuscript. The strict minimum amount of material that will be considered should include a detailed outline describing the planned contents of each chapter, a bibliography and several sample chapters.

   Authors should be aware that incomplete or insufficiently close to final manuscripts almost always result in longer refereeing times and nevertheless unclear referees' recommendations, making further refereeing of a final draft necessary.

   Authors should also be aware that parallel submission of their manuscript to another publisher while under consideration for LNM will in general lead to immediate rejection.

3. Manuscripts should in general be submitted in English. Final manuscripts should contain at least 100 pages of mathematical text and should always include

   - a table of contents;
   - an informative introduction, with adequate motivation and perhaps some historical remarks: it should be accessible to a reader not intimately familiar with the topic treated;
   - a subject index: as a rule this is genuinely helpful for the reader.

   For evaluation purposes, manuscripts may be submitted in print or electronic form (print form is still preferred by most referees), in the latter case preferably as pdf- or zipped psfiles. Lecture Notes volumes are, as a rule, printed digitally from the authors' files. To ensure best results, authors are asked to use the LaTeX2e style files available from Springer's web-server at:

   ftp://ftp.springer.de/pub/tex/latex/svmonot1/ (for monographs) and
   ftp://ftp.springer.de/pub/tex/latex/svmultt1/ (for summer schools/tutorials).

Additional technical instructions, if necessary, are available on request from lnm@springer.com.

4. Careful preparation of the manuscripts will help keep production time short besides ensuring satisfactory appearance of the finished book in print and online. After acceptance of the manuscript authors will be asked to prepare the final LaTeX source files and also the corresponding dvi-, pdf- or zipped ps-file. The LaTeX source files are essential for producing the full-text online version of the book (see http://www.springerlink. com/openurl.asp?genre=journal&issn=0075-8434 for the existing online volumes of LNM). The actual production of a Lecture Notes volume takes approximately 12 weeks.

5. Authors receive a total of 50 free copies of their volume, but no royalties. They are entitled to a discount of 33.3 % on the price of Springer books purchased for their personal use, if ordering directly from Springer.

6. Commitment to publish is made by letter of intent rather than by signing a formal contract. Springer-Verlag secures the copyright for each volume. Authors are free to reuse material contained in their LNM volumes in later publications: a brief written (or e-mail) request for formal permission is sufficient.

Addresses:

Professor J.-M. Morel, CMLA,
École Normale Supérieure de Cachan,
61 Avenue du Président Wilson, 94235 Cachan Cedex, France
E-mail: morel@cmla.ens-cachan.fr

Professor B. Teissier, Institut Mathématique de Jussieu,
UMR 7586 du CNRS, Équipe "Géométrie et Dynamique",
175 rue du Chevaleret
75013 Paris, France
E-mail: teissier@math.jussieu.fr

For the "Mathematical Biosciences Subseries" of LNM:

Professor P. K. Maini, Center for Mathematical Biology,
Mathematical Institute, 24-29 St Giles,
Oxford OX1 3LP, UK
E-mail : maini@maths.ox.ac.uk

Springer, Mathematics Editorial, Tiergartenstr. 17,
69121 Heidelberg, Germany,
Tel.: +49 (6221) 4876-8259

Fax: +49 (6221) 4876-8259
E-mail: lnm@springer.com